CELLS

An introduction to the
anatomy and physiology
of animal cells

by Ellen Johnston McHenry

SECOND EDITION

Ellen McHenry's Basement Workshop
www.ellenjmchenry.com

ISBN 978-1-7374763-4-4

Published by Ellen McHenry's Basement Workshop, Pennsylvania, USA
www.ellenjmchenry.com
Printed by Lightning Source, Ltd.
Retails can order through Ingram Content

Also by this author:
The Elements; Ingredients of the Universe (5th edition)
Carbon Chemistry (2nd edition)
Botany in 8 Lessons
Protozoa; A Poseidon Adventure
Mapping the World with Art
Mapping the Body with Art (video e-course)

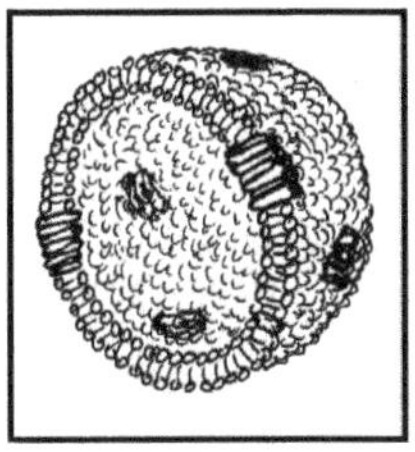
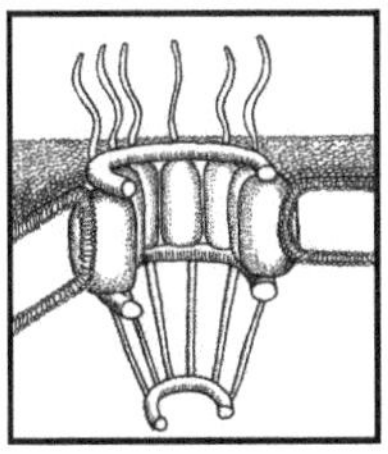
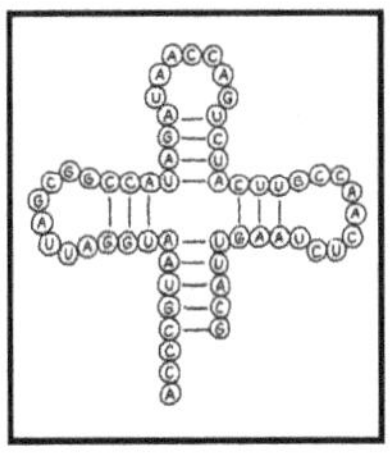
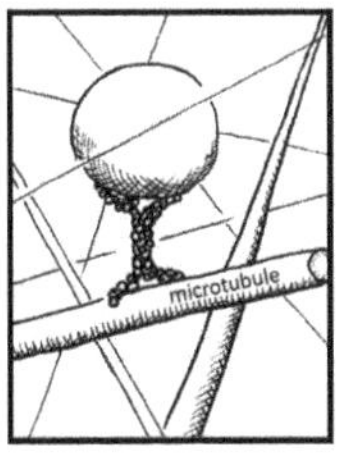

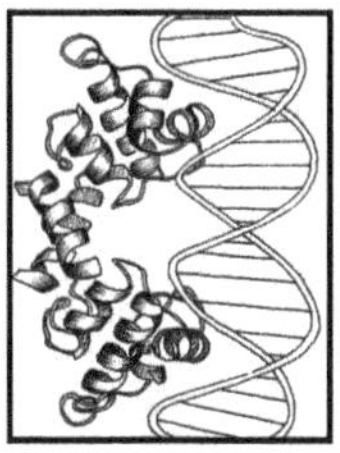

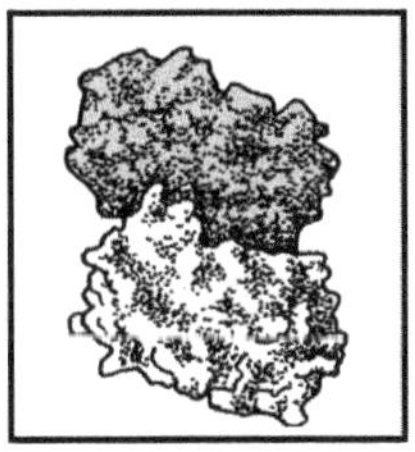
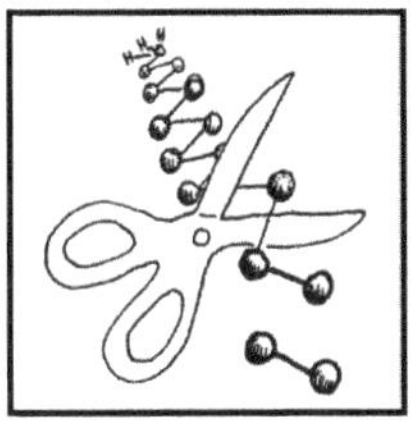
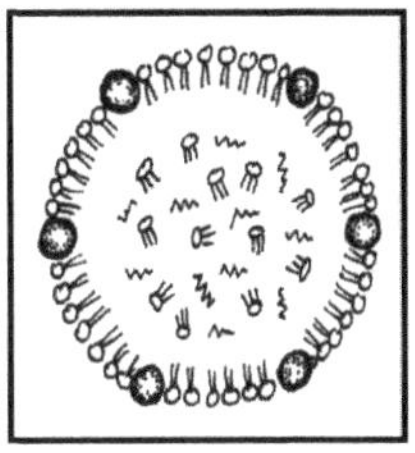

CELLS

An introduction to the anatomy and physiology of animal cells

Anatomy means "what the parts are"
Physiology means "how they work"

by Ellen J. McHenry

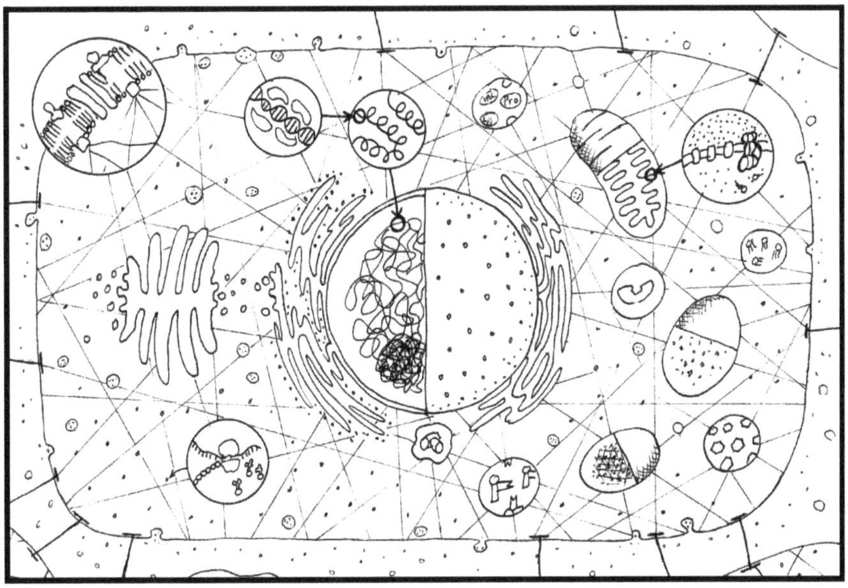

TABLE OF CONTENTS

CHAPTER 1: HOW DID WE FIND OUT ABOUT CELLS?

There was a time in the not-too-distant past when not a single person on earth knew that cells existed. Galileo, who used lenses to view distant planets, knew nothing of cells. It was in the decades following Galileo (the late 1600s) that someone figured out how to use lenses to make very small things visible. Two lenses were used, one at each end of a tube, forming a ***compound microscope***.

Englishman Robert Hooke (1635-1703) was probably the first person to observe cells. One day he sliced an extremely thin piece of cork and put it under his microscope. What did he see? Rows and rows of little box-like shapes that reminded him of the tiny rooms, or ***cells***, in monasteries (where monks live). Today we don't use the word "cell" when referring to a room, except when we talk about prison cells. But in Hooke's day the word "cell" was commonly used for a small room, so it was natural for him to use the word "cell" to describe these little compartments he saw in the cork. He didn't really know what these cells were made of or how they functioned, but the name he gave them has been used ever since.

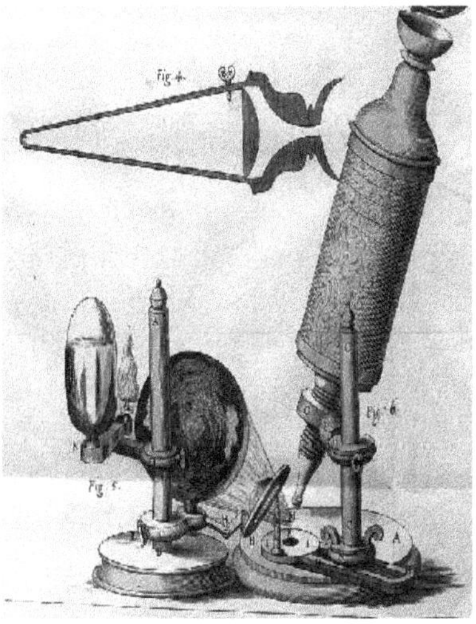

Hooke's compound scope

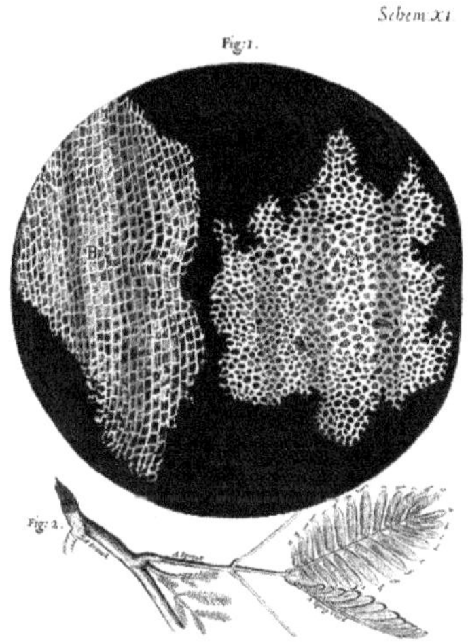

The famous cork cells

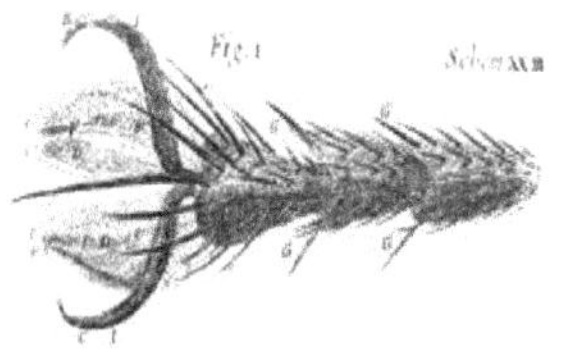

A fly's foot

Hooke eventually wrote a book called Micrographia telling about his amazing microscopic discoveries. He drew pictures of cells, parts of insects, hairs, specks of dirt and many other things that fascinated him. He discovered that no matter how sharp he made the point of a needle, the end of it still looked dull when viewed under his microscope! The only objects that still looked sharp when viewed under magnification were the tiny claws on the ends of insects' legs and the almost invisible "hairs" he found on the stems and leaves of plants.

Hooke was a brilliant man. He was also a surveyor, an architect, an astronomer and a physicist. He was working on the principles of motion and gravity at the same time that Isaac Newton was. He didn't really want to go down in history as the man who named cells. He would rather have been known for one of his other achievements: figuring out the laws of gravity and motion, helping to re-design London after the fire of 1666, or proposing the wave theory of light. But as history would have it, most people know him as the man who gave us the word "cell."

Hooke in a wig

Hooke without his wig

Then along came Antoni van Leeuwenhoek *(LAY-ven-hook)*, who lived his whole life in the Dutch town of Delft. He bought a copy of Hooke's Micrographia while on a trip to England in 1665, the only time in his life that he left Holland. Not long after reading Micrographia, Leeuwenhoek began making single-lens microscopes the likes of which have never been equaled. Leeuwenhoek perfected the art of making tiny lenses, but was careful to keep his technique a secret. He never wrote down his method, so we can only guess what he did. Modern glass making experts are fairly sure that Leeuwenhoek probably heated a glass rod and stretched it until it was a thin string. Then he would take

the very thin strand of glass and put it back into the flame and let the end melt until it formed a tiny round ball. This tiny round ball would be trimmed off and used as his lens. Other lens crafters of his day would spend hours grinding and polishing their lenses to get them into the right shape. Leeuwenhoek just took advantage of the natural physics of hot glass. He could make these tiny glass beads fairly quickly and easily; he managed to make over 500 of these little microscopes while keeping up with a full-time job as a cloth merchant. He mounted his lenses in silver panels and attached a screw mechanism on one side. With this simple magnifier, he was able to achieve magnification of at least 300 times larger than life size.

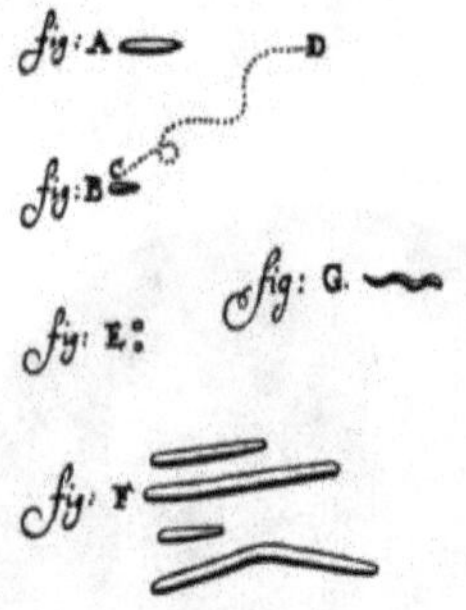

Leeuwenhoek observed bacteria

Leeuwenhoek was an incredibly patient person. He would sit for hours watching the specimens he had mounted on his microscope. He watched long enough to be able to observe the behavior and life cycles of microorganisms. He observed the microscopic food chain and knew what each little "animalcule" would eat. He saw eggs hatch. He saw blood cells circulate inside tiny circulatory systems. He observed sperm cells swimming. Once he kept a colony of fleas in a pouch inside his sock (to keep their eggs warm) and every hour or so he would check on them to see what changes had occurred. He spent several decades reporting all his findings to the Royal Society in London. At first, his descriptions of bizarre invisible creatures were almost too much to believe. The Royal Society had to send some of their members to visit Leeuwenhoek to verify that what he was saying was true, and he wasn't just imagining his microscopic "zoo." The visitors from the Royal Society looked through the little microscopes and were amazed to see exactly what Leeuwenhoek had written about. From then on, Leeuwenhoek's reports were treated as valid science. Prominent scientists and politicians began visiting Leeuwenhoek. Peter the Great of Russia put Delft on his European travel itinerary so that he could see Leeuwenhoek's little animalcules. Today, Leeuwenhoek is generally considered to be the "father" of modern microscopy.

The man in "The Geographer" by Verneer is probably Antoni van Leeuwenhoek.

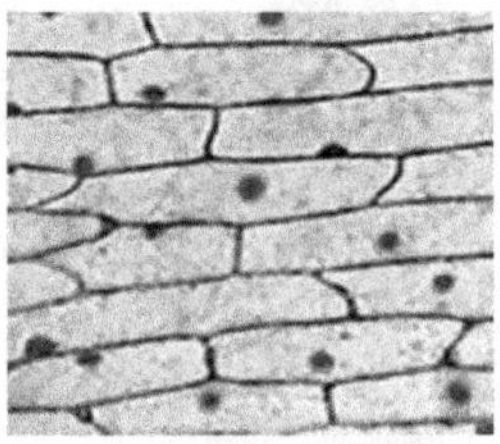

In the early 1800s, a Scottish botanist named Robert Brown made the next advances in our understanding of cells. Brown didn't have to make his own microscopes; by this time there were technicians who specialized in making optical devices such as microscopes. Since Brown was a botanist, it was plant cells he observed. He noticed that inside every cell there was a dark blobby thing. He called this the ***nucleus*** but he didn't have a clue what it did. Today we know that the nucleus contains the cell's DNA.

In 1827, Brown made another important microscopic discovery. While observing pollen grains under his microscope, he noticed that tiny particles inside the pollen grains were vibrating. He wondered if these particles were alive, since they were inside a plant cell. He tried a similar experiment with dust particles and saw the dust particles moving in the same way. He knew the dust particles were not alive, so he concluded that the motion must be due to a law of physics, not biology. He was right. Molecules are in constant motion and often collide. It is these molecular collisions that cause tiny particles to look like they are moving. We call this motion ***Brownian motion***, after Robert Brown.

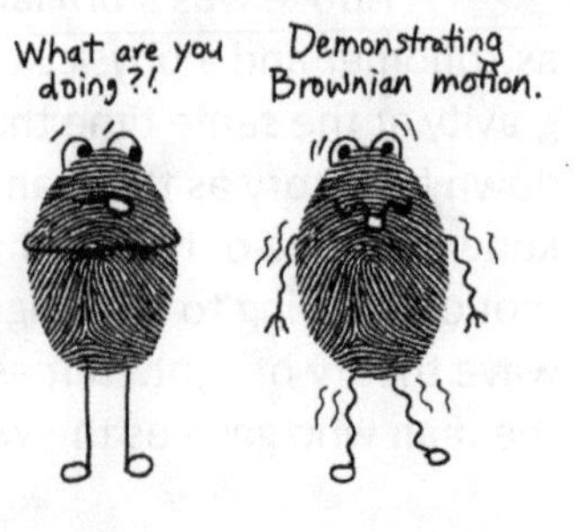

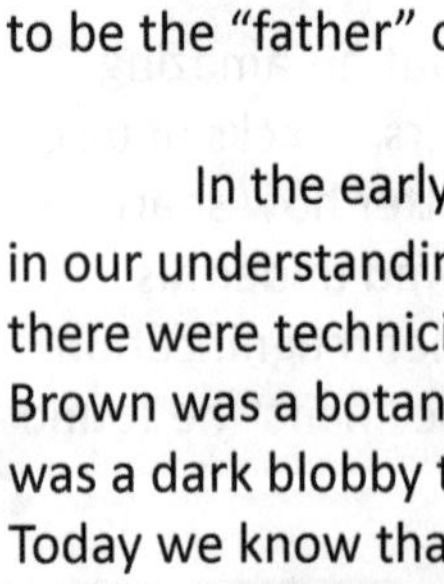

As an interesting historical side note, an ancient Greek named Lucretius was actually the first person to conceive of the idea of Brownian motion. In 60 BC, almost 2,000 years before Brown was born, Lucretius said this:

Observe the dust particles in sunbeams. You will see a multitude of tiny particles moving in a multitude of ways. Their motion is an indicator of underlying movements of matter that are hidden from our sight. It originates with the atoms which move of themselves. Their collisions set in motion slightly larger particles, and so the movement mounts up from the atoms and gradually emerges to the level of our senses, so that those particles we see in sunbeams are moved by blows that remain invisible.

In 1837, a German scientist named Theodor Schwann developed a theory that we now call "cell theory." Schwann came to realize that all living things are made up of cells that are very similar in basic structure. He also observed that cells only came from other cells. Cells could not come out of nowhere. This sounds obvious to us, but until Schwann's time many people still believed that living things could just suddenly appear. They saw flies appear seemingly out of nowhere when fruit or meat spoiled. Most people did not know that the flies had hatched from eggs because fly eggs are too small to see.

This is a drawing that Schwann made of different types of plant and anaimal cells he observed under his microscope.

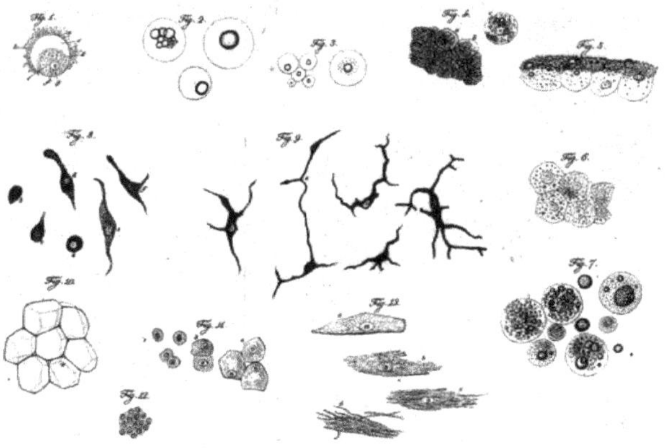

Schwann had a friend named Matthias Schleiden who was also a botanist. Together, they figured out that the nucleus played some role in cell division. They also observed the cytoplasm (fluid) inside the cell and saw that the organelles inside the cells moved around. Schleiden is considered to be the co-founder of ***cell theory***, along with Schwann. Cell theory says that cells can only come from other cells—they can't just pop into existence from nothing or from inorganic materials. (Ironically, Schleiden also accepted the theory of evolution, which suggested that cells did originally come from inorganic materials. He believed a theory that contradicted his theory?)

By the late 1800s, many different types of cells had been observed. There were fairly accurate pictures of plant cells, animal cells, single-celled organisms such as protozoa and bacteria. The big question now was how cells worked inside. Scientists knew that cells had some little "organelles" inside of them, but no one really knew what they did. The most obvious organelles were the nucleus (present in all cells) and chloroplasts (found only in plant cells). The chloroplasts were easy to spot because they were green. Other little spots and dots could be seen floating around inside the cell, but even the highest power on their microscopes could not enlarge them enough so that they could be studied. Another problem was that some of the little organelles were almost transparent. How can you study something you can hardly see?

A major breakthrough came when cell scientists learned how to stain cells before putting them under the microscope. The most famous "stain scientist" was Hans Christian Gram from Denmark. His technique of staining bacteria cells is still used today and bears his name: ***the Gram stain***. This stain will be absorbed by some kinds of bacteria but not by others. This helps to identify what kind of bacteria you are working with. Other stain experts developed stains that would penetrate the nucleus or other organelles, making them highly visible so they could be studied more easily. Then an Austrian scientist named ***Camillo Golgi*** discovered how to use a silver compound to stain nerve cells. His stains made possible many discoveries about nerve cells and how the nervous system works. Golgi's most famous discovery was another type of organelle found in almost all cells: the ***Golgi apparatus*** (or Golgi body).

An electron microscope from the 1930s

Then cell science "hit a wall," so to speak. Even the very best microscopes in the world could not magnify something beyond about 1000 times. Scientists knew that many mysteries of the cell would not be discovered until there was a way to achieve magnifications beyond 1000. Then, in the mid 1900s, a completely new type of microscope was invented: the ***electron microscope***.

Regular microscopes use light and lenses to make things look larger. Electron microscopes work on an entirely different principle; they use electrons instead of light. Electrons from a tungsten filament are "fired" at the sample being studied, and the electrons either go through it (in the case of transmission electron microscopes, or TEM) or they bounce off at various angles (in the case of scanning electron microscopes, or SEM). In both TEM and SEM, the electrons then hit a screen to form a visible image. Pictures from electron microscopes, which are known as ***micrographs***, are always in black and white. Color requires light, and electron microscopes don't use light. Colored micrographs are made by adding the color afterward. They use computer programs to adjust the graphics, just like you might use a program like Photoshop®.

An SEM microscope opened up to show you the vacuum chamber where the sample goes

Modern electron microscopes can provide images that are up to a million times larger than life. That's large enough to be able to see even the tiniest parts of the cell. However, electron microscopes have a big drawback. The samples being studied must be put into a vacuum chamber where there is not a single molecule of air (like outer space). Living cells need air. so basically, only dead specimens can be studied. The specimens can be killed and preserved only minutes before loading them into the machine, but nothing alive and moving can be viewed. Usually, the specimens have to be prepared by spraying them with an ultra-thin layer of gold, or some other metal. This means that you can't sit and watch little critters moving around under an electron microscope like you can with a regular (compound) microscope. You can't watch as a cell eats or grows or divides. You only get one picture of a cell at one moment in its life. Cell scientists must collect lots and lots of still pictures, then use "detective skills" to draw conclusions based on comparing all the pictures. Sometimes scientists can think of a way to test their theories about cells by "tagging" particular molecules with radioactive or fluorescent dyes that will show up on the screen. In the next chapter, we'll read about a cell part that was discovered in this way.

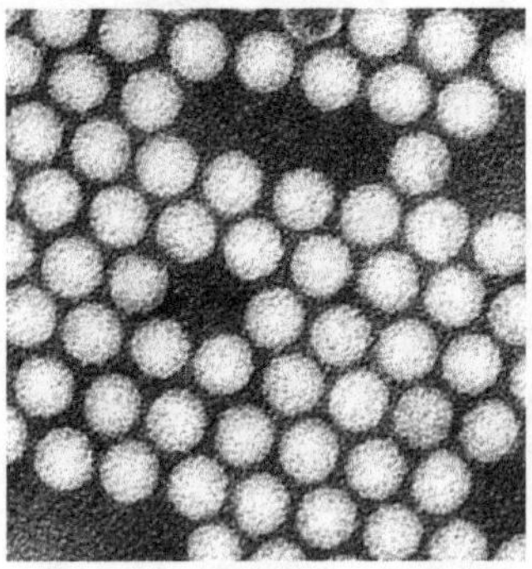

viruses

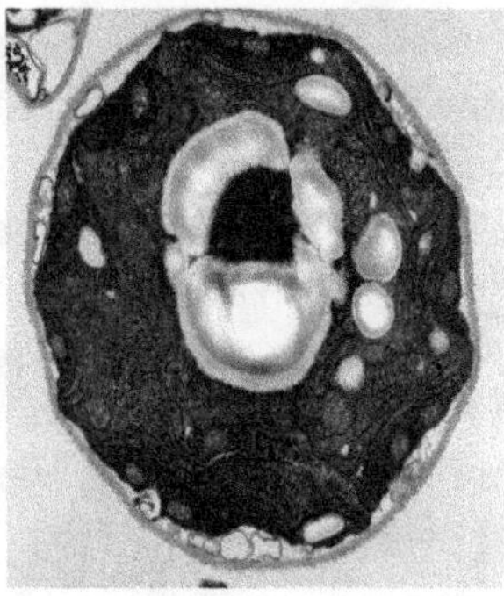

a single-celled organism

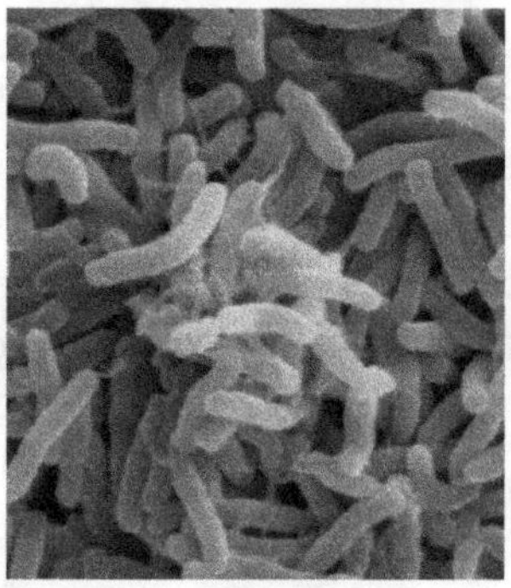

bacteria

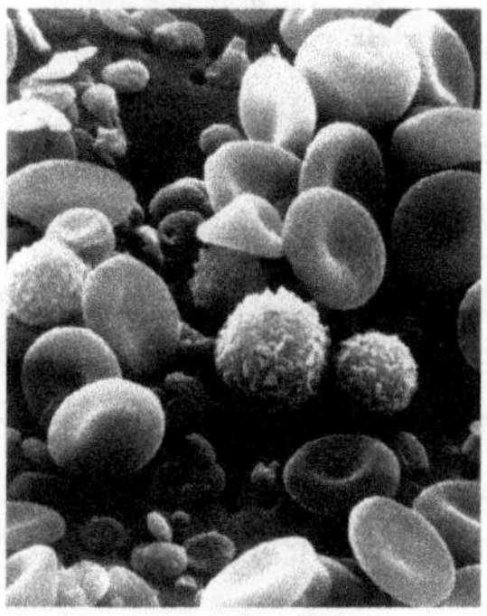

blood cells

TEM images look flat

SEM images look 3D

Images produced by TEM microscopes look flat. The electrons pass through the sample in much the same way that light passes through samples on a regular (compound) microscope. This type of image can be very good for studying the insides of cells. SEM electron microscopes produce 3D images. SEMs let you see textures and shapes. It takes both types of images to give us enough information to be able to understand what a cell is really like. Scientific illustrators try to create pictures that combine information gained from both types of images. Books about cells often contain many images made by scientific illustrators.

Electron microscopes are used for more than just biology. They can be used in the fields of material science (metals, crystals and ceramics), nanotechnology, chemistry, and forensics. They have become an essential tool for many branches of science.

<u>Comprehension self-check</u> If you can't think of the answer, go back and read that part of the chapter again until you find the answer. If you need to check your answers, check the answer key.

1) The first person to ever see a cell was:
   a) Galileo   b) Hooke   c) Lucretius   d) Leeuwenhoek

2) Which one of these did Hooke NOT do?
   a) develop theories about gravity and motion   b) propose a wave theory of light
   c) help to redesign London   d) develop cell theory

3) About how many microscopes did Leeuwenhoek make?
   a) less than 10   b) about 100   c) about 500   d) millions

4) TRUE or FALSE? The Royal Society immediately made Leeuwenhoek a member, as soon as they read his descriptions of "animalcules."

5) What is Brownian motion?
   a) a physical phenomenon caused by the constant motion of molecules
   b) the movement of dust particles in air   c) a biological phenomenon found only in living things
   d) the movement of cells under the microscope

6) When was the idea of tiny invisible particles (atoms) first proposed?
   a) 60 BC   b) 600 AD   c) 1827   d) early 1900s

7) TRUE or FALSE? Schwann and Schleiden proved that life could come from nonliving things.

8) TRUE or FALSE? By the late 1800s, scientists had seen many different types of cells.

9) TRUE or FALSE? Many cells, and their inner parts, are transparent.

10) What is the most noticeable object found inside a cell?
   a) Golgi body   b) nucleus   c) DNA   d) cytoplasm

11) Who has a staining method named after him?
   a) Antoni Leeuwenhoek   b) Theodor Schwann   c) Camillo Golgi   d) Hans Christian Gram

12) The staining method named referred to in question 9 is used to stain bacteria.

13) What is the maximum magnification you can get with most ordinary (compound) microscopes?
   a) 100x   b) 500x   c) 1000x   d) 100,000x

14) TRUE or FALSE? TEM images look 3D.

15) TRUE or FALSE? Electron microscopes can let you watch a cell as it divides.

16) For electron microscopy, what do the specimens have to be in?
   a) a vacuum   b) suspended animation   c) a frozen state   d) high temperature environment

17) TRUE or FALSE? There is a special kind of electron microscopy that can show you both a flat image and a 3D image at the same time.

18) What does SEM stand for? Scanning Electron Microscopes

19) What metal is common used to spray samples that will be observed with electron microscopes? Gold

20) TRUE or FALSE? Electron microscopes are used exclusively for biology.

Activity 1.1: Just for fun—can you guess what these are?

Here are some SEM and TEM micrographs. Try to figure out what they are. (Answers are in the answer key.)

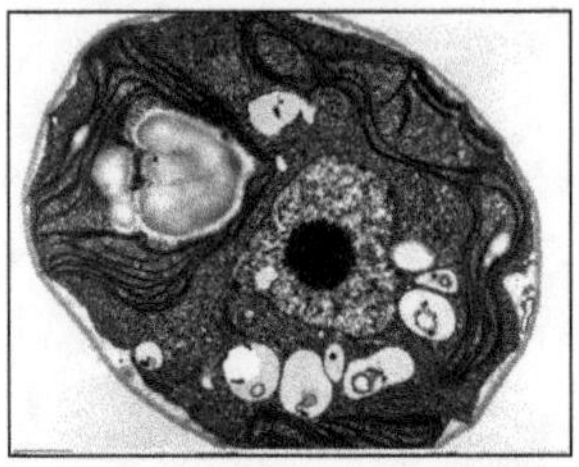
A

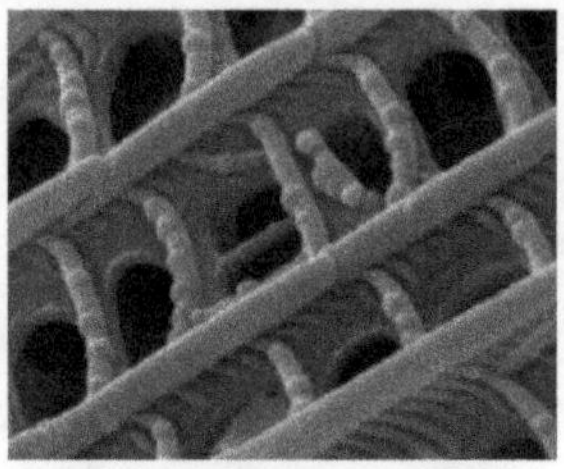
B

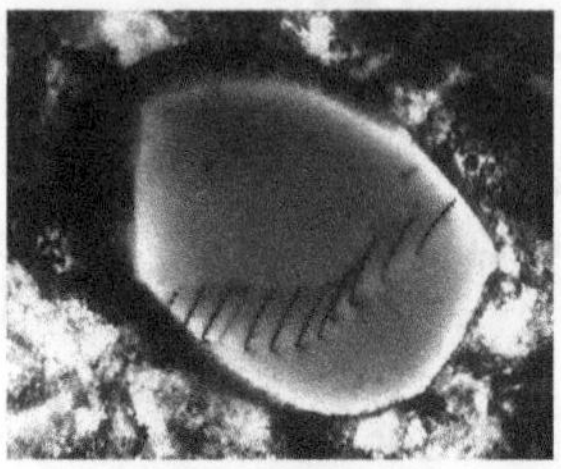
C

D

E

F

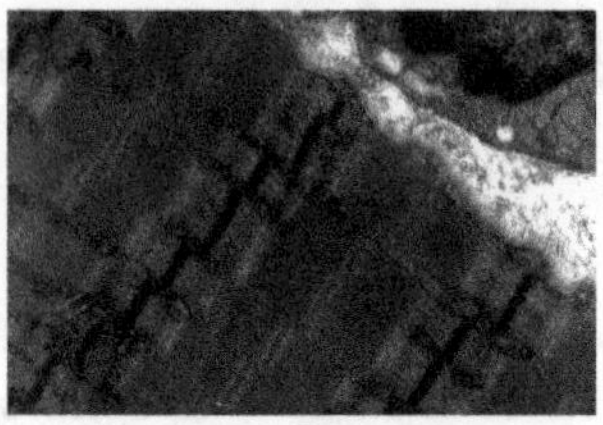
G

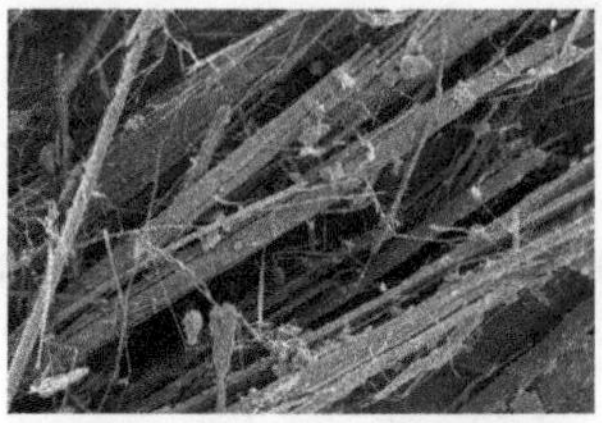
H

I

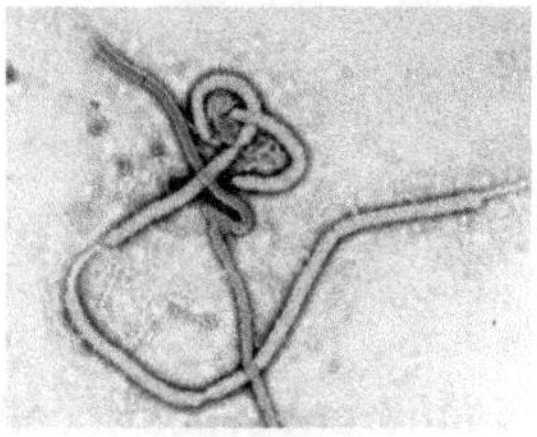
J

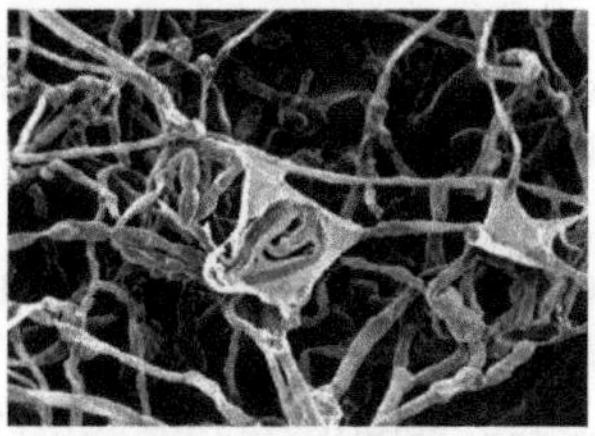
K

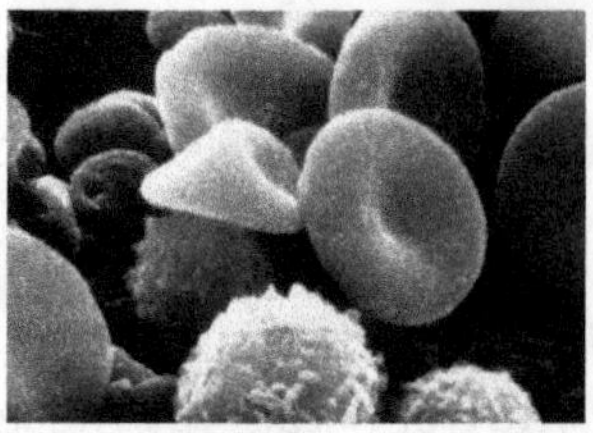
L

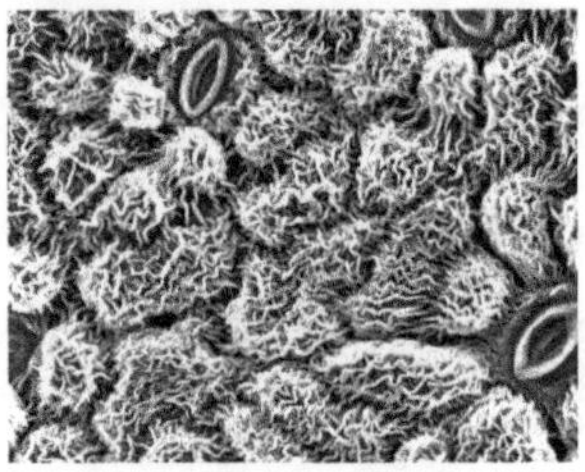
M

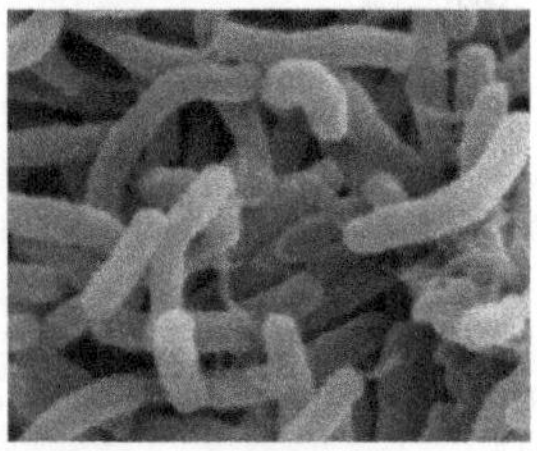
N

O

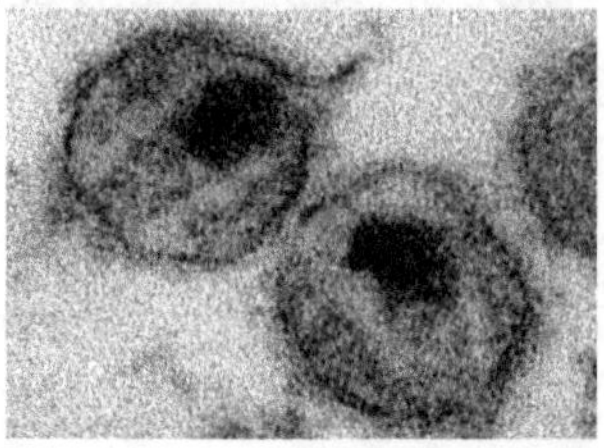
P

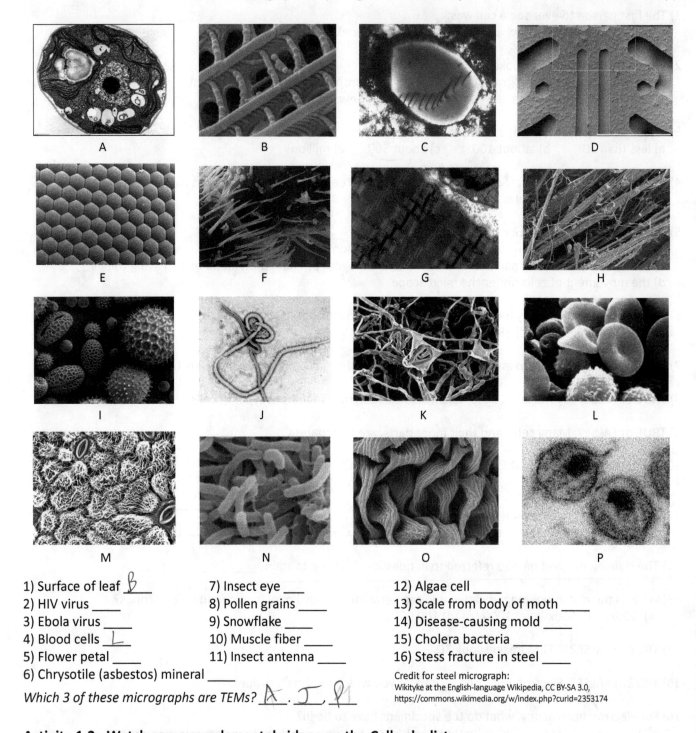

1) Surface of leaf B____
2) HIV virus ____
3) Ebola virus ____
4) Blood cells L___
5) Flower petal ____
6) Chrysotile (asbestos) mineral ____

7) Insect eye ___
8) Pollen grains ____
9) Snowflake ____
10) Muscle fiber ____
11) Insect antenna ____

12) Algae cell ____
13) Scale from body of moth ____
14) Disease-causing mold ____
15) Cholera bacteria ____
16) Stess fracture in steel ____

Which 3 of these micrographs are TEMs? A, J, P

Credit for steel micrograph:
Wikityke at the English-language Wikipedia, CC BY-SA 3.0,
https://commons.wikimedia.org/w/index.php?curid=2353174

Activity 1.2: Watch some supplemental videos on the Cells playlist

This curriculum has a YouTube playlist. Go to www.youtube.com/TheBasementWorkshop. The Cells playlist might not be visible at first glance. Sometimes you have to click on "See all playlists" and then click on arrows to advance the list to see all the titles. The playlist is made of videos posted by various people around the world, and they have the right to take down the videos at any time, so occasionally there will be a blank spot. The author of this book tries to keep the playlist updated, but it is not possible to check it daily or even weekly. The videos that do appear on this list have been previewed by the author so they don't contain anything offensive and they hopefully aren't too boring. YouTube does not provide a way to label the videos to indicate which chapter they go with, but the will be in approximately the right order, so you can go down the list as you read the book.

CHAPTER 2: THE CELL MEMBRANE

So now that we know a little bit about how cells were discovered, let's start learning about what cells are made of and how their insides work. We'll start with the outer surface, the ***plasma membrane***.

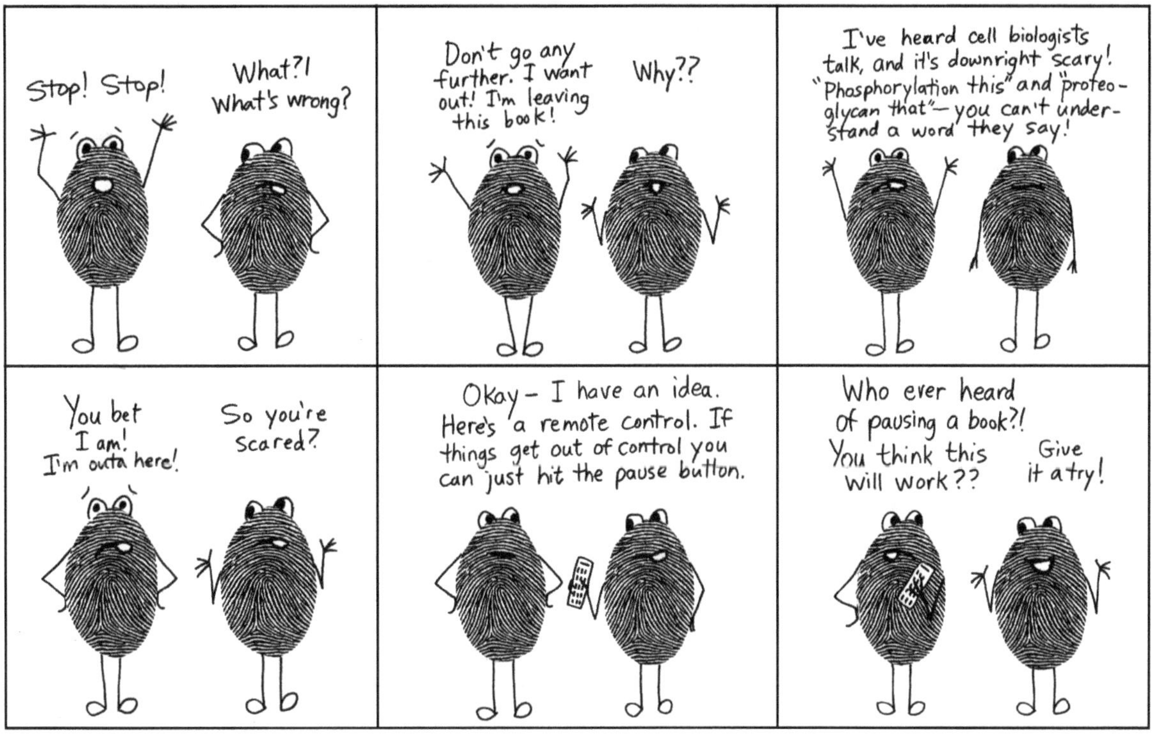

Okay, we're ready now?

Let's look at the outside of the cell first. The outer layer of a cell is called the ***membrane***, or, to be more precise, the ***plasma membrane***. (Don't worry about the word "plasma." We'll get to that later.) In plant cells there is an extra layer outside the membrane—a thick coating made of tough ***cellulose***. Bacteria, also, sometimes have thick outer walls. However, underneath those thick walls there is still a thin membrane. The membrane is so thin that it would take 10,000 of them stacked on top of each other to be as thick as a sheet of paper. It's barely visible even with one of those high-power electron microscopes, because it's only two molecules thick!
The molecules that form a cell membrane are called ***phospholipids***.

WAIT! DON'T PAUSE THE BOOK!

Let's look at this word and figure out what it means. The second part of the word, ***lipid***, means "fat." You know what fats are—those white streaks in your meat, the vegetable oil you use to fry your potatoes, the cream on top of fresh milk, even the grease that builds up on your scalp if you don't shampoo your hair for a few days. Lipids are greasy and oily and don't mix with water. If we look at one molecule of grease or fat, we'll see that it is made of a chain of carbon atoms with hydrogen atoms attached.

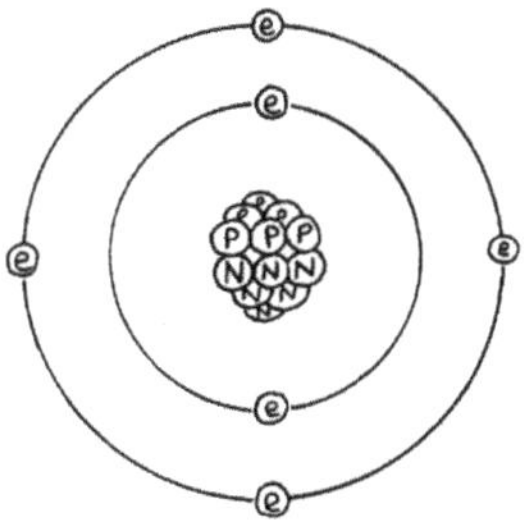

A single carbon atom is often drawn like this. Atoms have layers of electrons orbiting a nucleus made of protons and neutrons.

This circled part of the molecule is the "acid."

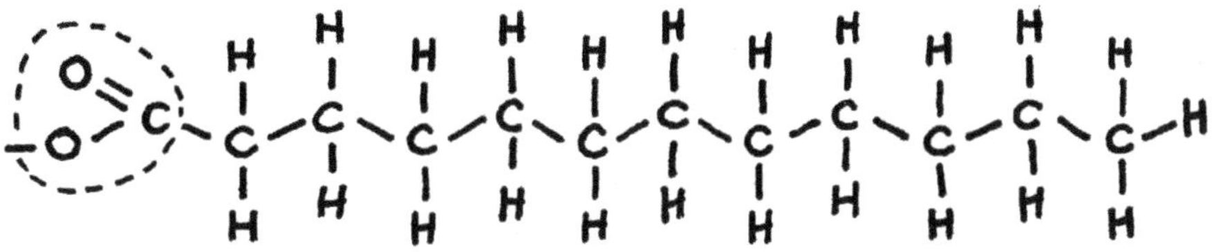

The simplest lipid molecule is called a ***fatty acid***.
The "fatty" part is the string of carbon atoms.

Drawing entire atoms is too difficult. In this book, we are going to simply use a letter to represent an atom. C is carbon, H is hydrogen, O is oxygen. If you want to know more about atoms and how they join together to make molecules, check out "The Elements" book by Ellen McHenry.

Phospho is short for "phosphate." A phosphate molecule is made of four oxygen atoms attached to an atom of phosphorus, P. (Phosphorus is a fascinating element that can glow in the dark and is found in matches and fluorescent light bulbs, but it is also found in many biological molecules, including phospholipids. We'll be meeting phosphorus again in future chapters.)

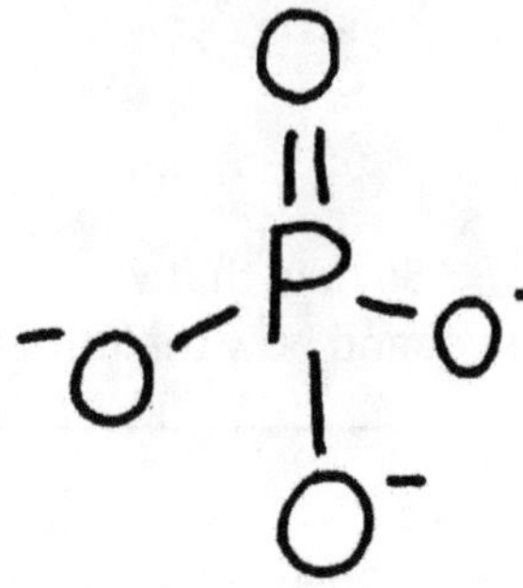

Phosphorus atoms can make five bonds. Imagine phosphorus having five arms so that it can shake or clasp hands with five people all at once. Oxygen atoms can make only two bonds. They are more like you, having two arms. In phosphate, one oxygen atom does a double handshake and uses two of phosphorus's five bonds. The three other oxygen atoms make one bond with phosphorus, but have their other "arm" unattached to anything. This is indicated by that minus sign next to them. (The minus sign represents a free electron that can make a bond.) Phosphate is written like this: PO_4^{3-}. The "4" tells you how many oxygen atoms there are. The "3-" tells you how many dangling, unattached arms there are.

PAUSE

PLAY

For our study of cell membranes, the most important difference between the phosphates and lipids is their reaction to water molecules. Phosphates are said to love water. Yes, scientists really do use the term "love," but they say it in Greek. They combine the Greek word for water, "hydro," with the Greek word for love, "philia" to make the word "hydrophilic," meaning water-loving.

Lipids, on the other hand, "hate" water molecules. They are said to be "hydrophobic." The Greek word "phobia" means "fear," so perhaps they feel more fear than hatred? It is silly, of course, to say that molecules can love or hate or fear or have any other emotion. So, if molecules aren't actually loving or hating, what makes them react the way they do to water molecules? Let's look at a water molecule to find out.

A water molecule consists of an oxygen atom holding on to two hydrogen atoms. Remember, oxygen has two "arms" and can hold on to two atoms. Hydrogen atoms are very small and have only one "arm" with which to make a bond. The water molecule is a pretty happy molecule, since each atom is making the number of bonds that it wants to make. However, it does have one issue. The oxygen atom is much larger, with a nucleus that is 16 times larger than a hydrogen's nucleus. The eight protons in the oxygen's nucleus have a very strong pull on the eight electrons that are being shared between the oxygen and the hydrogens. The result is that the electrons spend more time going around the oxygen than they do the hydrogens. The presence of the negatively charged electrons around the oxygen atom makes the molecule electrically lopsided. The side without the hydrogen atoms is slightly negative, and side where the hydrogens attach is slightly positive.

All molecules that are electrically lopsided are called ***polar*** molecules. "Polar" means that something has two ends, or sides, that are different. (Sometimes we think of the world "polar"as meaning "cold," because the north and south poles of the earth are located where it is cold, but the real meaning of "polar" is "opposite.")

WATER MOLECULE

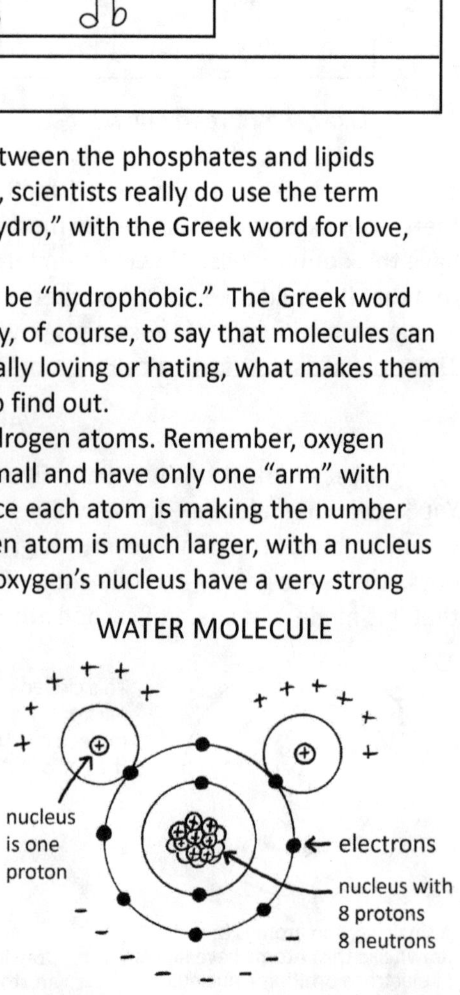

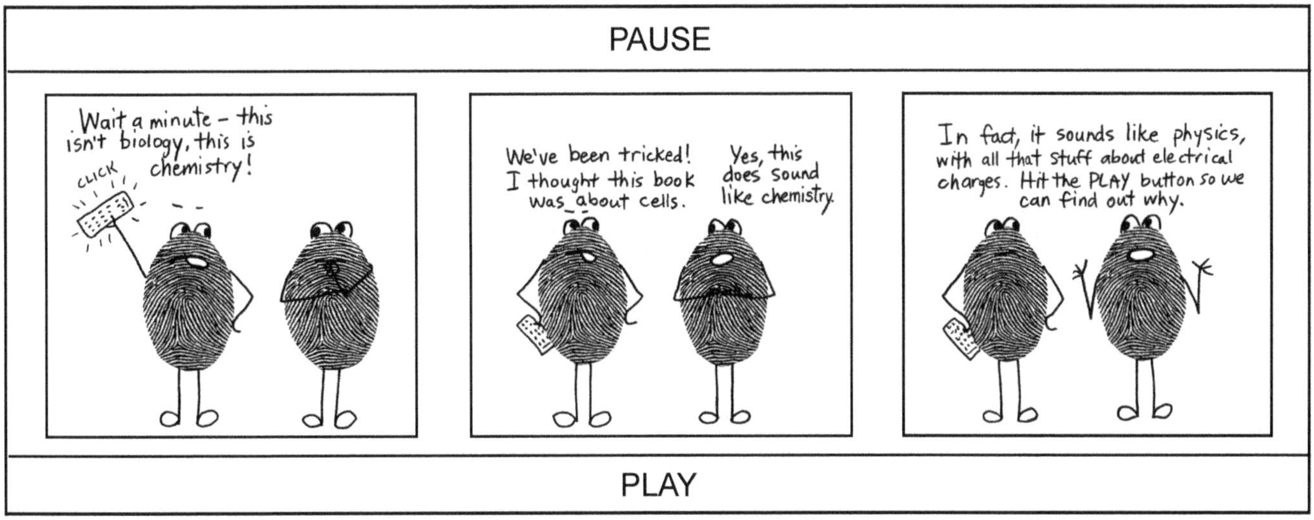

Chemistry is the foundation on which biology rests. The answers to many biology questions involve chemistry. And yes, chemistry is largely about the electrical interactions between atoms. So, in a way, biology boils down to physics. Let's continue on and see what role polarity plays in a membrane.

You may have heard the phrase, "Opposites attract." The context in which you heard the phrase might not have been physics, but the origin of this saying is rooted in physics. Positive and negative charges attract. The reason why this is true can only be explained by studying quantum physics, which is far beyond the scope of this book. All we need to know is that those dangling minus signs on molecules will want to be close to a molecule, or a part of a molecule, that has a positive charge. Those minus signs on the phosphate will be attracted to the positive side of a water molecule. But what about lipid molecules? If you look at the fatty acid that our friend is holding, you can see that there aren't any minus or plus signs around it. Therefore it doesn't interact with water molecules.

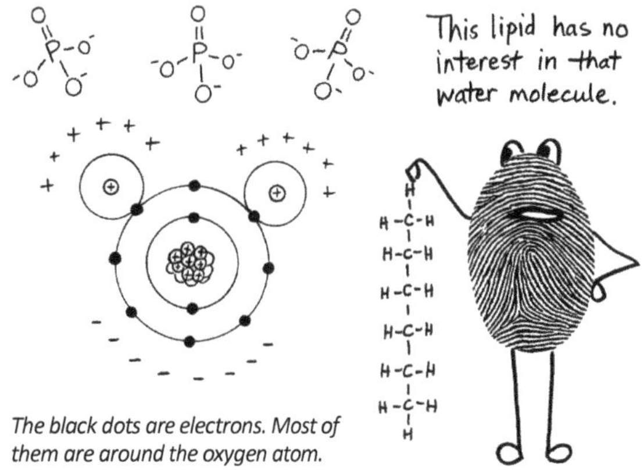

The black dots are electrons. Most of them are around the oxygen atom.

And, now, finally, we are ready to look at a phospholipid molecule. The name tells us that it is a phosphate connected to a lipid (a fatty acid carbon chain). A little clump of atoms called ***glycerol*** *(GLISS-er-ol)* holds them together. Glycerol is like a 2-sided clip that can hold on to both a phosphate and two carbon chains.

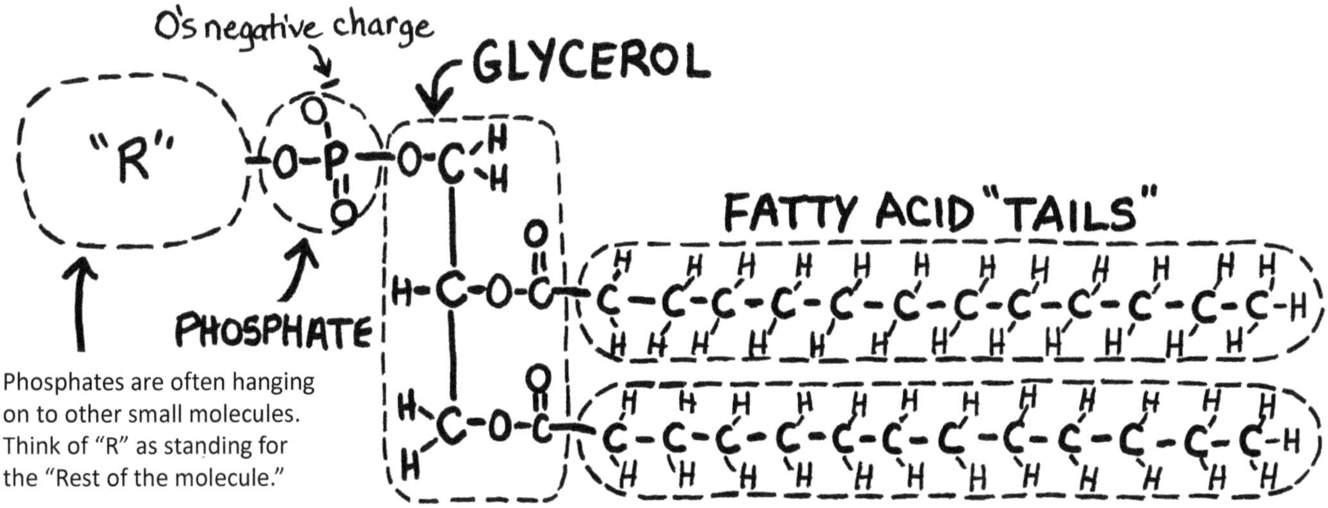

Phosphates are often hanging on to other small molecules. Think of "R" as standing for the "Rest of the molecule."

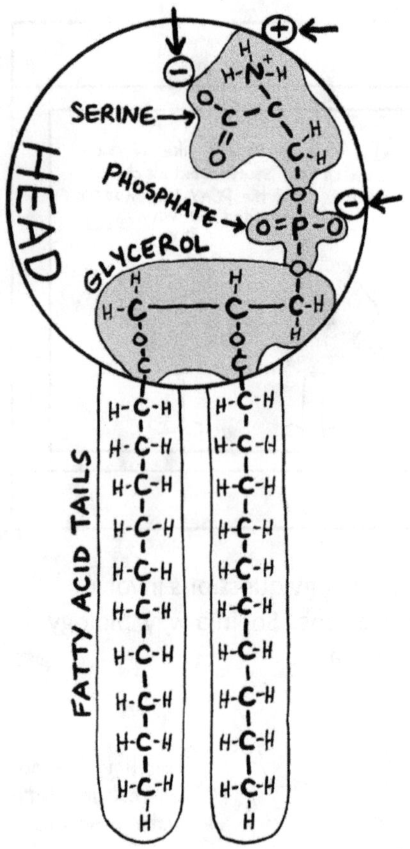

This is the most complicated phospholipid diagram you will see in this book. It shows that the "head" area is made of a glycerol, a phosphate, and another small molecule (here we see a simple protein called "serine"). The head isn't really round, but we can imagine that these odd-shaped molecules are contained in a circular area. The arrows are pointing to the electrical charges of the head area. We see two negative charges, and one positive charge, so the negative charges win by one. The overall charge of this head is (-1). This negative charge will be attracted to a water molecule's positive side. Some phospholipid heads have one positive and one negative charge, which makes them electrically neutral. They are still considered "hydrophilic" molecules, though, even without a negative charge.

Now that we've seen the complete structure of a phospholipid molecule, we are going to learn a helpful shortcut. We'll need to draw lots of these molecules to make a cell membrane, and it's obvious that this molecule is far too complicated to draw repeatedly. Scientists use a simple diagram that is easily recognizable as a phospholipid. The tails can be thick or thin, and are sometimes zig-zaggy or slightly bent.

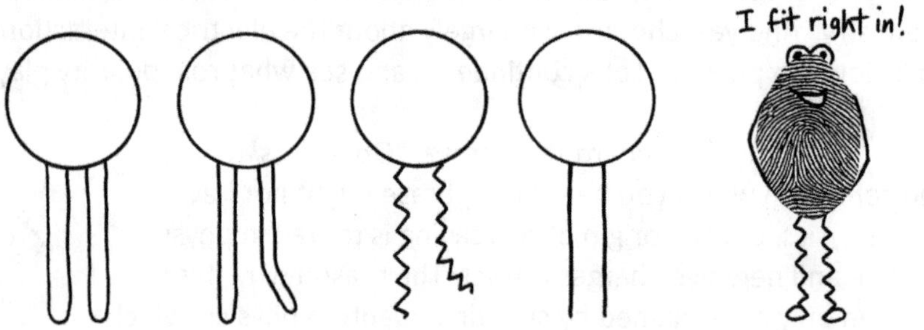

Now that we can easily draw a lot of phospholipids, we are ready to see how they behave as a group. What would happen if you took a bunch of these phospholipids and tossed them into a bucket of water? The heads would feel quite at home among the water molecules because they are hydrophilic, loving water. The hydrophobic tails, however, would be freaking out. It would be a nightmare for them to be surrounded by water molecules. They would need to cooperate to create a "NO WATER" zone where they can feel comfortable and safe. This is accomplished by having the phospholipids create a sphere where the tails are all pointing to the inside.

This shape is called a ***micelle*** (mie-SELL). (Though shown as a circle in this diagram, a micelle is actually a sphere.)

BONUS INFO: Micelles are found throughout your body, especially in your blood. They allow hydrophobic molecules (such as digested fats and some vitamins) to be transported in the bloodstream. Since blood is 90% water, the bloodstream is a very uncomfortable place for a molecule that hates water! The hydrophobic molecules hide inside the micelle.

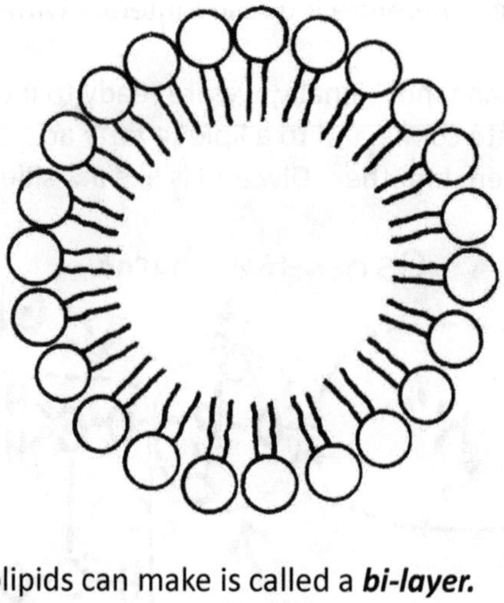

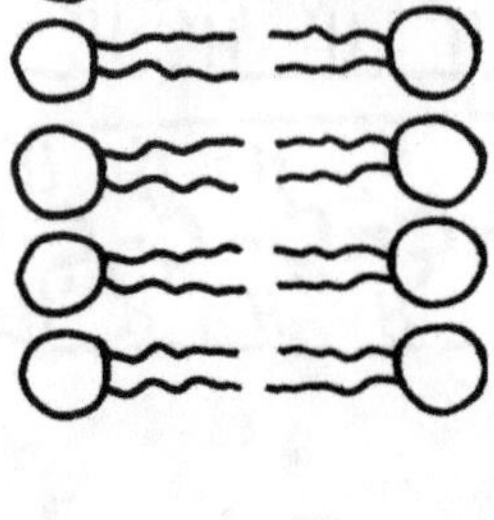

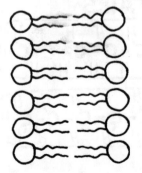

Another arrangement that phospholipids can make is called a ***bi-layer.*** ("Bi" means "two.") This shape is sort of what you would get if you took a micelle and flattened it. The diagram here on the left shows the phospholipids lined up perfectly, with the tails of one molecule exactly opposite the tails of another. In reality, they don't always line up so perfectly. (Notice the optical illusion: a white line down the middle.) If we had a large, three-dimensional sheet of bi-layer we might be able to fold it in such a way that it formed a hollow sphere.

If we use hundreds, or perhaps even thousands, of phospholipid molecules, we can make a sizable sphere. Look at the cut-away edge. Can you see all the individual phospholipid molecules lined up tail to tail, forming a bi-layer? If we had not cut the sphere in half, you would not be able to see the tails. The surface of the sphere would look like a "sea" of balls.

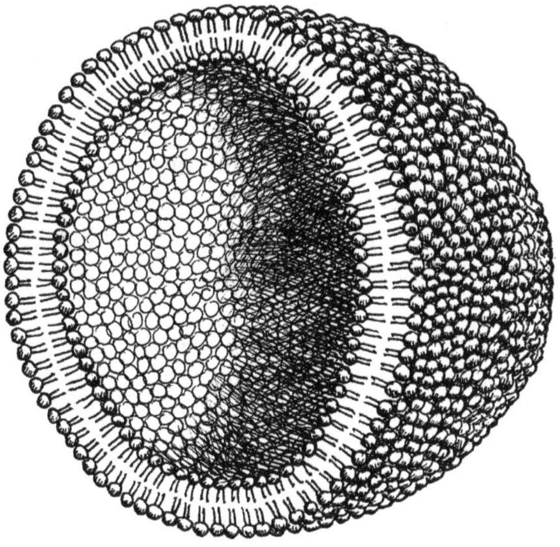

Now we have a nice, tight, almost leak-proof ball. Very small molecules, such as oxygen or carbon dioxide, might sneak through the cracks, but large molecules don't stand a chance of getting through. This is what the outer layer, or ***plasma membrane,*** of a cell is made of. Most of the organelles inside a cell are also wrapped in a bi-layer membrane. Additionally, the cell makes and uses spheres like this for storing things, almost as if they were plastic bags.

PAUSE

PLAY

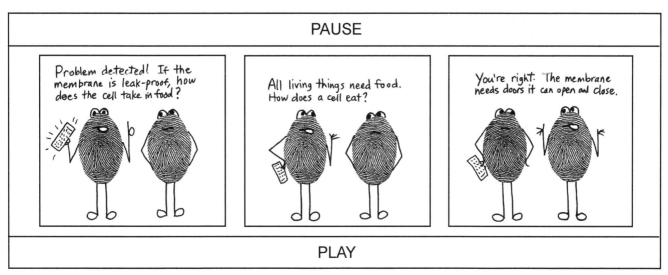

Yes, the membrane needs doors and portals that it can control. Now that we understand the structure of the membrane itself, we can look at the microscopic "gadgets" that are in it.

Many of the structures that are embedded in a cell's membrane are made of ***protein***. They are called (no surprise here) ***membrane-bound proteins***. In future chapters we'll learn what proteins are made of and see how the cell manufactures them. For now, we are just going to take a brief survey of the general types of proteins we find and the jobs they do. Membrane-bound proteins can be found on the outside, on the inside, or going all the way through the membrane.

Proteins that go all the way through the membrane are called ***transmembrane proteins***. ("Trans" is Latin for "across.") They act as **tunnels**, or **portals**, letting some types of molecules pass through and keeping others out. A cell wants to take in food molecules (sugars, fats, proteins) and get rid of waste molecules. It might also want to take in a chemical message sent by another cell. Some transmembrane proteins act like **pumps**, pushing water or salt molecules in or out of the cell.

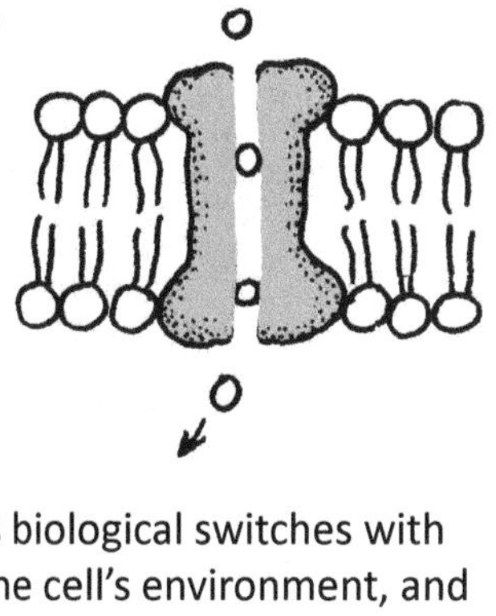

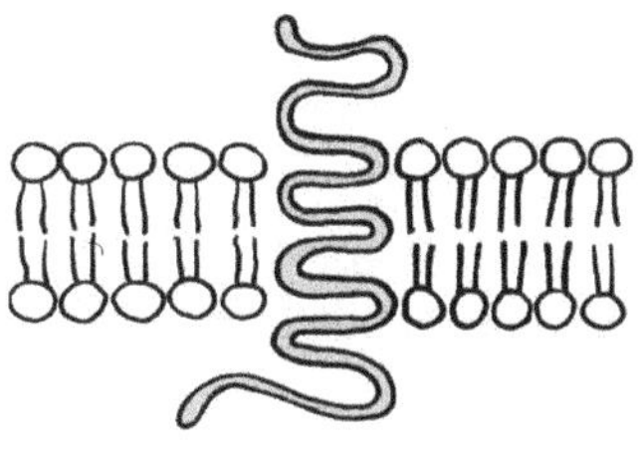

Another type of transmembrane protein acts like a **switch**. When you flip a light switch, it moves internal parts that are hidden in the wall, and the result is that electricity flows into a light bulb. A cell has biological switches with an external part that can be triggered by something in the cell's environment, and an internal part that reacts to the stimulus and makes a change inside the cell.

Proteins attached to the outer surface of the cell are properly called ***peripheral membrane proteins.*** (Peripheral means "on the outside.") Here are some examples of jobs they might do:

1) Act as a **flag**, identifying the cell as belonging to the organism it is part of, so that it doesn't get attacked by immune system cells who are out looking for foreign invaders. Cells don't have eyes and can't see; they rely on their sense of touch to identify things around them. Immune system cells feel the surface of any cell that crosses their path. If they feel this ID flag, they know not to attack the cell. If this flag isn't there, the battle is on!

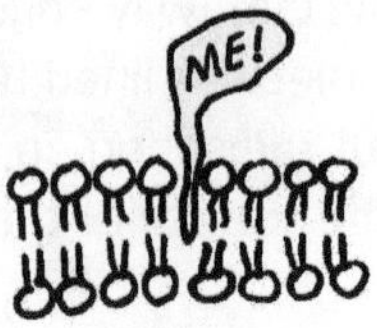

2) Act as a **mailbox**, receiving messages from other cells. Cells are usually part of a larger organism and they must all work together to keep themselves alive. They communicate by sending chemical messages to each other. These messenger molecules have a particular shape that will only fit into a receptor that has a complimentary shape, like a key into a lock.

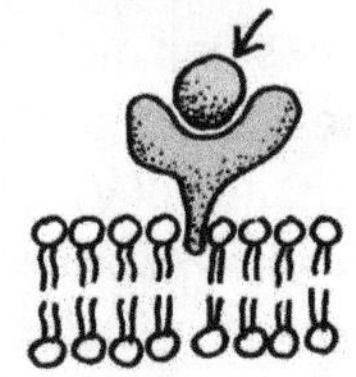

3) Act as an **anchoring hook**, allowing the cell to stick to other cells. These proteins can grab and hold on to the corresponding anchor proteins of other cells. Some anchoring mechanisms are designed to hold the cells tightly together. Other mechanisms allow for a looser, more flexible attachment. The anchoring hooks between skin cells (shown here) are called ***desmosomes***. Desmosomes allow your skin to be very strong yet very flexible.

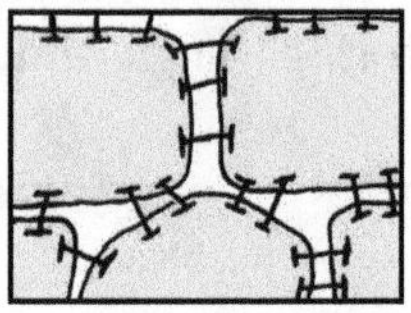

The proteins on the inner side of the membrane most often function as a place to attach things to, sort of like a hook or clip stuck into a wall. The most common cell part that needs to be anchored to the membrane is the cell's skeleton.

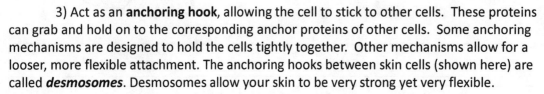

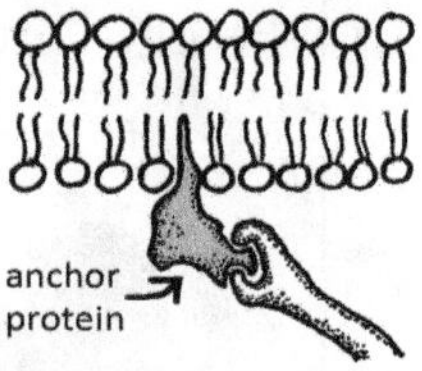

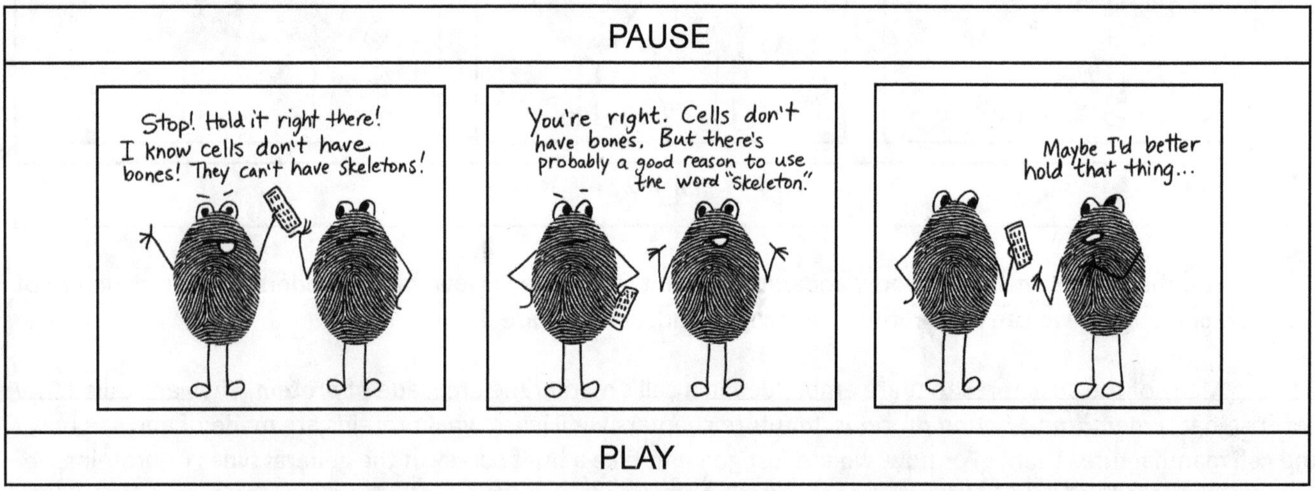

We'll learn about the cell's skeleton in the next chapter. No, cells don't have bones; but they do have rafts...

Some cell processes require more than one of the protein structures that are embedded in the membrane. Think of how many tools a carpenter needs to build something. All the tools and materials must be right there within reach so that he can do his job. Imagine if the carpenter's tools kept wandering off all the time because the floor was in constant motion. He turns around to grab the saw and it's not there. He must go and search for it, and by the time he gets back his lumber is missing. And when he is ready to nail something, his hammer is gone. He spends all his time trying to keep his tools in one place and never actually gets the job done. Fortunately, this would never happen because we live in a world where gravity and friction keep objects (on flat surfaces) firmly in place.

The surface of a membrane, however, is quite unlike the firm floors on which we place our furniture and our tools. The phospholipids are not tied together and are probably in constant motion. How much they move is still being researched, but most scientists think that the membrane forms what they call a ***fluid mosaic***. The word "mosaic" is used by artists to describe a picture or pattern that is formed by many small objects such as colored pebbles or pieces of colored glass. The word "fluid" means "in motion." (Imagine a small pond completely covered by floating ping-pong balls. A ball would not be able to travel across the pond but would still be able to shift its position quite a bit.)

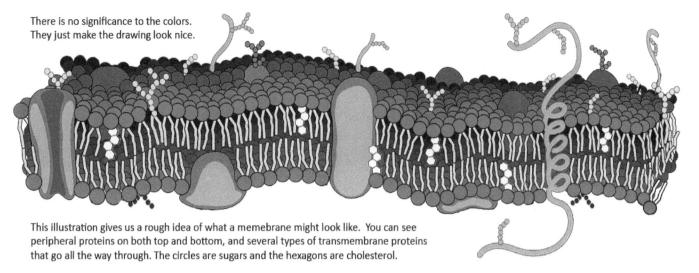

There is no significance to the colors. They just make the drawing look nice.

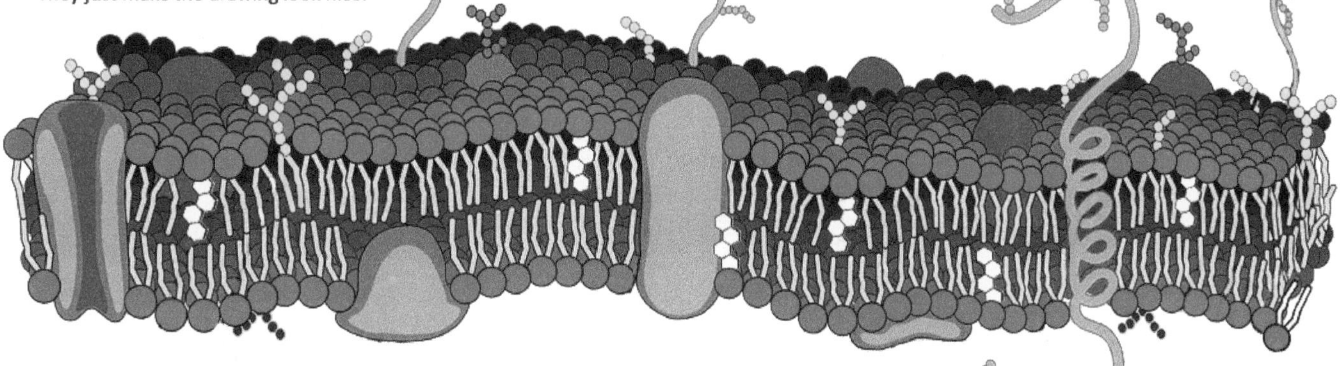

This illustration gives us a rough idea of what a memebrane might look like. You can see peripheral proteins on both top and bottom, and several types of transmembrane proteins that go all the way through. The circles are sugars and the hexagons are cholesterol.

If the phospholipids in the membrane can shift their position and move about, this creates a problem for structures that must work together (like the carpenter's tools). How will they stay together so they can work together? One answer is to secure all of them into a structure called a ***lipid raft***. The raft is made of a certain type of phospholipid that is very good at sticking to another molecule that is present in all membranes: ***cholesterol.*** You may have heard the word "cholesterol" used during a discussion about foods that are bad for you. Cholesterol is a lipid molecule that your body makes, but you can also consume it in your diet. It is most often found in food that contain animal fats. Eating too much cholesterol can sometimes be a problem, but the molecule itself is not "bad." Cholesterol helps to hold fatty acid tails together. Lipid rafts are areas that contain many cholesterol molecules. The protein structures embedded in these rafts stay in place. The rafts themselves seem to be able to move about, but the movement of the entire raft does not interfere with the ability of the protein structures to do their job. Protein tools that need to be together stay next to each other.

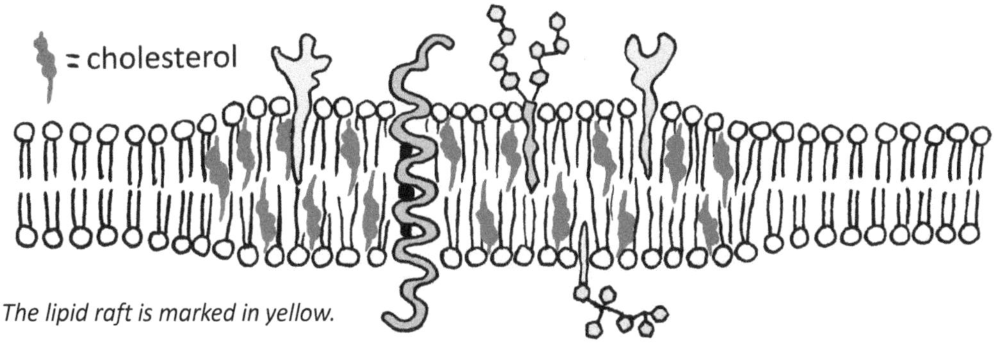

The lipid raft is marked in yellow.

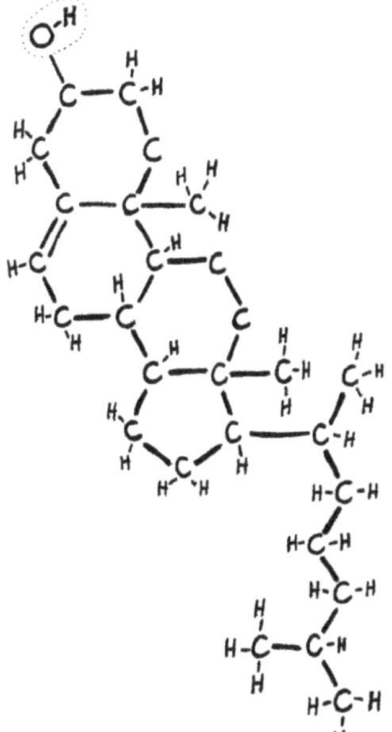

The cholesterol molecule is a member of a group of molecules that are based on hexagonal (6-sided) rings of carbon atoms. Other members of this group include vitamin D, and hormones such as testosterone, estradiol, progesterone, and cortisol. (You don't need to remember these names.)

You can see that the cholesterol molecule has three hexagonal rings of carbon, one pentagon, and a "tail" of carbons that reminds us of a fatty acid. The entire molecule, except for the O-H part at the top, wants to be tucked into the fatty acid tails of the phospholipids. The O-H is hydrophilic like the heads, so it stays as closes to the heads as it can. Since this molecule is so complicated, you'll see a variety of short-cuts, some of them with neat hexagons, and others very blobby.

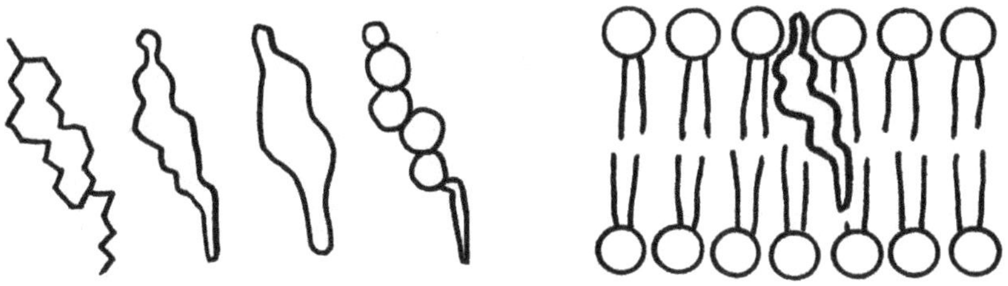

Before we end this chapter, there's one last feature of the membrane that must be mentioned. If you looked carefully at the illustrations on the previous page, you may have noticed strings of little circles or hexagons. They represent sugar molecules. As funny as it might sound, cells are "sugar coated."

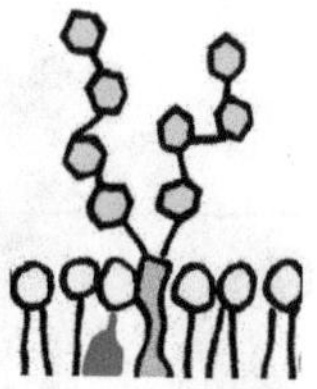

The smallest and simplest sugar molecules are hexagonal in shape and made of carbon, oxygen and hydrogen atoms. (The last one is a special sugar that includes a nitrogen atom.)

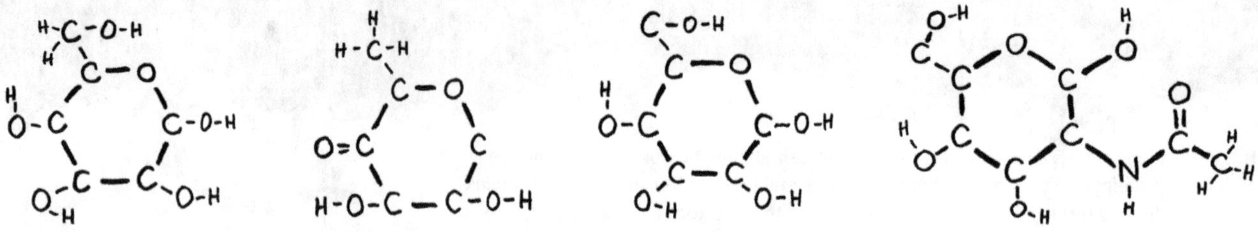

GLUCOSE FRUCTOSE MANNOSE "GlcNAc

(*You might be wondering where "table sugar" fits into this scheme. Table sugar, or "sucrose", is made of a glucose molecule attached to a fructose molecule. Your digestive system breaks them apart and puts them into your blood. Your cells take in the glucose molecules and then break them apart to harvest energy.)*

Sugars can be used as a source of energy, but your cells can also use sugars for many other purposes. Just like wood can be either burned for energy or used to make furniture, sugars can also be either "burned" for energy or used to build things. The sugars that the cell uses to build things include these simple sugars shown above, but also include more complicated variations. It isn't necessary for us to go into the chemistry of these more complicated sugars to appreciate what they do. Here are some examples of ways that various sugar-based molecules are used by cells. (Let's called them by their official name: ***glycans.*** The root "glyc-" means "sugar.")

1) Glycans can function as "**mailing labels,**" helping manufactured parts to get to their destination inside the cell. (We will mention this again in the chapter on Golgi bodies.)

2) Your red blood cells have short glycans sticking out of their membranes, and, depending on what the string looks like, your blood will be "typed" as **A, B, AB, or O**. (There are other blood types, as well, though we don't hear much about them because a mismatch is not life-threatening like the ABO types.)

3) A sugar called "O-GlcNAc" is a key player in **cell division**. Scientists used to think that cell division (the cell cycle) was controlled by proteins called cyclins. Then new research revealed that these proteins were being controlled by glycans, especially O-GlcNAc. Too much of this sugar was as bad as too little—either way, things went terribly wrong when the cell duplicated itself. A new cell might end up with two nuclei, or with a wrinkled, weird-looking nucleus where all the DNA was crumpled on one side.

4) A dense coat of glycans can act as a **protective coating**. Bacterial cells are the most striking example of this "sugar-coating," but all cells make at least some protective sugars. One way this protection can happen is by "hiding" the cell's peripheral proteins from the natural, protein-dissolving chemicals that roam throughout the body looking for bits of protein "garbage" that need to be recycled.

5) Certain glycans on key molecules help to control the **growth of embryos**.

6) Glycans can act as **clips** that hold molecules in a storage area, so the molecules will be ready to go if the need arises.

7) The cells of our immune system use a "sugar code," using glycans to **communicate** with each other.

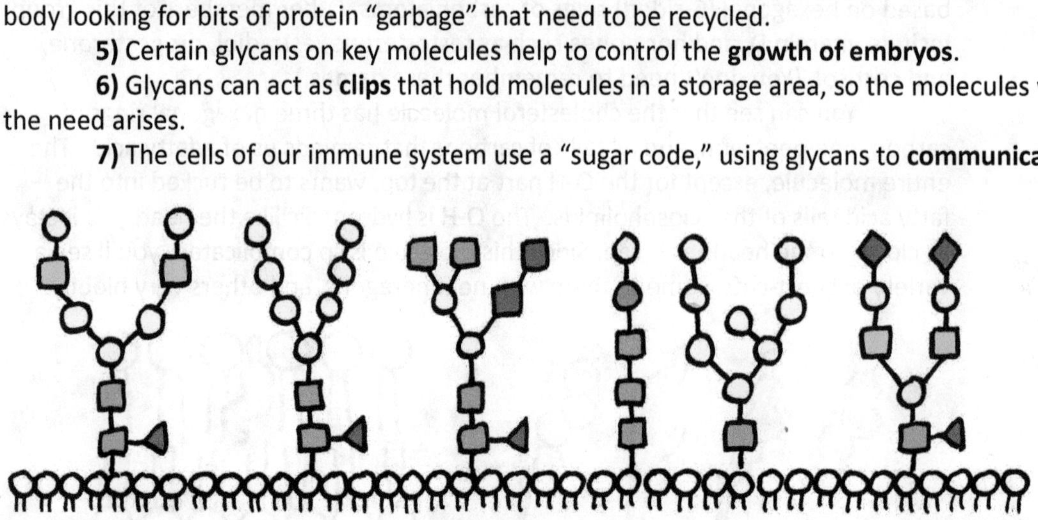

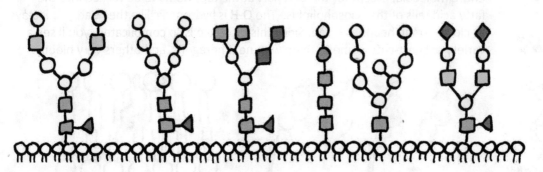

Diagrams always give you a key to let you know which shapes represent which types of sugars. For example, here we have squares for GlcNAc, circles for mannose, diamonds for glucose.

Don't forget to check the "Cells" YouTube playlist for animations of what you just read!

<u>Comprehension self-check</u> If you can't think of the answer, go back and read that part of the chapter again until you find the answer. If you need to check your answers, check the answer key.

1) The word "lipid" means: a) protein b) fat c) sugar d) membrane

2) The most natural shape for a group of phospholipid molecules to make is:
a) flat surface b) ball c) long line d) cytoskeleton

3) How many bonds (to other atoms) can an oxygen atom make? 2 (Look at the water molecule, for example.)

4) How many bonds can a phosphorus atom make? 5 (You can count the lines coming out of it.)

5) What does glycerol do in the phospholipid molecule?
a) keep the phosphate and the lipid together b) push the phosphate toward water
c) push the lipid away from water d) allow the cell to stick to other cells

6) Which hates water? a) the phosphate head b) the lipid tail

7) Where would you find a membrane-bound protein?
a) all the way through the membrane b) stuck to the inside of the membrane
c) stuck to the outside of the membrane d) all of the above

8) Which of these elements would you NOT find in a phospholipid molecule?
a) carbon b) potassium c) nitrogen d) hydrogen e) oxygen f) phosphorus

9) Which one of these would you NOT find as part of a phospholipid molecule?
a) glycerol b) fatty acids c) water d) phosphate e) serine

10) TRUE or FALSE? Micelles are made and used in your body to transport fats through the blood.

11) TRUE or FALSE? A micelle is a made of a bi-layer of phospholipids.

12) TRUE or FALSE? The only place you find phospholipid bi-layers is the outer layer of a cell.

13) TRUE or FALSE? Tiny molecules, such as oxygen, O_2, might be able to go through a phospholipid bi-layer, but larger molecules cannot.

14) TRUE or FALSE? The correct name for the outer membrane of a cell is the "plasma membrane."

15) Which one of these is *least likely* to be an example of a transmembrane protein function?
a) tunnel b) portal c) anchor d) pump

16) What happens to cells that do not have an ID flag on their surface?
a) Nothing. b) They take one from another cell. c) They are killed by immune system cells.

17) Which one of these is NOT a function that a protein on the outside of the membrane might perform?
a) act as an ID flag b) act as a mailbox to receive messages
c) act as an anchor for the cytoskeleton fibers d) act as an anchor for "cables" that attach to other cells

18) The word "mosaic" means:
a) a moving picture b) pattern made with tiny pieces c) surface layer

19) This substance is found through the membrane but is particularly concentrated in lipid rafts:
a) cholesterol b) phosphate c) glycerol d) water e) phospholipids

20) Which one of these is something that sugars do NOT do?
a) act as protective coating around the cell b) allow cells to communicate c) control growth
d) act as mailing labels for products made inside the cell e) act as channels that allow molecules to enter the cell

ACTIVITY 2.1 Review crossword puzzle

19 ACROSS

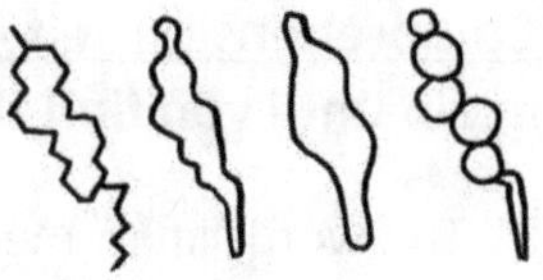

8 ACROSS

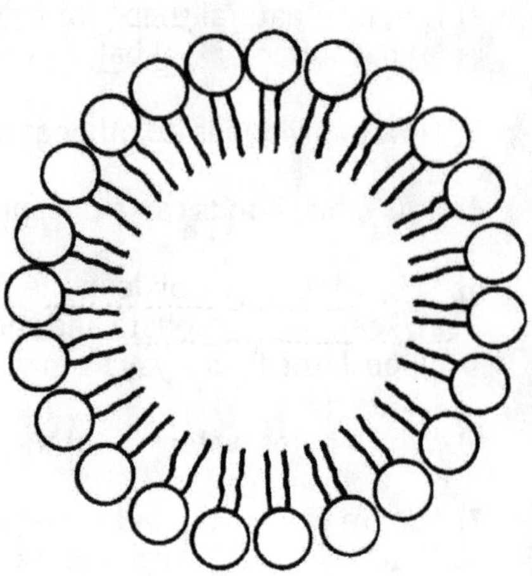

17 ACROSS

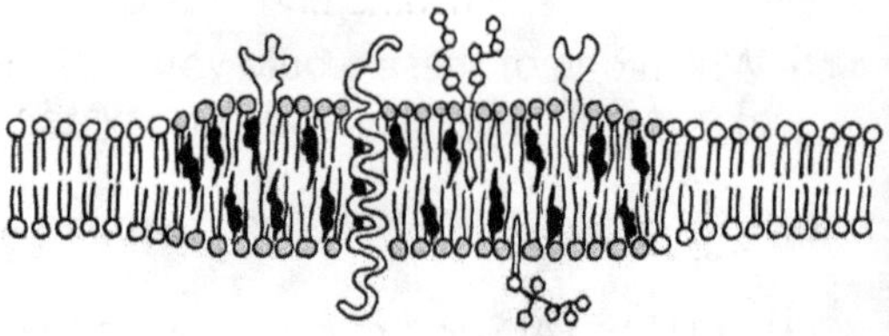

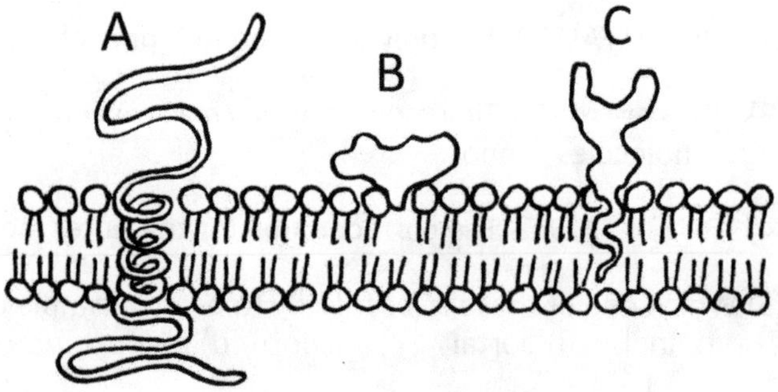

ACROSS:

2) Glucose and fructose are examples of ____ molecules.
3) The appearance and texture of cell membranes is often described as a fluid ______.
4) The person who gave us the name "cells."
5) The scientific name for fat is ____.
8) A <u>single</u> layer of phospholipids forming a ball shape.
9) Sugars most often take this 6-sided geometric shape.
11) Sugar tags known as A, B and O are found on ____ cells.
16) Proteins that go all the way through a membrane are called ________ proteins.
17) An area of membrane dense with cholesterol that keeps phospholipids and proteins together is called a ____.
19) This molecule has 3 hexagonal rings and a tail made of a string of carbons, and helps to keep phospholipids together.
20) This molecule is made of 1 phosphorus and 4 oxygens.
21) A molecule that "hates" water and won't interact with it.

DOWN:

1) This scientist worked with Schleiden to develop "cell theory."
2) Protein A, shown above, probably acts as a ____.
6) Protein B, shown above, is a ______ protein.
7) Protein C, shown above, probably acts like a ______.
9) Cholesterol is similar to both vitamin D, and a group of molecules called ______.
10) Water is a _____ molecule because electrons spend more time circulating around the oxygen atom.
12) This scientist is called the "father of microscopy."
13) Sugar codes can allow immune system cells to ____.
14) One oxygen atom and 2 hydrogens make ____.
15) This type of electron microscope gives us 3D images.
16) This type of electron microscope gives us flat images.
18) This scientist gave us the word "nucleus."

CHAPTER 3: THE CYTOSKELETON AND MOTOR PROTEINS

Inside their membranes, cells are filled with a watery fluid called ***cytosol***. The word "cyto" means "cell," and the ending "-sol" is short for "solution." The term ***cytoplasm*** is also used to describe the watery insides of a cell, but this term includes not only the fluid but also small things that are floating around in it. It can seem like these words mean the same thing because science writers use them almost interchangeably. To be fair, in many cases it doesn't really matter which word the writer chooses because either word will fit the meaning of the sentence. A helpful analogy might be to think of a can of soup. If the can is the membrane, the soup is the cytoplasm. If the soup is chicken noodle, it has noodles and bits of chicken floating in broth. The word 'cytosol" would correspond to the broth. The words "soup" and "broth" are very similar, but we can use the more specific word "broth" when we want to talk about only the liquid element of the soup, not the things floating in it. In many conversations, the word "soup" is good enough and we don't have to be any more specific than that.

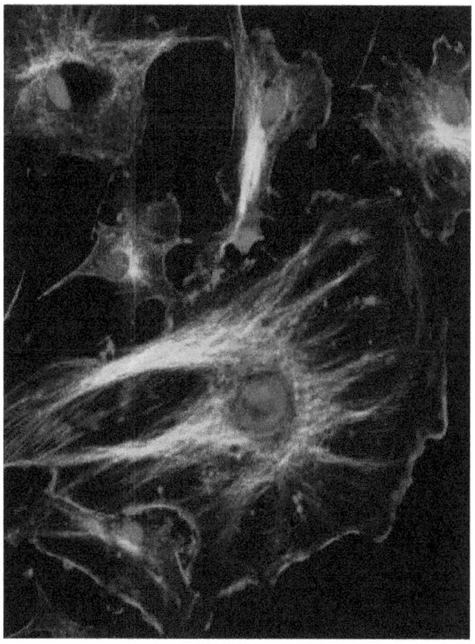

In 1903, a Russian scientist proposed that cells must have some kind of network inside that helps them keep their shape. He reasoned that if cells were nothing but a bag of liquid, they would be too easily flattened. The cells he saw in his microscope were anything but flat. This proposed network of fibers was nothing but a theory until the 1970s, when scientists discovered a way to detect fibers too thin to be seen under a microscope. Trying to see these filaments in the cytosol is like trying to see fishing line underwater. Both are transparent, and the fishing line is very thin. (That's the whole idea—the fish can't see it!)

The breakthrough came in the 1970s. Someone discovered a natural molecule, an "antibody," that would stick to these invisible fibers. (We'll read about antibodies in chapter 10.) The antibodies could be tagged with fluorescent dyes that glowed green or red. If you put these stained antibodies into a cell, they would stick to the fibers and make them show up as brightly colored lines. The images produced in these experiments were stunning. They showed organized networks of filaments, like a three-dimensional system of roads and highways. It was obvious that this network of fibers acted as a structural support, so it was named the ***cytoskeleton***.

With further research, scientists discovered that the cytoskeleton not only helps the cell to maintain its shape, but also functions as a transportation system. It really does act like a system of roads and highways. The roads come in three sizes: small, medium and large. The scientists who discovered them gave them these names: ***microfilaments, intermediate filaments and microtubules***. If we were to make a model of a cytoskeleton we might use thread, yarn, and drinking straws to represent these fibers.

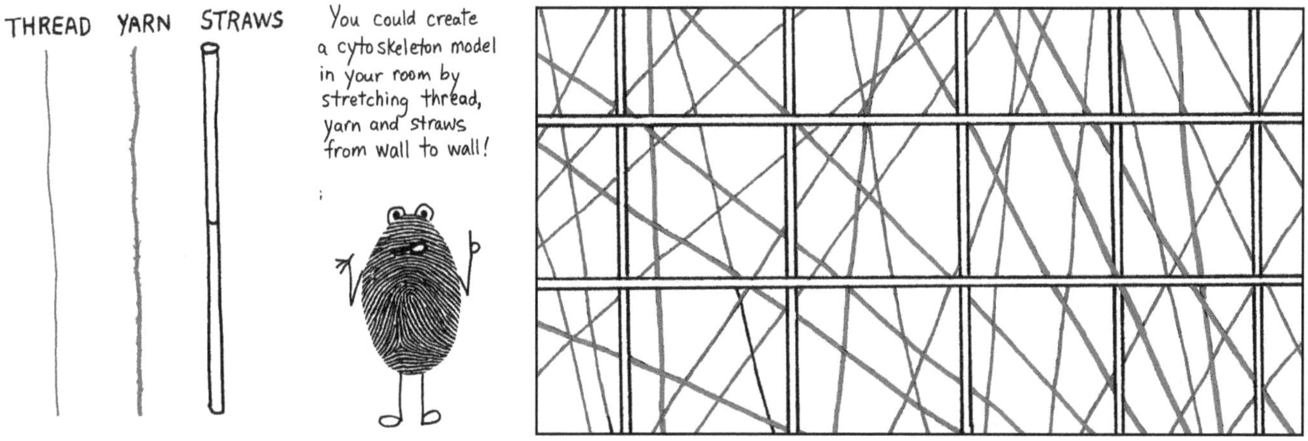

The smallest ones, the ***microfilaments*** (the thread in our model) are made of protein molecules called ***actin***, so the filaments are also known as ***actin filaments***.

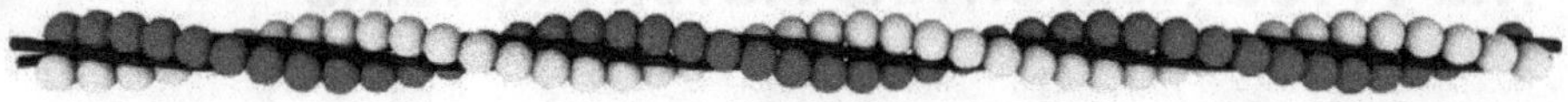

The balls represent little units of actin protein. More about protein later in the book.

The microfilaments are very important to the overall shape of the cell, and can also help the cell to change its shape. If a cell wants to move, it quickly builds a whole bunch of new microfilaments in that direction. The cell can build these at the rate of thousands per second. As the new microfilaments are built, they push the flexible membrane outward. Cytoplasm flows along with the microfilaments. Together they create what is called a ***pseudopod***, or "false foot." (Can you see the tiny microfilament lines in the pseudopods in this diagram?) We have white blood cells in our bodies that form pseudopods in order to surround and capture bacteria and viruses. Single-celled organisms like the ameba (old spelling: amoeba) also use pseudopods to move through their environment. In fact, this type of movement is often called "ameboid motion."

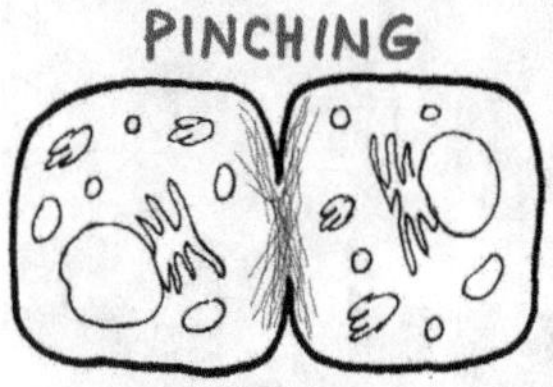

Microfilaments are also very important when it is time for the cell to reproduce by splitting itself in half. The microfilaments cause the cell to "pinch" in the middle, in preparation for the splitting process.

In some types of cells, such as muscle cells, we find actin filaments being used as a track along which another protein, ***myosin,*** can travel. The interaction between actin and myosin is what allows you to move your muscles. Most anatomy courses cover this topic quite thoroughly, so we'll just give it a brief mention in the last chapter when we look at various types of cells.

The medium-sized ***intermediate filaments*** (the yarn in our model) are especially abundant in nerve cells, skin cells, and muscle cells. They form a stretchy lattice inside the cell that help to give it strength. In skin cells, an intermediate filament called ***keratin*** forms a very strong, stretchy network that gives skin its flexibility and durability. It adheres to several types of anchor proteins in the membrane. One type allows skin cells to make strong connections to each other.

What would happen if something went wrong with the manufacturing process and the cell accidentally made these protein anchors the wrong shape? The intermediate filaments would not stay anchored. When this happens in muscle cells, it can cause a condition called muscular dystrophy. A person with this condition has very weak muscles. When intermediate filaments (keratins) in skin cells are the wrong shape, it causes life-threatening skin diseases.

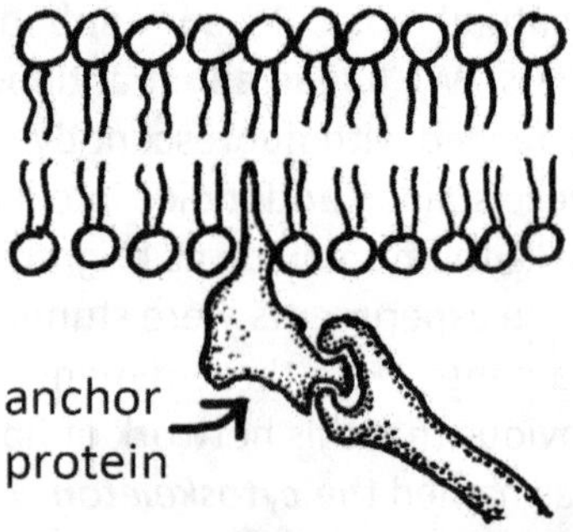

The shapes of the anchor proteins match the ends of the filaments a bit like like jigsaw puzzle pieces that fit together.

The largest filaments, the ***microtubules*** (the drinking straws in our model), really do look like tubes. These tubes are the "highways" that the cell uses to move things about. What does a cell need to move?

Some of the cell's organelles act like little factories and make proteins, fats and enzymes for their own use or to ship out to other cells. These products are then "packaged" into vesicles made of phospholipid membrane. The packages can't move on their own, so something has to take them to where they need to go. When it was discovered how this transport happens, scientists could hardly believe their eyes—they saw little proteins with "feet" that were "walking" along the microtubule roads! These ***motor proteins*** move like tightrope walkers at a circus, putting one foot in front of the other along the narrow rope. They carry their cargo as if it was a huge sack resting on their shoulders. How the motor proteins know where to go is still being studied.

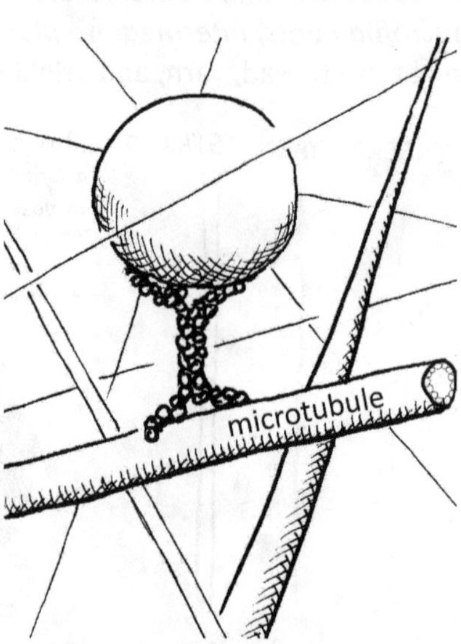

There are several types of motor proteins. The ones that are most abundant (and therefore the most studied) are called ***kinesin*** *(kin-EE-sun)* and ***dynein*** *(DIE-nin)*. As a general rule, they seem to "walk" in only one direction. Kinesin carries things away from the central nucleus and towards the outer membrane. Dynein goes the opposite way, carrying things from the outside towards the center. When they get to the end of the line and have fulfilled their mission, they often drop off the tubule and eventually float back towards their starting point. Then they receive new instructions and are off on another mission. They probably won't attach to the same tubule they used for the previous mission. Their missions can be as short as a few seconds or as long as a few minutes. If a kinesin and dynein run into each other, one of them will probably fall off, or dynein might side step long enough to let kinesin pass.

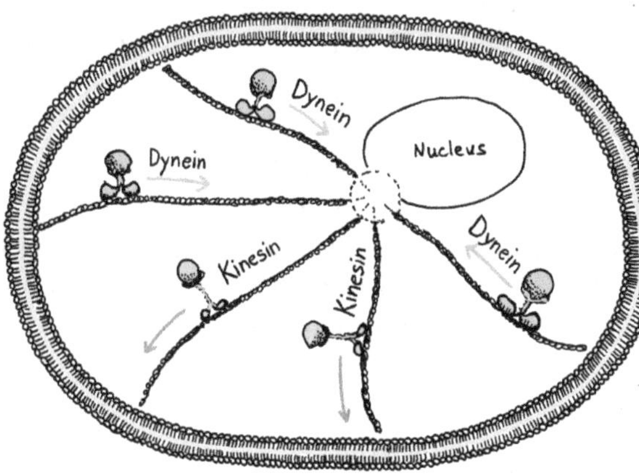

Now, here comes the tricky part. The things that look like feet on a motor protein are actually called ***heads***. And the things that look like its hands (holding the load) are actually called its ***tails***.

PAUSE

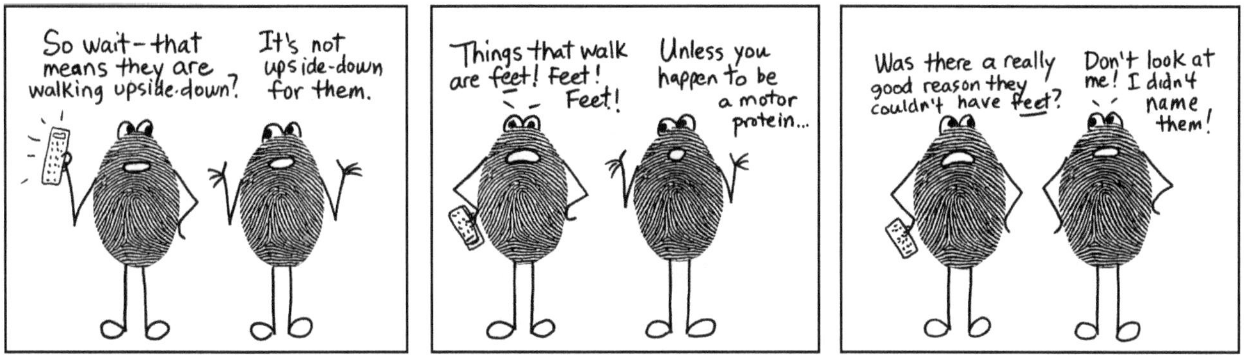

PLAY

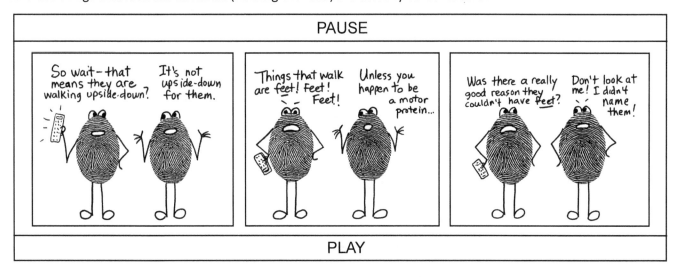

Yes, it is unfortunate. Sorry. But let's learn a little more about these heads and tails.

The shape and structure of motor protein heads is pretty standard because they all have to walk along the same tubule roads. Their tails can be quite different, though, because a particular shape is needed to connect to a particular type of cargo. The part of a motor protein that connects to the cargo is called the ***binding site***. (This term is also used for the place on each head that touches the microtubule.) So far, scientists have identified about 40 types of cargo binding sites. This picture shows four different kinesins. The one on the left is considered "standard" (the most common kinesin) and the others are variations.

Make sure you watch the recommended videos listed at the end of this chapter so you can see these guys in action!

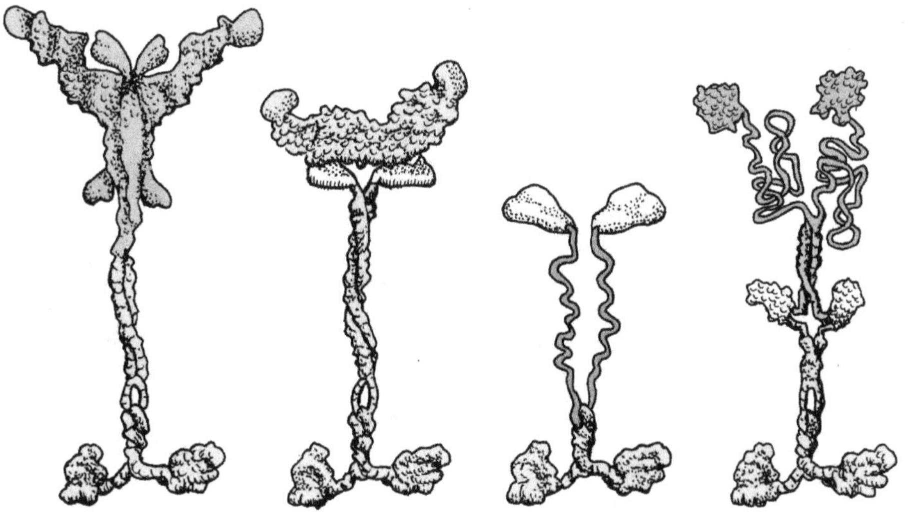

Notice that the heads (feet) all look the same. The tails (top) are specialized for different cargo. They don't have any natural color, but scientists often add color to diagrams.

With each step a motor protein takes, it will use a tiny energy molecule called ATP, which we will study in the next chapter. These steps occur very quickly, much faster than you can walk. One researcher saw a motor protein talking 100 steps per second. This would be like you running as fast as a car on a highway! They can also carry loads much larger than themselves, the equivalent of you towing a house. For exceptionally large cargo, several motor proteins can work together.

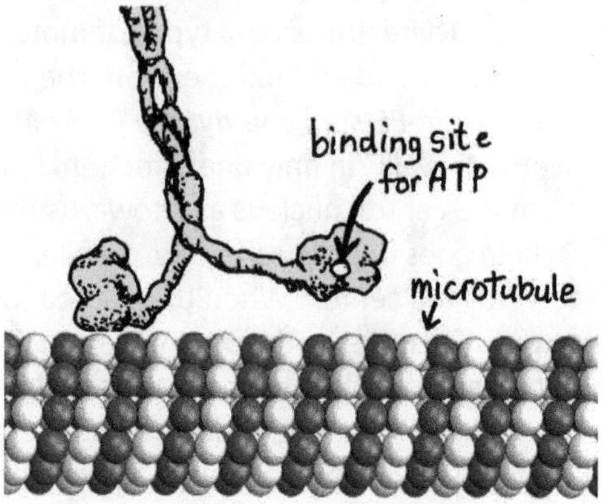

When a motor protein wears out (after only a day of action) it will be taken apart and its parts recycled, as if it was a used car. There are tiny factories inside the cell that are constantly using the atoms and molecules from the old motor proteins (and from other recycled cell parts) to make new ones.

How important are these motor proteins to the life of a cell? Well, just imagine what would happen to your town or city if many of the cars and trucks stopped working. Mail might not get delivered, food would not be shipped to grocery stores, hospital workers might be stuck at home, and construction materials might never arrive at construction sites. Everyone in the town or city would eventually be affected by the failure of the transportation system. When things go wrong with a cell's transport system, the result is often a very serious disease.

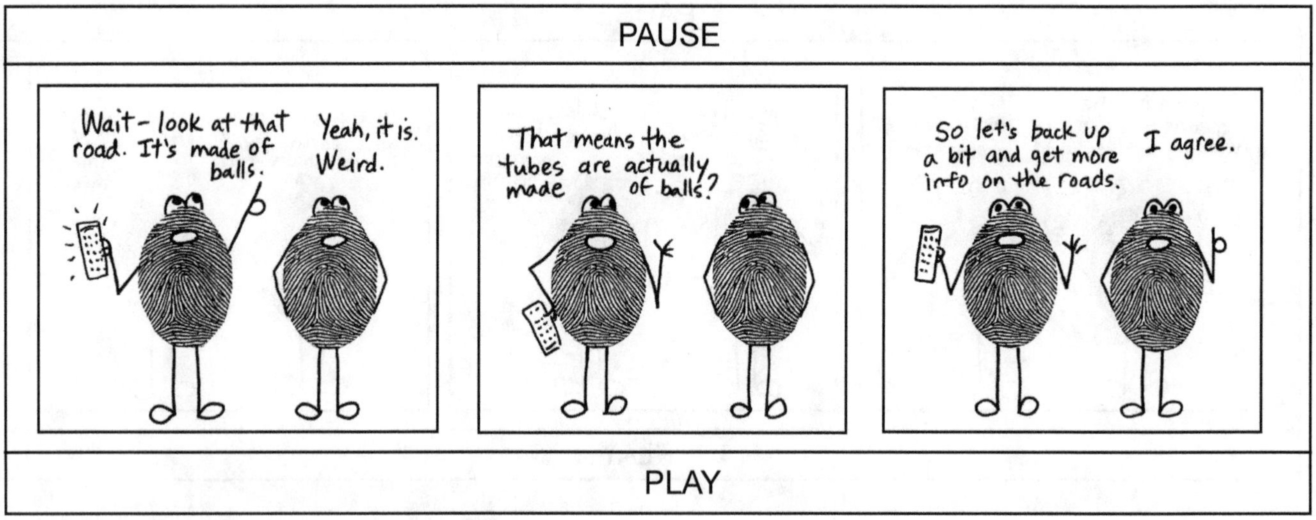

Microtubules are made of tiny individual units of protein called ***tubulin***. (As mentioned previously, we'll learn exactly what protein is in a later chapter.) There are two types of tubulin: *alpha* tubulin and *beta* tubulin. You will see the words "alpha" and "beta" used a lot in biology. They are Greek words for the letters A and B.

One alpha and one beta tubulin snap together to make a pair that stays together. There is a special word for a pair of molecules, a word that is easy to pronounce and to spell: ***dimer***. ("Di" is Greek for "two.") The cytosol of the cell is FULL of these tubulin dimers. They are simply everywhere!

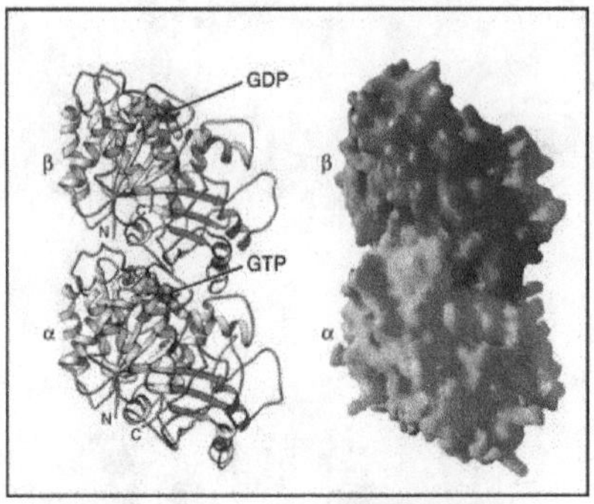

Most accurate drawing of tubulin.

Slightly simplified way to draw it.

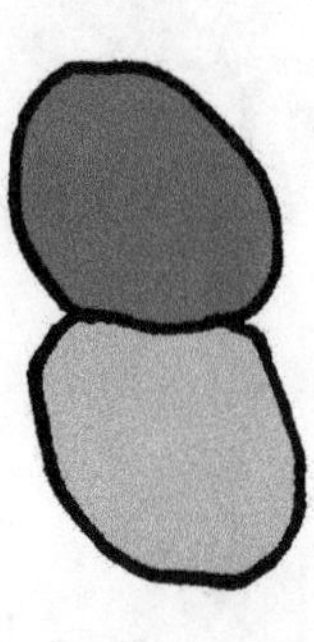

Simple enough for us to draw!

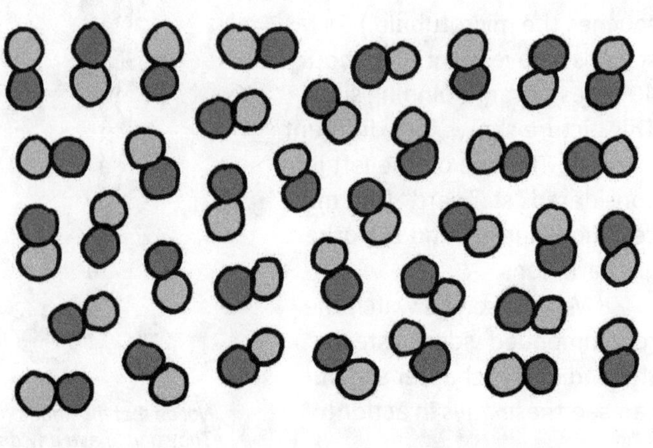

When two dimers bump into each other, they can stick together if a tiny molecule called GTP is present. GTP is similar to ATP, the energy molecule that motor proteins use. GTP fits into a little pocket in the tubulin molecule. When it sticks, it changes the shape of the dimer just slightly, making it a little more straight. Once it has been straightened, the tubulin dimer will be able to stick to other dimers. When there are many straightened dimers, they will forms long lines, then the lines will stick together to form sheets. The sheets will curl up to form tubes. (Video animations can show you this process in action. Be sure to check out the videos on the Cells playlist, or search for some on your own.)

Often there are "helper proteins" in the area, too, that will come alongside the forming tubes and help to add more dimers to the growing ("positive") end.

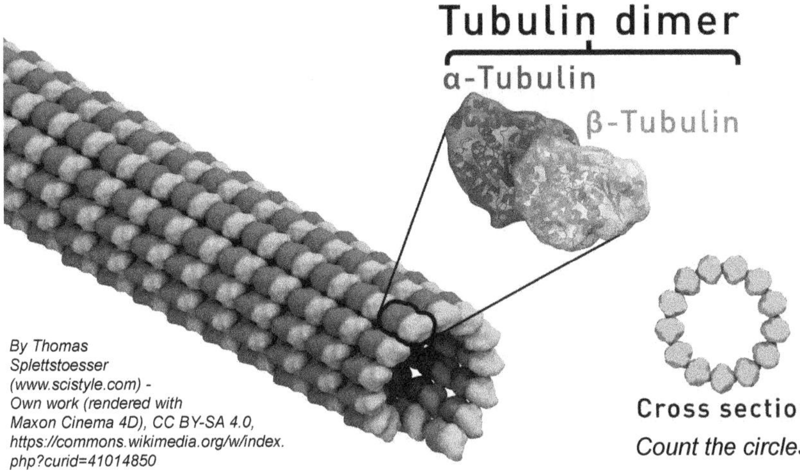

By Thomas Splettstoesser (www.scistyle.com) - Own work (rendered with Maxon Cinema 4D), CC BY-SA 4.0, https://commons.wikimedia.org/w/index.php?curid=41014850

Cross section
Count the circles.

If we look at the end of a tubule, we can count exactly how many dimers it takes to form the tube: 13. Isn't that an odd number to find in something biological? Although it is one of the Fibonacci numbers (1, 2, 3, 5, 8, 13, 21, 34, 55, etc.) it only shows up occasionally in a flowering plant.

Microtubules are not scattered randomly around the cell. Like a well-designed city, the cell's roadways are very organized. There is a central "hub" for the microtubules, like a railway station from which all the train tracks branch out. This central station is called the ***centrosome***. ("Som" or "soma" is Greek for "body," but it is often used in the same way that we use the word "thing.") The centrosome is also called the ***microtubule organizing center*** of the cell. It is made of two ***centrioles*** surrounded by a blob of protein gel. The little places where the microtubules are attached are cone-shaped structures (made of another type of tubulin) that acts like a foundation platform on which a microtubule can start to grow. The attached ends of a tubule are called the "negative" ends. Dynein walks towards them.

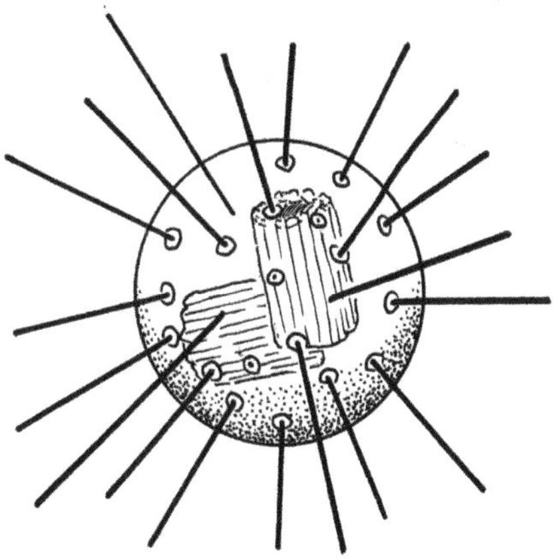

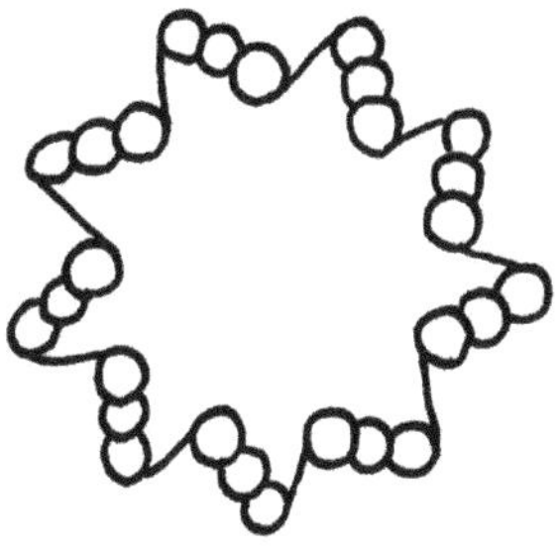

Cross section showing the "end view" of the centriole

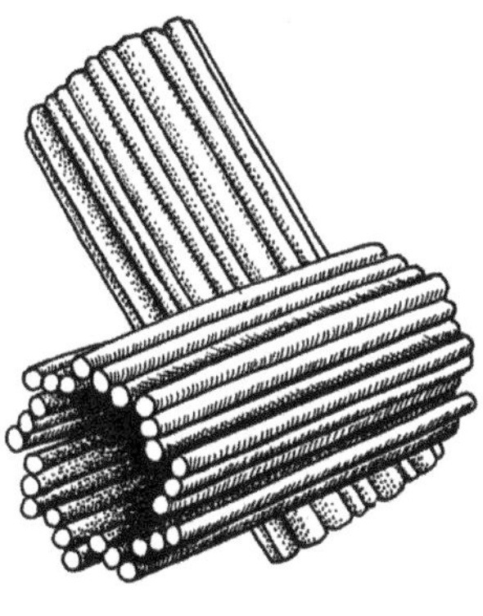

If we take a close-up look at those barrel-shaped centrioles inside the centrosome, we can see that they are made of microtubules arranged in a very precise pattern. Three microtubules get bundled together in a straight line. Then nine of these flat bundles are arranged into a circle. Some thin protein fibers hold them in place.

Notice that the centrioles are perpendicular to each other, making an "L" shape. They always stay in this position.

Don't pause the book—we are almost done learning about the cytoskeleton! Before we end the chapter, however, we need to mention two other important jobs that microtubules do, besides their role as a road system: they help a cell to make a copy of itself, and can form structures that allow the cell to move.

All living organisms must be able to grow, and growth occurs through the process of cell division. Cells are able to split in half, forming two identical copies of themselves. During this duplication process, the cell must make a second copy of its DNA. (We'll learn about DNA in a future chapter, but you probably already know that DNA is like a library, containing all the information the cell will ever need.) The two sets of DNA must be pulled apart and taken to opposite sides of the cell so that when the split occurs, each side will have a full set of DNA.

First, the cell duplicates its centrosome so now it has two of them. The centrosomes go to opposite sides of the cell, as shown in diagram (1). Meanwhile, the DNA duplicates itself inside the nucleus. During the duplication process, the DNA coils up into long sticks called chromosomes. The covering around the nucleus dissolves so that the chromosomes are sitting out in the open in the middle of the cell. The rest of the diagrams have all the other cell parts removed so that we can focus only on what happens to the chromosomes.

The two sets of chromosomes must be separated and pulled to opposite sides of the cell. In diagram (2), we see the centrosomes form a shape called a ***spindle,*** which is made from microtubules and looks a bit like an American football. In the middle of the spindle, the microtubules attach to the chromosomes. The microtubules then begin to pull the chromosomes apart, as shown in diagram (3). The pulling is caused by the microtubules beginning to disassemble themselves so they get shorter and shorter. Finally, the two sets of chromosomes are completely separated and arrive at opposite ends of the cell, as shown in diagram (4). The cell is now ready to split in the middle and form two independent cells. Microfilaments will help do the pinching in the middle.

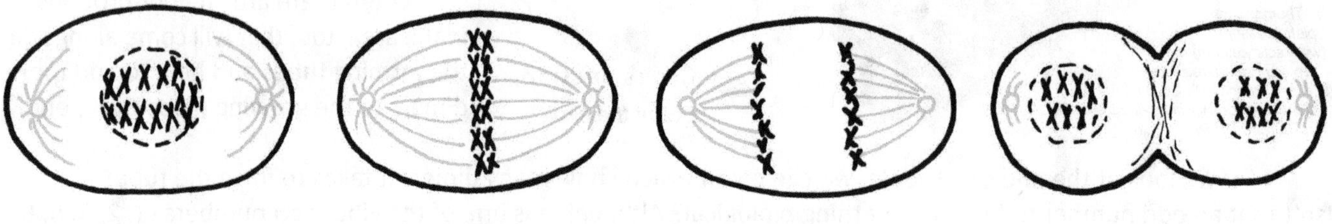

(1) (2) (3) (4)

We will study this more in a later chapter. The main point we need to learn now is that microtubules play an important role in cell division by forming a spindle that pulls chromosomes apart.

PAUSE

PLAY

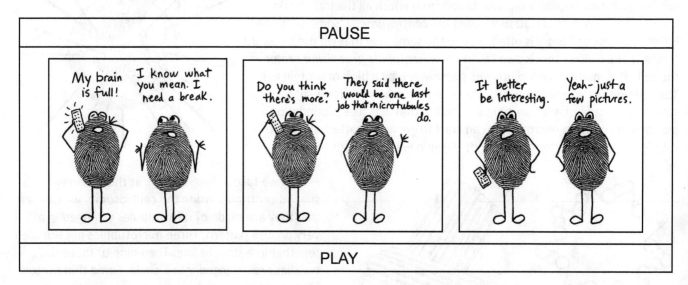

Microtubules are used in some specialized cells to make things that look like hairs or tails. The hairs are called ***cilia*** and the tails are called ***flagella***. These structures allow the cell to move by acting like paddles (cilia) or propellers (flagella). Cilia are used by single-celled animals like the paramecium, but are also found in our own bodies in the cells that line our trachea (the tube that goes down into our lungs). Flagella are used by many single-celled animals, but are also found in sperm cells made by most living organisms.

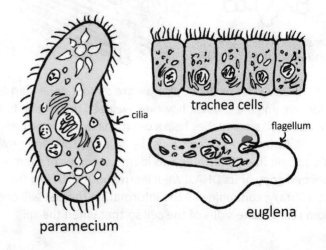

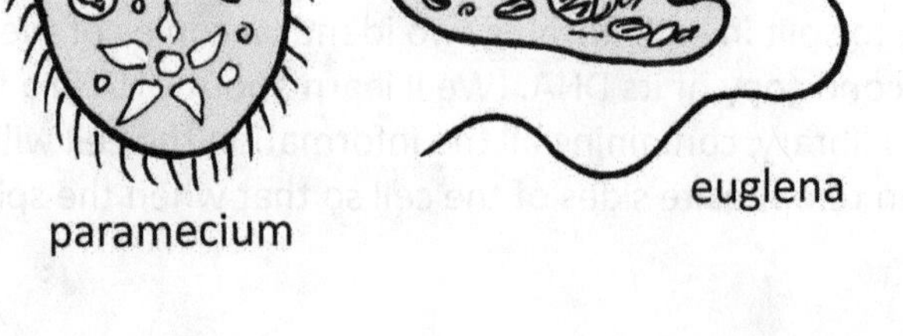

Can you remember what you read? If you can't think of the answer, go back and read that part of the chapter again until you find the answer.

1) What is the fluid inside a cell called? cytosol or cytoplasm

2) Which one is made of a protein called actin? a) microtubules b) microfilaments c) intermediate filaments

3) Which one of these (as far as we know) is NOT something a microfilament might do?
a) cause a muscle fiber to move b) act like a road for motor proteins
c) help a cell to divide in half d) allow the cell to move using pseudopods

4) TRUE or FALSE? The only cells that move using "ameboid" motion are single-celled organisms like the ameba.

5) Which one of these is attached to the membrane using anchor proteins in the membrane?
a) microfilaments b) intermediate filaments c) microtubules d) all of these e) none of these

6) Which of these motor proteins walks outward, towards the edge of the cell? a) kinesin b) dynein

7) Which part of a motor protein touches the microtubule highways? a) tails b) feet c) heads

8) How long (as far as we know) is the lifetime of a motor protein? a) days b) weeks c) months d) years

9) How fast (as far as we know) can a motor protein travel?
a) one step per second b) one step per minute c) 100 steps per second d) 100 steps per minute

10) What tiny energy molecule does a motor protein need in order to move? (three letters) A T P

11) Which one of these does the cytoskeleton NOT do?
a) form new phospholipid membranes b) help the cell maintain its shape
c) transport things across the cytoplasm d) form pseudopods

12) The protein unit that microtubules are made of is called: tubulin.

13) A pair of molecules that is bound together is called a dimer.

14) How many protein units are need to form a microtubule circle? a) 8 b) 9 c) 11 d) 12 e) 13

15) Which of these word roots means "body"? a) cyto b) soma c) pseudo d) pod

16) A centrosome is made of two Centrioles in a blob of protein gel.

17) What does the centrosome do? a) acts as a gathering point for all the proteins floating around the cell
b) builds microtubules c) organizes microtubules d) All of these e) b and c

18) What does the spindle do? a) pulls chromosomes apart b) makes centrosomes
c) gives shape to the cell d) duplicates chromosomes

19) Which type of fibers help the cell to pinch in the middle and form two new cells?
a) microfilaments b) intermediate filaments c) microtubules

20) Microtubules are used by specialized cells to make structures (cilia and flagella) that allow them to:
a) divide in half b) be identified by other cells c) use oxygen d) move

<u>ACTIVITY 3.1</u> "Must-watch" videos

NOTE: The original posters of the videos can decide to take down their videos. In some cases, someone else will put the video back up again, but the address will have changed. Check the Cells playlist for these videos, but if they are not there, try searching for them using key words.

1) "A Day in the Life of a Motor Protein"

This video has over a million views on YouTube. (If you can't access YouTube, check for it on other videos streaming services.) A biology lab at the University of Utrecht, in the Netherlands, put together an informative and marvelously funny animated film about kinesin and dynein. The cell in which these particular motor proteins live is a nerve cell, so the film starts out with a little information about this type of cell.

2) "The workhorse of the cell: Kinesin"

This video also has a lot of views, so it should pop right up in a search. The animation is fabulous and will bring to life everything you read about kinesins in this chapter.

3) "White blood cell chases a bacteria"

Another video with millions of views and posted from multiple sources. White blood cells use their cytoskeleton to change their shape very quickly. In this video clip you can see a white blood cell chasing some bacteria. It does catch one at the end.

WHILE YOU ARE AT THE PLAYLIST, CONSIDER WATCHING THE OTHER VIDEOS ABOUT THE CYTOSKELETON.

<u>ACTIVITY 3.2</u> Kinesin vs. Dynein strategy game

This is a two player game. It takes only a few minutes to play. The point of playing the game is to reinforce the fact that motor proteins go in only one direction. **Kinesin *(kin-EE-sin)* goes away from the nucleus, and dynein *(die-nin)* goes towards it.** The board represents a cell with a very simple microtubule arrangement. (Technically, we maybe should have added the centrosome because it is the microtubule organizing center, but the centrosome is very close to the nucleus, so for the sake of simplicity it is easier to show just the nucleus.)

You will need eight coins. Mark four of them with the letter K, and the other four with the letter D. You can use a permanent marker, or you could write the letters on paper circles and tape them on. If you have "sticky notes," you could cut pieces from the sticky strip of the note. (Using coins will give the tokens some weight so that if someone coughs or sneezes, the tokens won't blow off the board.)

Put the K tokens on the circles that are close to the nucleus. The goal of the K player will be to get the tokens out to the circles near the plasma membrane.

Put the D tokens on the circles that are close to the outer membrane. The goal of the D player will be to get all the token into the circles close to the nucleus.

Players will take turns making a move. The K pieces can ONLY go either <u>away</u> from the nucleus, or side to side, around the circle that they are currently on. The D pieces can ONLY go either <u>towards</u> the nucleus or side to side, around the circle that they are currently on. Players MUST make a move on each turn. Remember, the tokens can't go "backwards." Once they have advanced to another ring, they can't go back. They can go around that ring, but they can't go back to the previous one they were on.

The first player to get all their tokens to their destination wins the round.

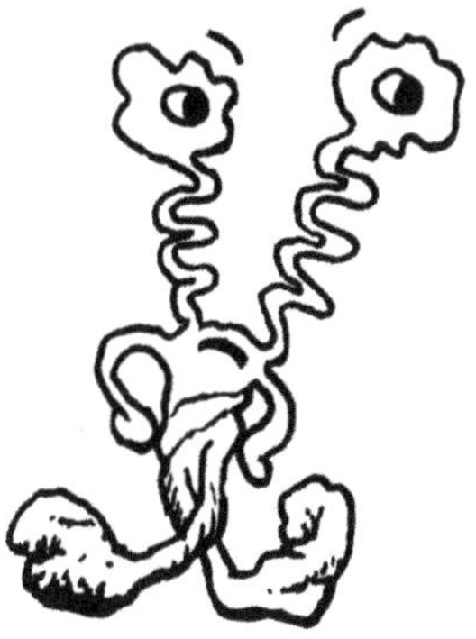

KINESIN VERSUS DYNEIN

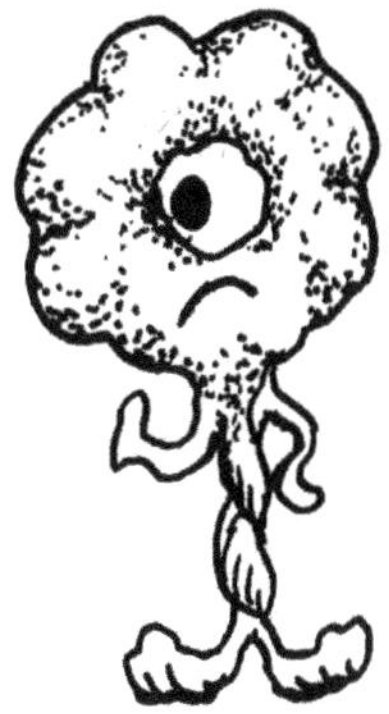

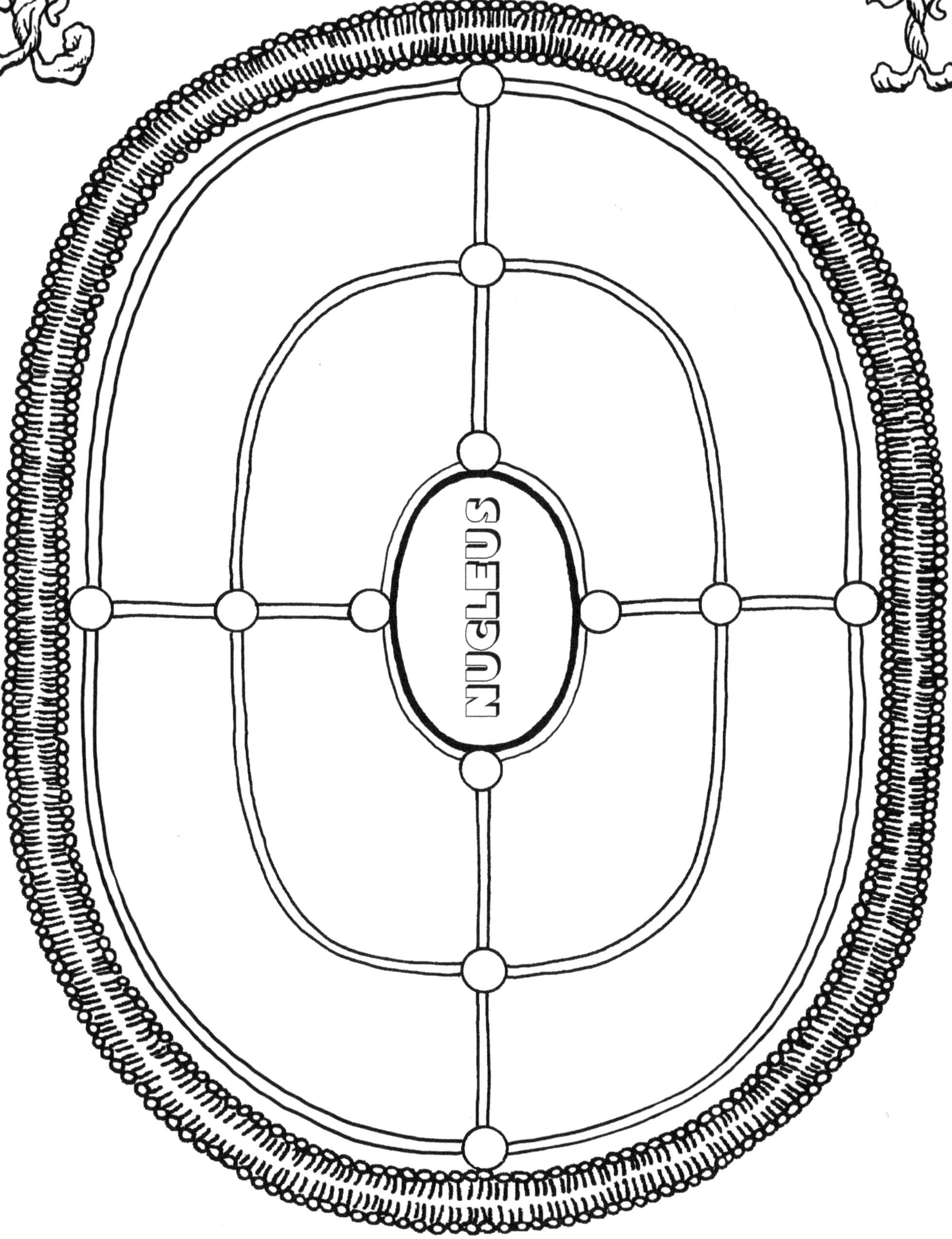

CHAPTER 4: ATP AND THE MITOCHONDRIA

Cells use a lot of energy. It takes energy for those motor proteins to walk along the cytoskeleton highway. It takes energy to build molecules. It takes energy to create pseudopods. Everything the cell does requires energy. Where do your cells get energy? The process begins with the food you eat. You know that your food is digested in your stomach and intestines, so that its molecules can be absorbed into your blood and distributed to your cells. But exactly how do your cells process the food molecules and harvest energy from them?

Surprisingly, human bodies are not that different from cars. Engines that run on gasoline use combustion to break apart the long carbon chains found in petroleum molecules. A molecule of gasoline looks something like this:

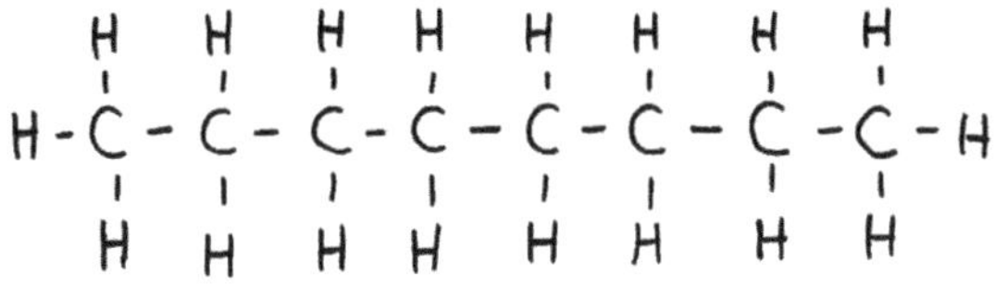

or this:

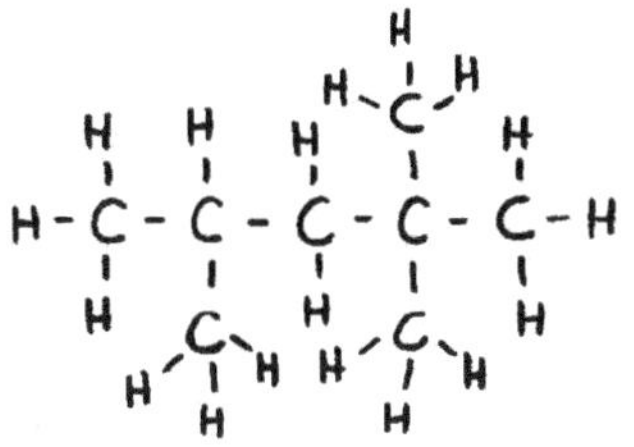

Both of these molecules have eight carbons with hydrogens attached to them. Compare these gasoline molecules with two types of molecules your body uses for energy: fat and sugar:

A LIPID MOLECULE

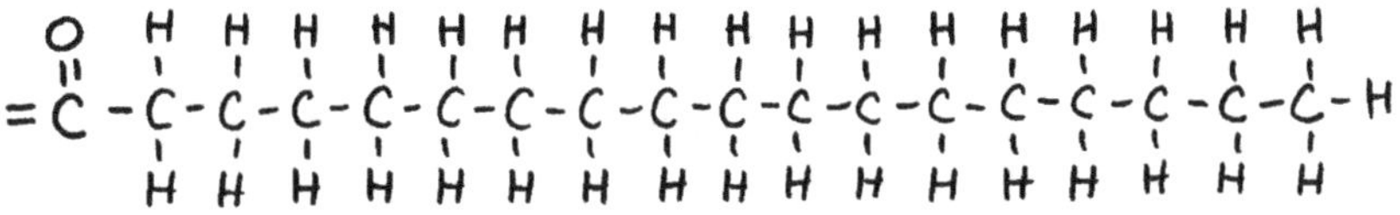

This is actually the lipid tail from our phospholipid molecule.

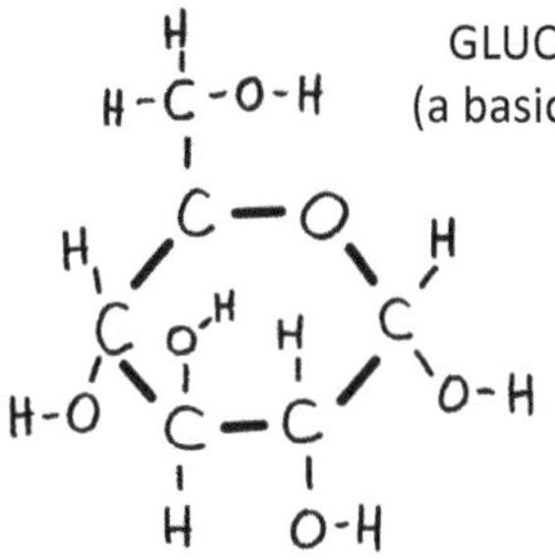

GLUCOSE
(a basic sugar)

Can you see how this explains why plants we eat can also be used to make "biofuel" for cars?

The similarity is striking, isn't it? All of these molecules are basically strings or clumps of carbon atoms with hydrogens attached to them. The edible molecules have a few oxygens thrown into the mix, too.

Basically, our food is not that different from gasoline. The way an engine gets energy out of a fuel molecule is to apply a spark to it, causing it to explode. The explosion happens inside a metal cylinder that contains a movable piston. The motion of the piston is then transferred, by various shafts and gears, to the wheels. The chemical waste products created by this explosion are carbon dioxide and water (plus some miscellaneous carbon chain molecules that somehow escaped being torn apart). Both carbon dioxide and water come out the exhaust pipe. Do you want to guess what waste products your body creates as it "burns" your food? Yep, carbon dioxide and water!

Now, you'd be in big trouble if your body used sparks and rapid combustion to get the energy out of your food. You wouldn't survive past your first meal. Instead, your body uses a very gradual method, a process that involves numerous small steps, so that there is never a harmful release of too much energy all at once. Your body breaks down the large energy molecules into smaller and smaller molecules, and finally into molecules so small that they are safe for a cell to use. The tiny energy unit that a cell uses is called ***ATP***.

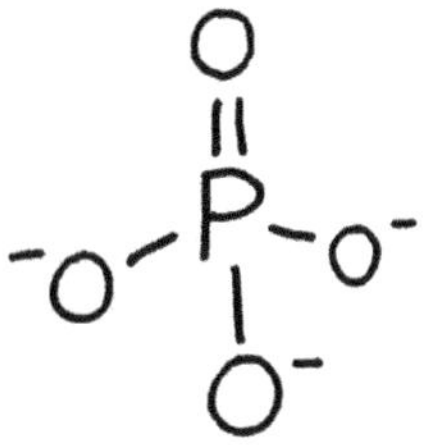

ATP is short for "**a**denosine **t**ri-**p**hosphate." *(a-DEN-o-zine)* The word "tri" tells you that there are three of something. Does the word "phosphate" look familiar? (If it doesn't, you might want to go back and read chapter 2 again.) A phosphate is a phosphorus atom with some oxygens attached to it. "Tri-phosphate" means that the ATP molecule has three of these phosphates. What about the adenosine part? Well... we'll show you what it looks like, but some of you may want to close your eyes because this looks a bit complicated.

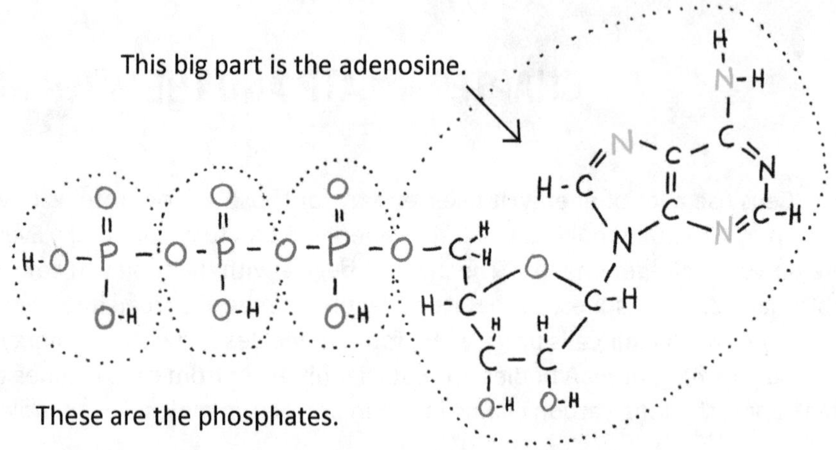

This molecule is way too complicated for scientists to draw every time they want to show ATP. So instead, they just draw it as circles, with one big circle representing the adenosine and three little circles for the phosphates. Often, they don't even put the letters in. Scientists just recognize this shape as ATP.

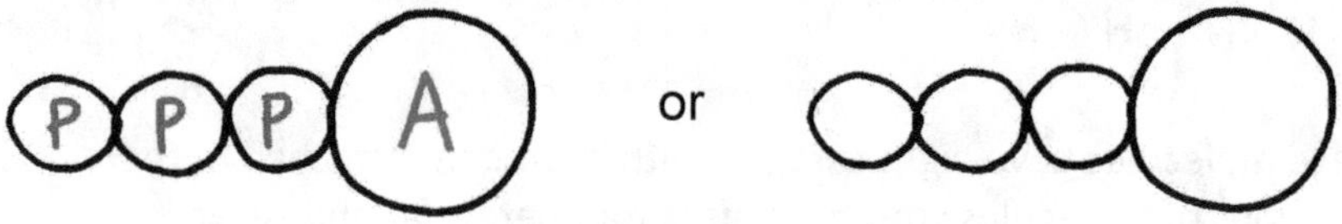

The way ATP releases energy is by popping off the phosphate on the end. There is energy stored in the bond that keeps that third phosphate on, and when it comes off, energy is released. Once the third phosphate is gone, the molecule is no longer called ATP because it no longer has three phosphates. It is now ***ADP***: adenosine **di**-phosphate. (You might wonder why ADP can't pop off its other two phosphates and provide three energy units instead of just one. The short, over-simplified answer is that those two other phosphates don't come off easily.) For the ATP molecule to be "recharged," the third phosphate must be put back on, and putting it back on takes energy. That's where your food comes in. The energy from your food is used to pop that third phosphate back onto ADP and recharge it into ATP. The primary reason your body needs food calories is for recharging ATPs. Once the ATPs are made, they can float around freely inside the cell, available to any cell part that needs them.

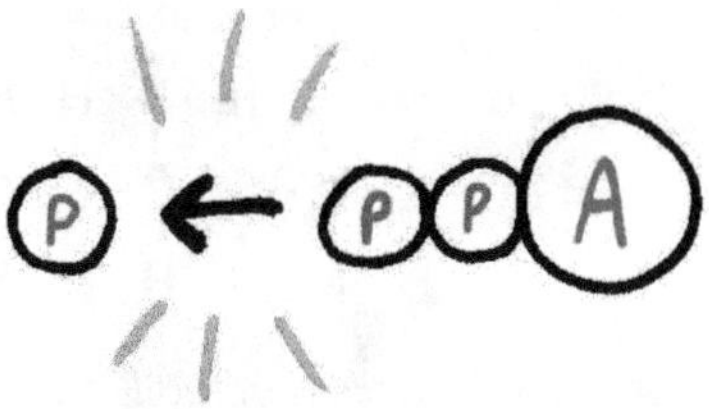

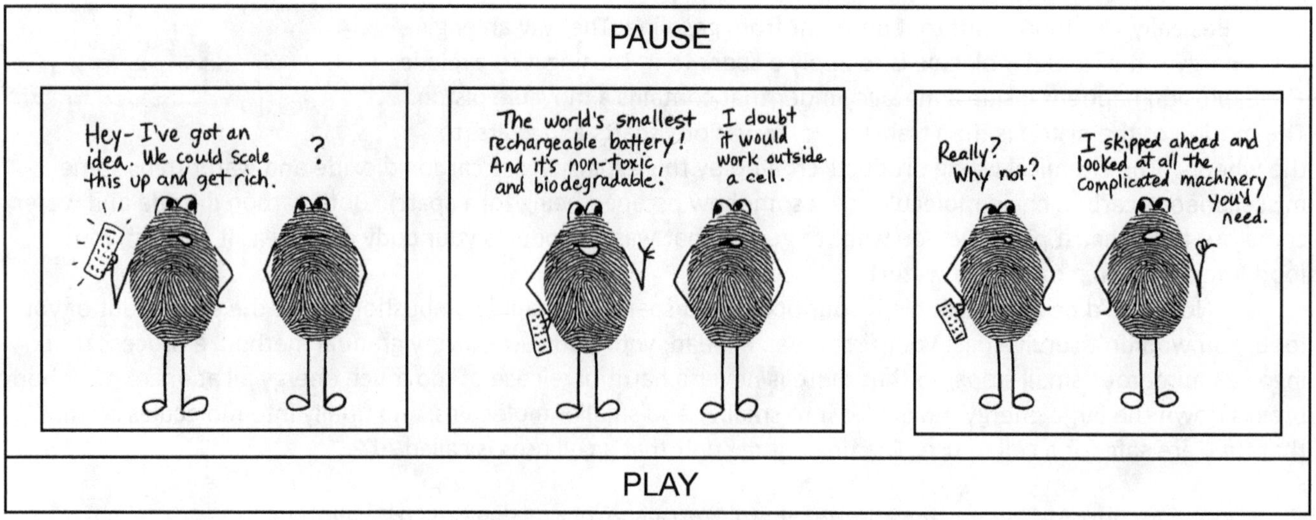

Cells have little recharging "machines" specially designed to put those third phosphates back onto ADP, turning them back into ATP. The scientific word for "make" is "***synthesize***." So scientists (who love Latin and Greek words) decided to name this machine ***ATP synthase***. (The "-ase" on the end of the word is the ending they always use for this type of protein. It's sort of like a last name.) This little machine looks like a cross between an old-fashioned telephone and an old-fashioned egg beater.

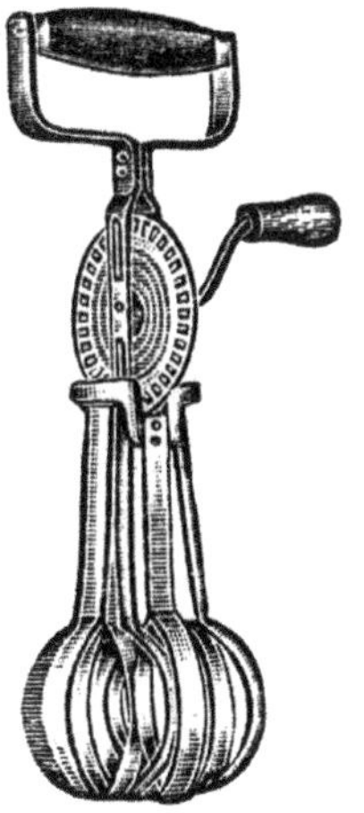

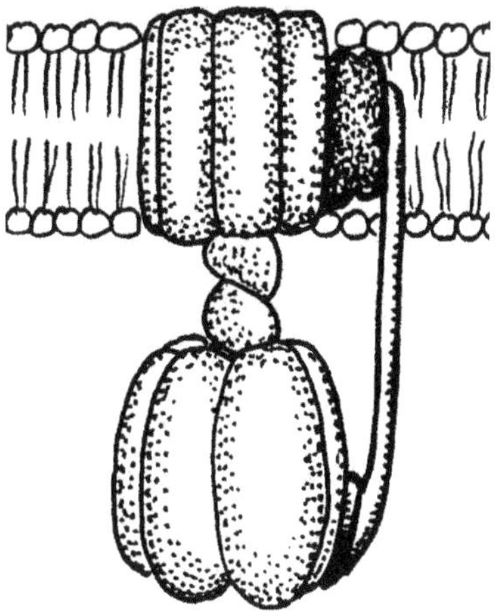

The ATP synthase machine sits in the middle of a phospholipid membrane, held in place by the phospholipid molecules. As we will see, this machine is very similar to a motor, except that it runs on protons, not electrons. The electricity that comes out of the outlets in our walls is made of a continuous stream of moving electrons. ATP synthase uses a stream of moving protons. (We usually find protons locked inside the nucleus of an atom, but here we will see some floating around in the cytoplasm.) A stream of protons will travel through the ATP synthase machine and cause that lumpy "rotor" on the bottom to turn. -

There is more to say about how this little machine works, but first, let's find out exactly where it is located. It's <u>not</u> in the outer membrane of the cell. It's in a special membrane inside organelles called ***mitochondria***.

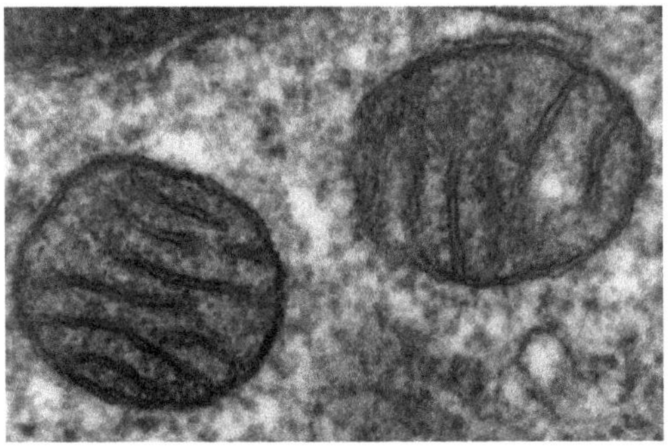

Every cell has at least one mitochondrion, and some cells, such as those found in our liver, have thousands of mitochondria. In this micrograph (a photograph taken with a microscope) we see two mitochondria from a cell taken from the lung tissue of a mammal (probably a lab rat or mouse). The dark circle around the outside of the mitochondria is a phospholipid membrane. Because the mitochondria are "bound" around the outside by a membrane, they are called ***membrane-bound organelles***. In future chapters we will meet other membrane-bound organelles.

When we look at a diagram of a mitochondrion, we might be a bit confused because it doesn't seem to match the micrograph pictures. In the micrographs, we see stripes. In diagrams, we see a plump, wiggly structure.

The dark area inside the wavy lines is called the ***matrix***. The matrix is thicker than the cytosol. Imagine the cytosol as water and the matrix as syrup.

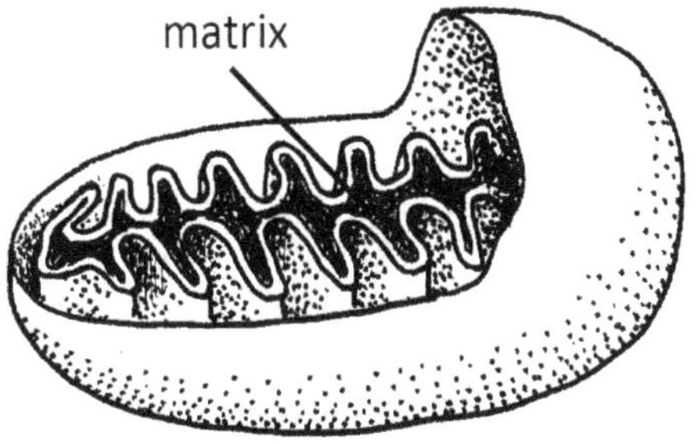

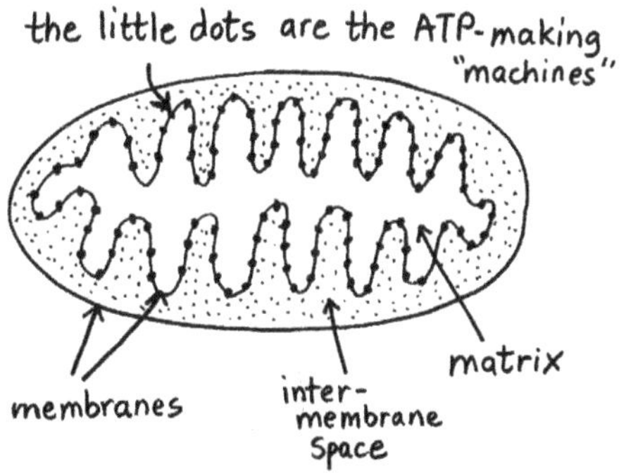

The diagram on the left shows a "cut-away" view, revealing the inner structure. The diagram on the right is a cross section, or slice, taken right through the middle, as if the "knife" in the left hand drawing had kept on going and cut the top completely off. This view is less interesting to look at since it is flat, but it shows us where these little ATP synthase machines are located. They are embedded in a wavy, complicated-looking membrane. The area inside the membrane is called the ***matrix***. The area outside the membrane is called the ***intermembrane space*** because it is "inter" (meaning "between") the matrix membrane and the outer membrane. The matrix membrane is made of exactly the same thing as the cell's outer membrane: a double layer of phospholipid molecules.

To learn more about what happens in the mitochondria, we will have to zoom in and take a very close-up view of the membrane surrounding the matrix.

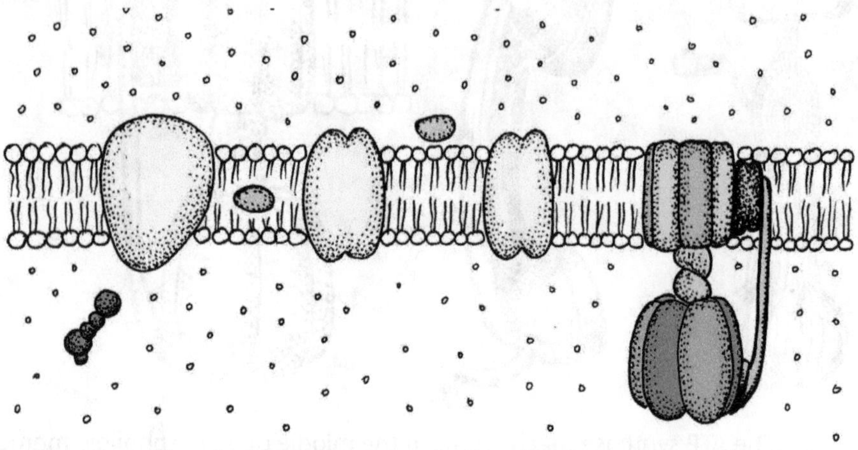

Remember, these things don't really have any color. We add color to make them less boring.

In this diagram we can see that the ATP synthase machine isn't alone. You can see other little machines labeled as pumps, plus three tiny "carriers." (All the dots are protons.) This assembly line of machines is called the ***electron transport chain***. This chain of machines transports electrons through the assembly line to power the pumps. The pumps pull protons from below the membrane and put them above it. Like people, molecules usually don't like to be in a place that is too crowded, so as soon as there are more protons above than below, the protons will want to go back down to where it is less crowded. The only way to go back down is through a channel in the ATP synthase machine. As they funnel down through the machine (headed for the less-crowded side of the membrane), they cause the "beaters" of the ATP synthase machine to rotate.

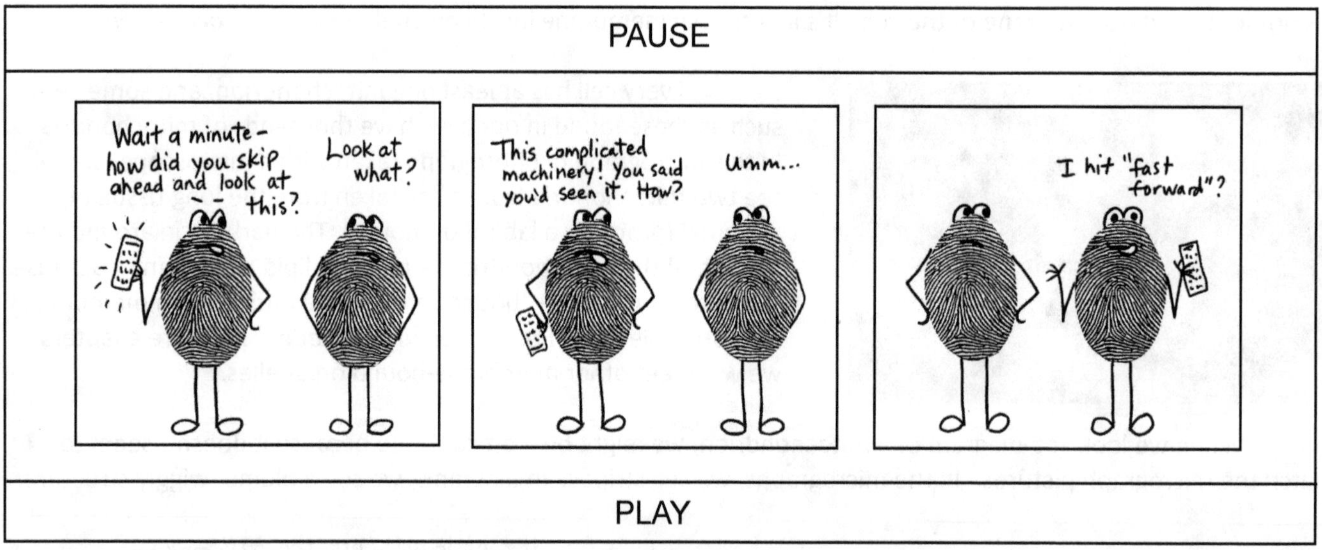

Before we continue with the details of how these machines work, let's do the overview one more time. The goal of this assembly line is to turn those beaters on the ATP synthase machine. The ATP synthase machine will be powered by protons going down through it. The protons will be able to go down through the machine because they were pumped up above the membrane by pumps. The pumps will be powered by electrons.

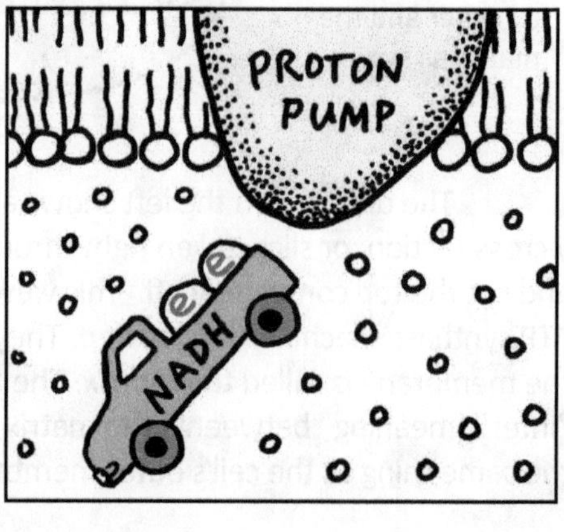

Okay, now for more details! To get things started, a pair of electrons is brought over to the assembly line by a molecule called NADH. Never mind what NADH means—all you need to know is that these electrons came from a process where sugars from food were digested. Think of NADH as a little pickup truck carrying two electrons. These electrons have a lot of energy and will be used to power the three proton pumps.

Now we will look at the whole assembly line again and add some arrows to show the path of the electrons.

The electrons will go through all three pumps, but will need to ride on a carrier molecule to get from pump to pump. You could think of these carriers as ferry boats or shuttle buses.

The electrons are headed for an oxygen atom waiting at the bottom of the third pump. This is their final destination.

Every time an electron goes through a pump, the pump pushes a proton from below the membrane up into the space above the membrane. As the electrons go through the pumps, their energy gets used up.

ELECTRON TRANSPORT CHAIN

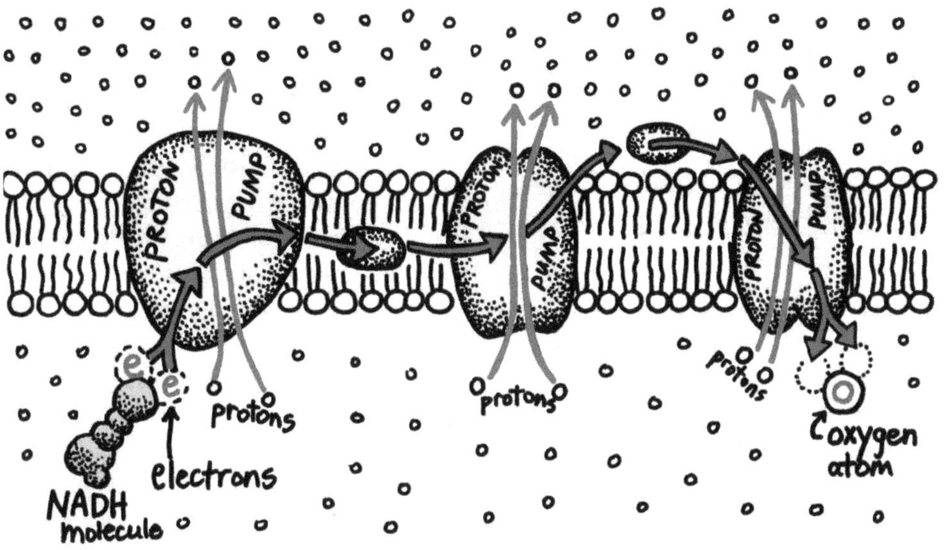

By the time they get to the third pump, the electrons have "slowed down" and lost much of their ability to do work. The fate of these "worn out" electrons is actually an important feature of this assembly line. It just so happens that there are oxygen atoms hanging out near the bottom of this third pump. An oxygen atom would love to turn itself into a water molecule; all it has to do is acquire two hydrogen atoms. A hydrogen atom is nothing more than an electron stuck to a proton. So all the oxygen has to do is grab two of those protons that are floating around, then take the two tired electrons as they come off the assembly line. The electrons stick to the protons to make hydrogen atoms and presto—a molecule of water (H_2O) is made! Your body can use this water molecule for other cell processes, or it can get rid of it as waste. The water vapor in your breath when you exhale contains many water molecules that were made by this assembly line.

As strange as it may seem, the only reason you need to breathe is to provide oxygen molecules for the end of this electron transport chain. (If there aren't any oxygen atoms waiting near that third pump, your body immediately shuts down the assembly line.) Because this process is directly tied to breathing, it is called ***cellular respiration***. It's the use of oxygen ("respiration") at the cellular level.

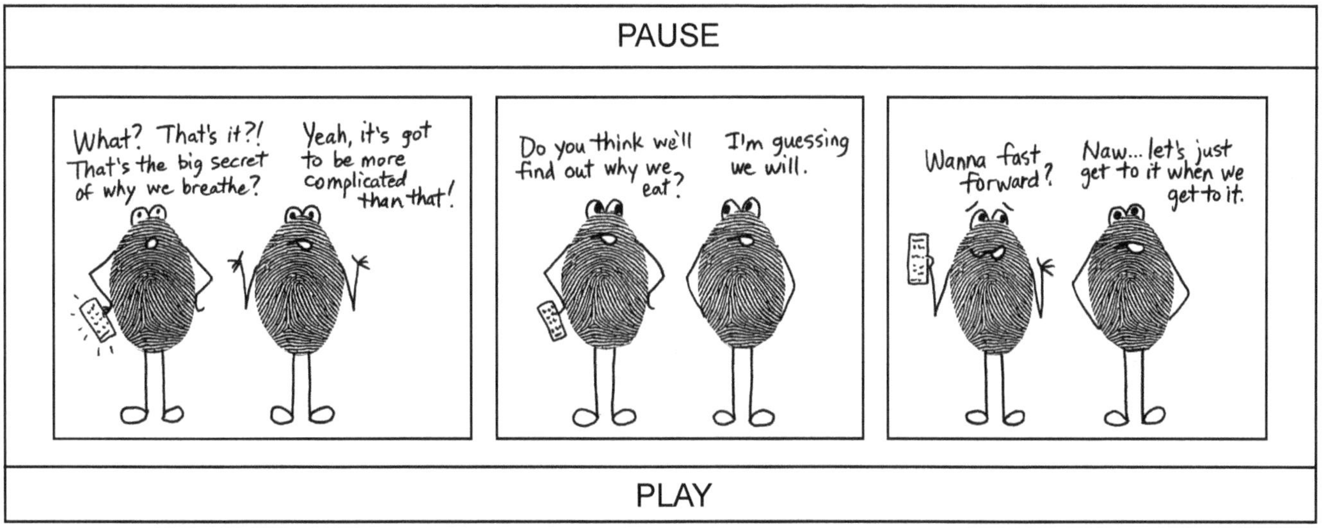

Now let's go back and follow the protons and see what happens to them. As the pumps push the protons up into the space above the membrane, the protons start to accumulate up there. Soon, there are more protons "upstairs" than "downstairs." Imagine someone who lives in a two-story house inviting about 50 friends to come and visit. During the visit, all 50 friends are required to go upstairs. First, it was too crowded downstairs, and now it is too crowded upstairs! Then someone is able to quietly sneak down the staircase. What a relief to be back downstairs—so much empty space! Then someone else comes downstairs. With only two friends downstairs, it

is still a relief to be in an uncrowded space. Then another friend comes down, then another, and another. When will they stop coming down? As soon as there are an equal number of friends upstairs and downstairs, there will be no advantage to going downstairs. Unless you force them to go down, (or tell them that a new batch of cookies has just come out of the oven downstairs), the migration downwards will stop because going down would mean going to a <u>more</u> crowded, not less crowded, space.

When atoms and molecules behave like this, going to a less crowded place, we call it ***diffusion***. The ATP synthase machine takes advantage of the diffusion of protons. The synthase machine is like the staircase in our imaginary scenario, allowing the protons to sneak down to a less crowded place. (However, this analogy quickly breaks down because unlike the staircase, the ATP synthase machine can only be used to go down. To make our house analogy work, you would have to imagine one-way pumps going up through the floorboards, into the upstairs. Too weird.)

Another analogy that scientists sometimes use is a water reservoir and a water wheel. When the reservoir starts to overflow, the water goes over a spillway and down into the wheel. The mechanical motion of the turning wheel is then used to turn machinery that grinds grain or runs a saw. The ATP synthase machine is like the water wheel, and it really does turn. Those things on the bottom that look like egg beater blades actually rotate.

There are lots of ADP molecules floating around near the beaters, as well as individual phosphate molecules. Both ADP and phosphates will be taken up into the beaters. They snap into little pockets in the beaters that are exactly the right shape to hold one ADP and one phosphate. The turning motion of the beaters will squeeze them together and the phosphate will be pressed back onto the ADP, turning it into ATP. With the next turn of the beater, the ATP is released and falls out. The ATP is then ready to be used by any cell process that needs it.

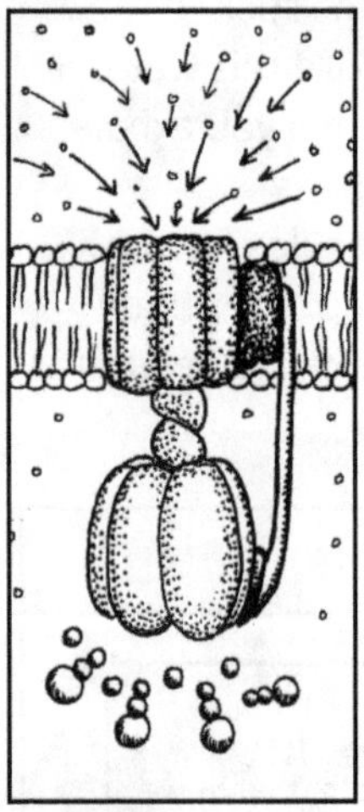

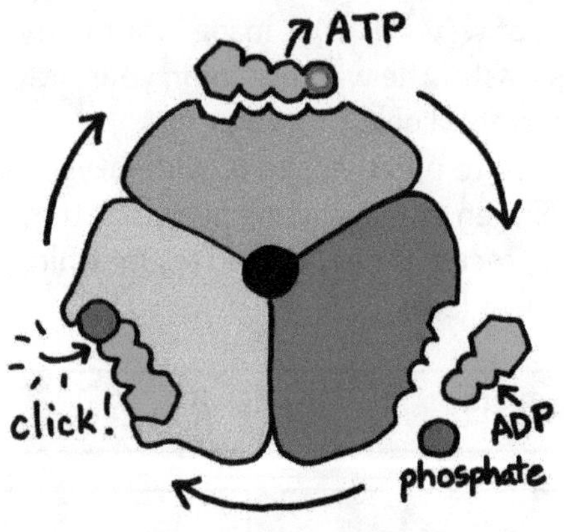

Bottom view of ATP synthase "beaters"

The Wikipedia article on ATP synthase has a helpful animated graphic that shows the bottom view. You can watch ADP and phosphate going into a pocket, snapping together, then being released as ATP.

NOTE: You might want to watch the videos suggested in Activity #1 before answering these questions.

If you can't remember these answers, go back and find them in the chapter.

1) What is the basic unit of energy used by cells? ___ ___ ___

2) Why can plants be made into fuel for vehicles?
   a) because plants contain petroleum b) because plants contain protein
   c) because plants contain strings of carbon atoms with hydrogens attached d) trick question — they can't

3) What is the molecule shown here? ____________________ 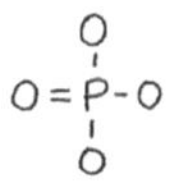

4) TRUE or FALSE? Your body uses a slow form of combustion to burn your food.

5) What happens when the third phosphate is popped off ATP? a) a molecule of water is formed
   b) energy is released c) energy is used up d) a proton is released

6) The word "synthesize" means "to ____________."

7) What is the inside of the mitochondria called? a) matrix b) membrane c) cytosol d) synthase

8) TRUE or FALSE? The electron transport chain assembly line sits in the outer membrane of the mitochondria.

9) What happens when electrons pass through the pumps in the ETC assembly line?
   a) ATP is formed b) water is formed
   c) two protons are pumped upward d) two electrons are pumped upward

10) What does the NADH molecule do? a) pumps protons b) shuttles electrons c) makes ATP

11) What is the ultimate goal of the electron transport chain assembly line? a) to make an oxygen atom happy
   b) to pump protons upwards c) to make you study science d) to recharge ATP molecules

12) TRUE or FALSE? Atoms and molecules like to be packed tightly together.

13) How many pumps are in the electron transport chain? ____

14) What happens to the electrons after they go through the third pump?
   a) They stick to an oxygen atom. b) They are pumped back into the matrix.
   c) They go down through the synthase machine. d) They escape through gaps in the membrane.

15) If you join an electron and a proton, what do you get? a) water b) hydrogen c) oxygen d) nothing

16) What is it called when a lot of something goes to a place where there is less of it? ________________

17) TRUE or FALSE? The electron transport chain produces both ATP and water molecules.

18) What does the word root "di" mean? _____

19) REVIEW: Which of these motor proteins walks outward, towards the edge of the cell? a) kinesin b) dynein

20) REVIEW: Which part of a motor protein touches the microtubule highways? a) tails b) feet c) heads

ACTIVITY 4.1 "Must-watch" videos

1) Animations showing the ATP synthase machine in action

There should be several animations posted on the Cells youtube playlist. If they have disappeared, just use the search feature with key words "ATP synthase." If you don't want to use youtube, use these key words in your preferred video posting/streaming site.

2) Animations showing how the electron transport chain (ETC) works

You might hear the word "gradient" in some of these videos. A "gradient" means that there is an area of higher concentration and an area of lower concentration (like the upstairs and downstairs in our house analogy). Also, you might hear the term "hydrogen ion," which is another word for a proton. The word "ion" means any atom or molecule that carries an electrical charge. Protons are positively charged, so they are ions. Phosphate is also an ion because it carries a negative charge. A proton pump is a type of ion pump. Don't be too concerned about understanding every word of the narration. Just enjoy the animations and notice all the parts that we discussed in this chapter.

ACTIVITY 4.2 Crossword puzzle with wordless clues

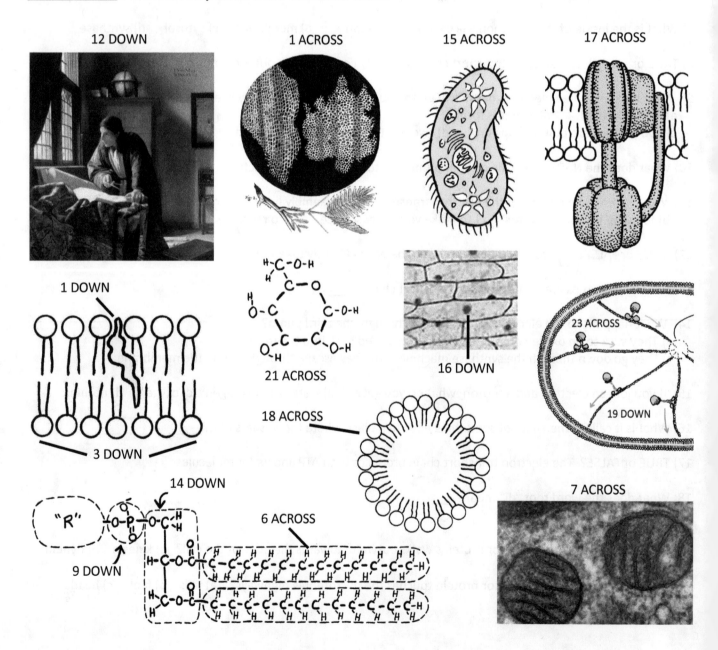

12 DOWN

1 ACROSS

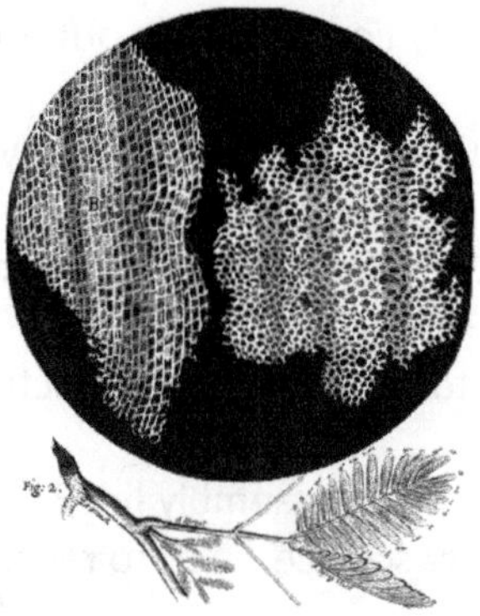

15 ACROSS

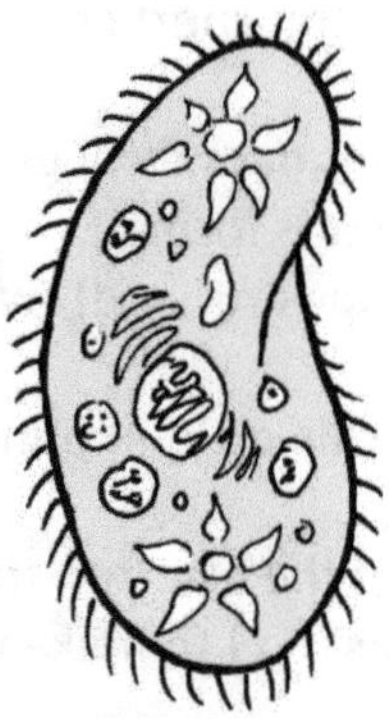

17 ACROSS

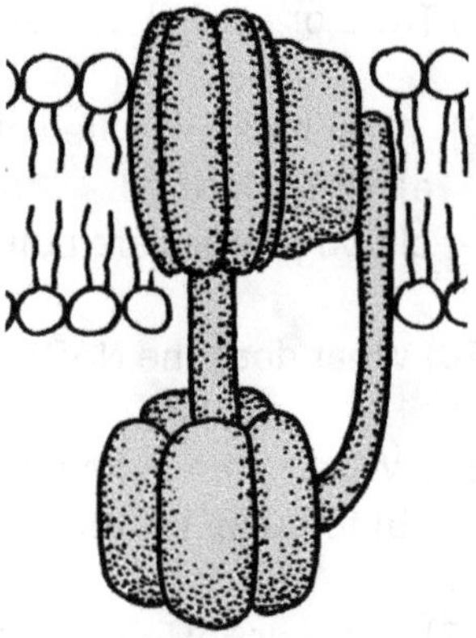

1 DOWN

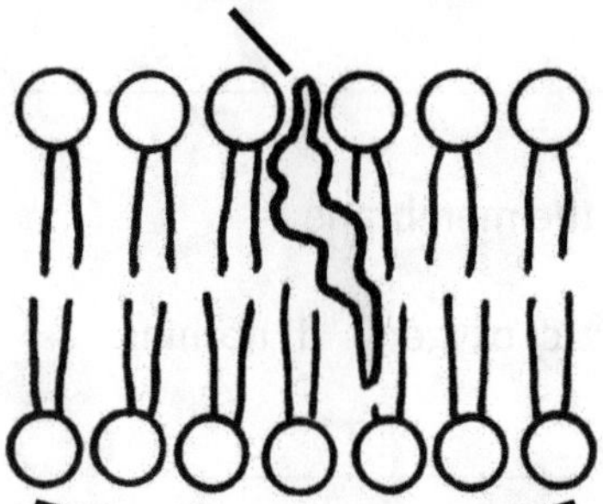

3 DOWN

21 ACROSS

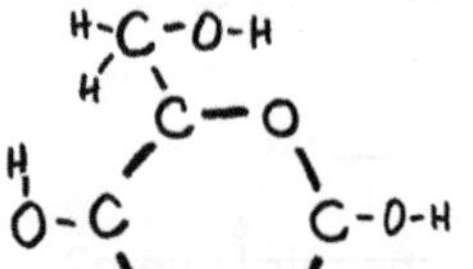

16 DOWN

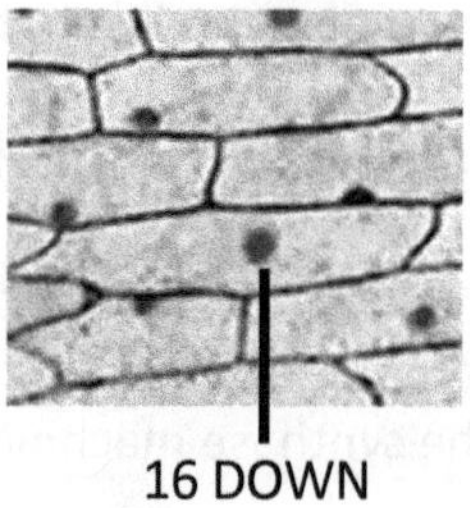

18 ACROSS

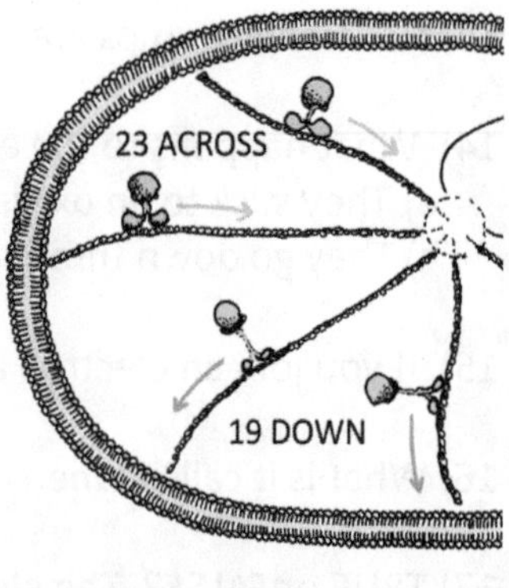

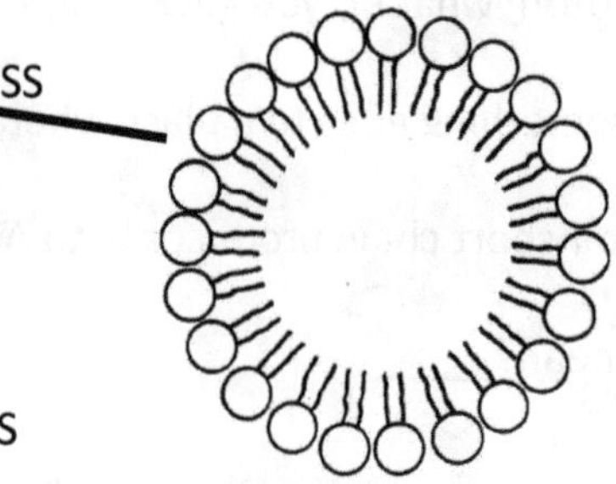

7 ACROSS

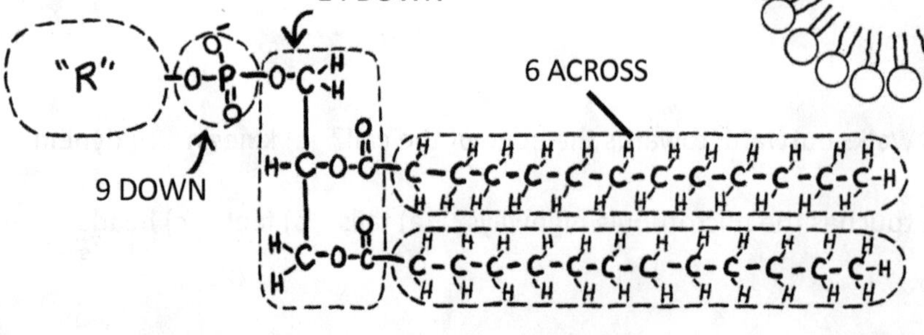

CROSSWORD REVIEW CHAPTERS 1-4

13 ACROSS

5 DOWN

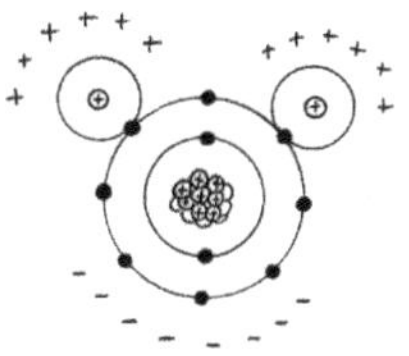

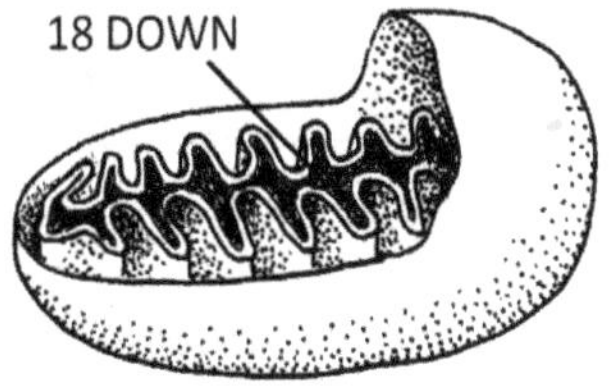

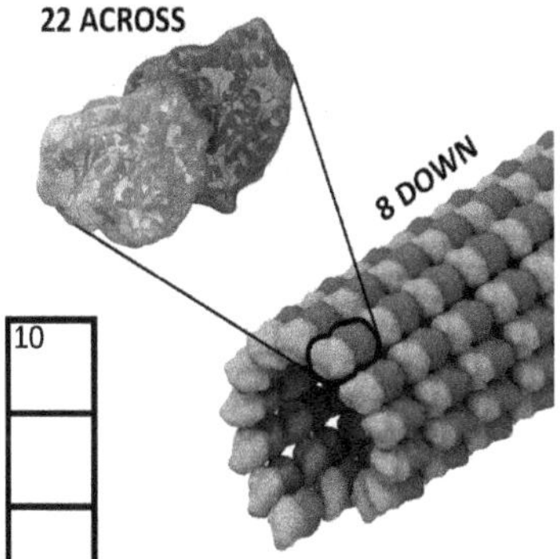

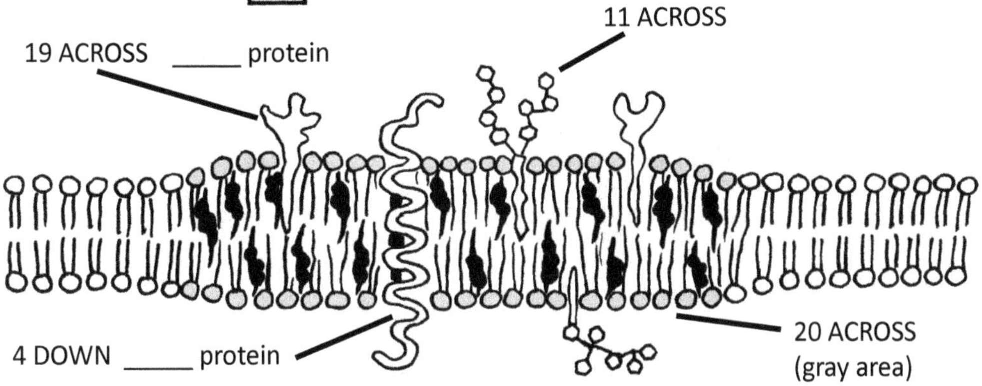

10 DOWN

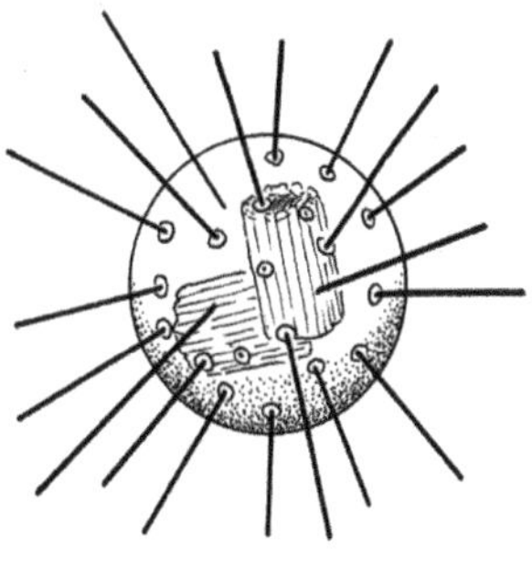

ACTIVITY 4.3: What do these little cell parts really look like?

People who draw diagrams of cell processes are aware that the viewer can easily be overwhelmed by too much information. One way they can make diagrams less complicated is by simplifying the shapes of the little cellular machines. Our diagrams of the electron transport chain used very simplified shapes. What do these little pumps and shuttles really look like?

The ATP synthase machine is made of proteins curled up into a spiral shape called the ***alpha helix***. We will learn more about this shape in the next chapter. No individual atoms are shown.

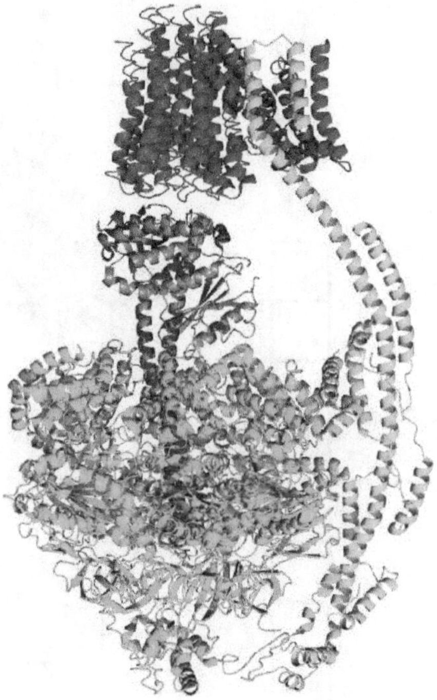

By BiochemEkaterina - Own work, CC BY-SA 4.0, https://commons.wikimedia.org/w/index.php?curid=59979304

This is also ATP synthase. The artist has chosen to show the atoms as little balls. The spiral shapes are harder to see. The colors show you individual sections that fit together to make the whole machine.

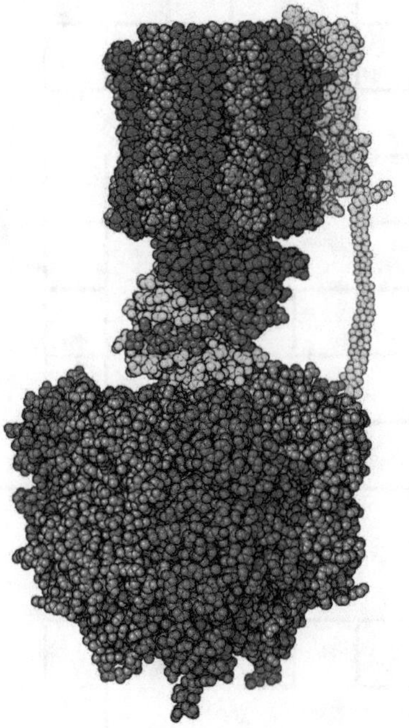

By Alex.X - enWiki (PDB.org for coordinate), CC BY-SA 3.0, https://commons.wikimedia.org/w/index.php?curid=1618502

This protein machine isn't in our chapter, but we are looking at it anyway because it shows you another common shape for transmembrane channels: the "beta barrel." (The yellow things are called "beta sheets.")

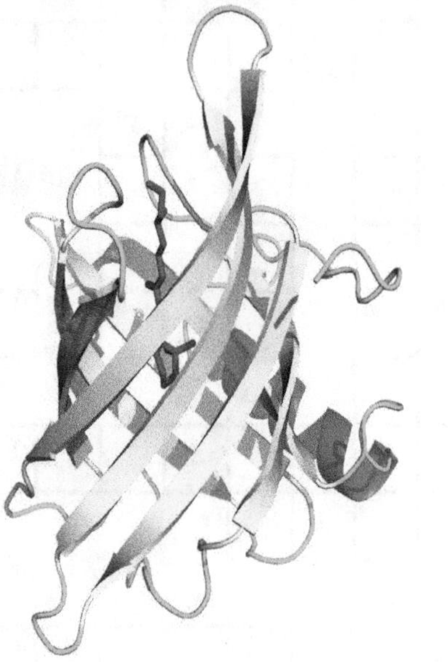

By Opabinia regalis - Self-created from PDB ID 1BRP using PyMol, CC BY-SA 3.0, https://commons.wikimedia.org/w/index.php?curid=1775129

This is the first "shuttle" that carries electrons through the membrane to the second pump. It has two names: "coenyzme Q10" and "ubiquinone"

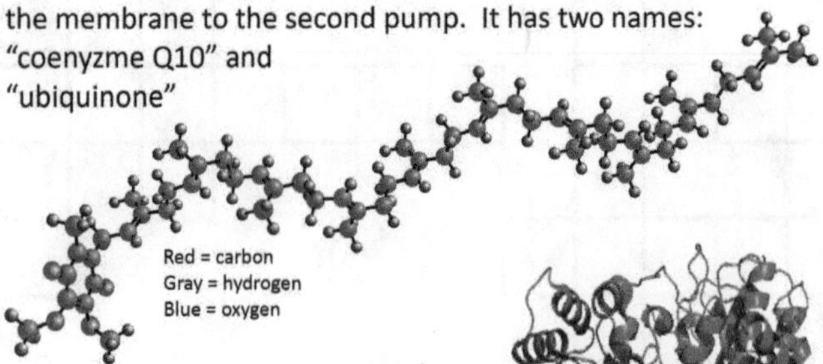

This is "cytochrome C," the second "shuttle" in the transport chain.

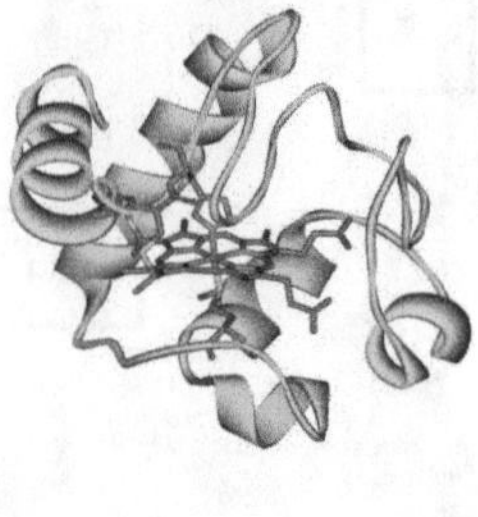

This is the middle pump in the electron transport chain.

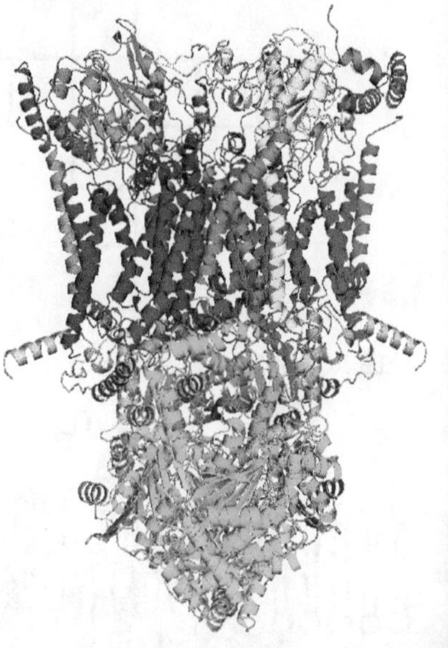

By C31004 at English Wikipedia, CC BY-SA 3.0, https://commons.wikimedia.org/w/index.php?curid=24133439

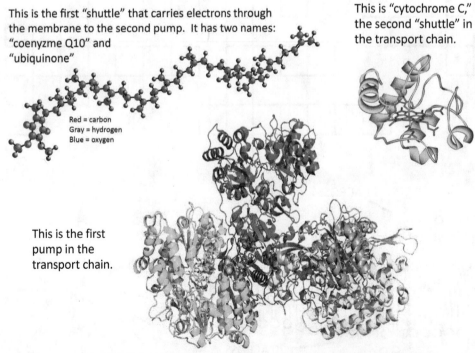

This is the first pump in the transport chain.

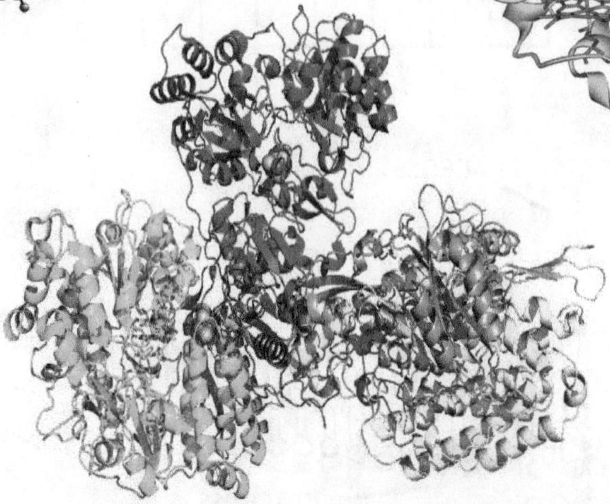

By A2-33 - Own work, CC BY-SA 4.0, https://commons.wikimedia.org/w/index.php?curid=41107517

CHAPTER 5: PROTEINS, DNA, AND RNA

So now we have a cell with an outer membrane, an organized cytoskeleton, motor proteins to get things to where they need to be, and an assembly line that can make energy molecules. We're slowly building a cell!

PAUSE

PLAY

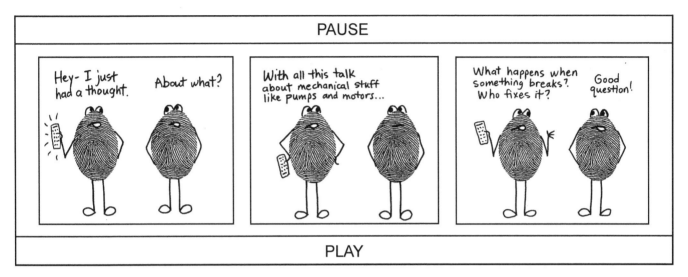

Before we can discuss the repairs, we need to learn more about the substance that most biological machines are made of: ***protein***. Proteins are made of individual units called ***amino acids***. Like lipids and sugars, amino acid molecules have many atoms of carbon, oxygen, and hydrogen. But they also have one type of atom that lipids and sugars do not have: ***nitrogen***.

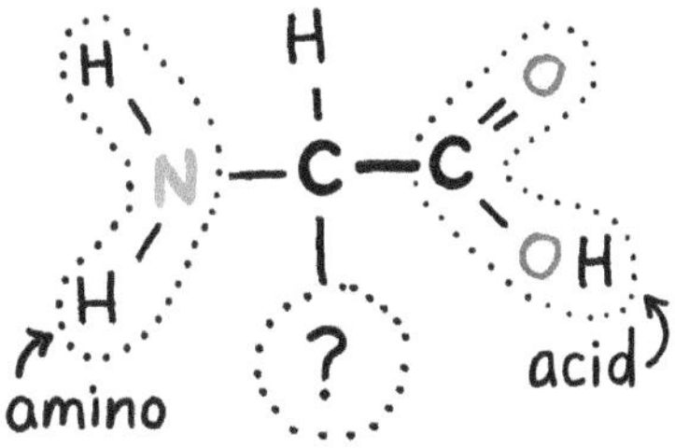

AMINO ACID STRUCTURE

An amino acid molecule is built around two carbon atoms. The nitrogen (N) and its two hydrogens make the "amino" part of the molecule. The COOH at the other end is what makes it an acid. COOH is a special group of atoms that you can stick onto any molecule to turn it into an acid. (We will learn a little more about acids in the next chapter.) You might be wondering whether an amino acid tastes sour, like a lemon (which is very acidic). No, an amino acid would not taste sour. Meat, eggs, and beans are loaded with amino acids and they don't taste sour. (We'll meet another molecule in this chapter that is called an acid but is not sour.)

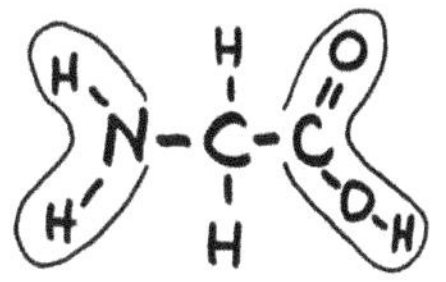

GLYCINE

The (?) at the bottom of the molecule is the "wild card," meaning that there are many options for what might go here. The simplest thing you could do is make that carbon happy by giving it a hydrogen to hang on to. When this happens, you get the smallest of all amino acids, ***glycine*** *(GLIE-seen).* Glycine comes in handy in places where you need a small building block that fits into tight spaces and won't interact much with its neighbors. Another simple and easy option for replacing that (?) is to add another carbon and give it three hydrogens (CH_3). This amino acid is ***alanine*** *(AL-uh-neen).* We could keep going on this theme and add more carbons. If we add four carbons (and give them hydrogens), we'd have ***leucine***. *(LU-seen).*

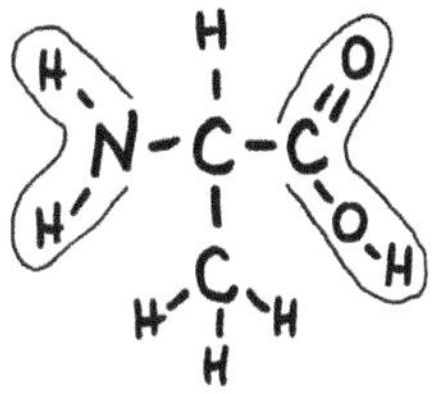

ALANINE

Each amino acid has something attached to it that makes it just what you need in certain situations. Some aminos have a positive or negative electrical charge. Some are hydrophobic and love to hide out amongst the fatty acid tails of phospholipid molecules, making them good "anchors" for transmembrane proteins. The amino acid named ***cysteine*** *(SIS-teen)* contains a sulfur atom which gives it the ability to form "cross links" making proteins very tough and durable. Hair contains a lot of cysteine, which is why it is so strong. You can smell the sulfur atoms when you burn hair. Burning sulfur stinks!

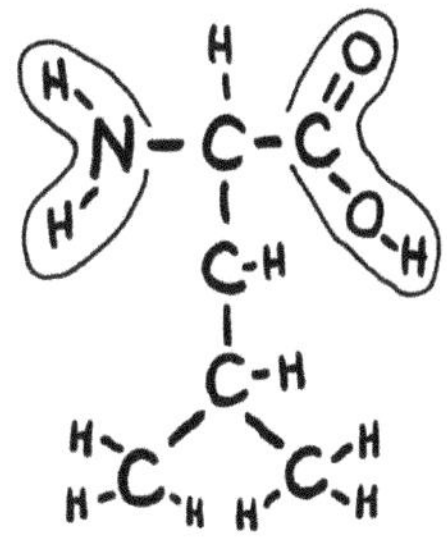

LEUCINE

Surprisingly, there aren't that many different kinds of amino acids. There are just **20** of them. That's it. Not hundreds or thousands or millions, just 20. The chart opposite this page shows all 20.

Like phospholipids, amino acid molecules are too complicated to draw frequently. Unless it is really important to see all the atoms in an amino acid molecule, scientists usually represent amino acids as circles with a three-letter abbreviation written inside. If the circles are really tiny, they can abbreviate even more and use just one letter. These abbreviations have been standardized; all scientists must use the same abbreviations. In the chart on the next page, you can see the three-letter abbreviation written below each circle. The one-letter abbreviation is inside the colored dot next to the name.

Amino acids link together to form long chains, like a string of beads. A typical chain has about 300 amino acids, but very long chains can have more than 2,000. On the next few pages we will see how the cell knows which amino acids to connect together, and in what order to connect them. But first, let's see what happens to these long chains after they are assembled.

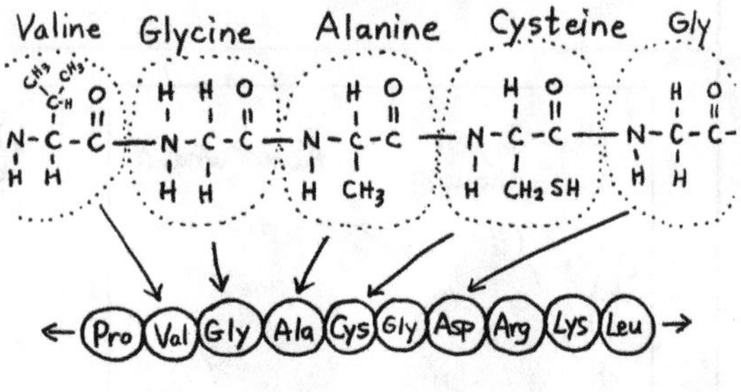

The long chains of amino acid "beads" get folded into compact shapes, then into complex shapes. This process is called ***protein folding.*** (For once, scientists didn't give something a hard name!) We've already seen some of the complex shapes that proteins get folded into. The ATP synthase machine and the proton pumps are made of folded proteins.

There are two levels of folding that can happen to a chain of amino acids. The first level of organization is often to coil the chain into a shape called the ***alpha helix***. Another option is to bend the chain back and forth to make "pleats," forming a shape called the ***beta sheet***. The dotted lines in these diagrams (between oxygens and hydrogens) represent an attraction between atoms called "hydrogen bonding." It is a relatively weak bond, but strong enough to form these shapes.

ALPHA HELIX

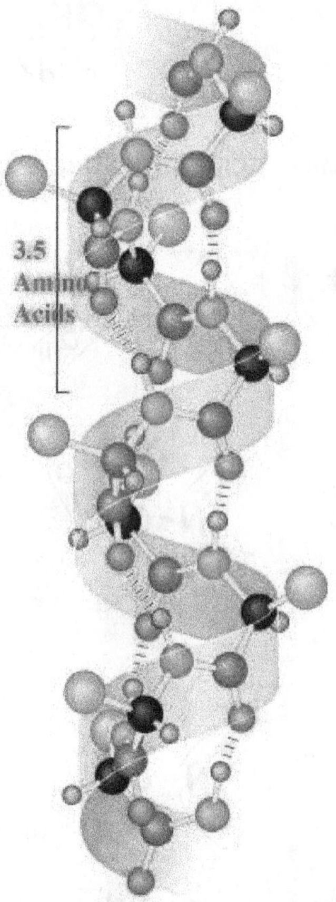

Red= oxygen Blue= nitrogen
Black and gray= carbon
Small gray= hydrogen

*BETA SHEETS**

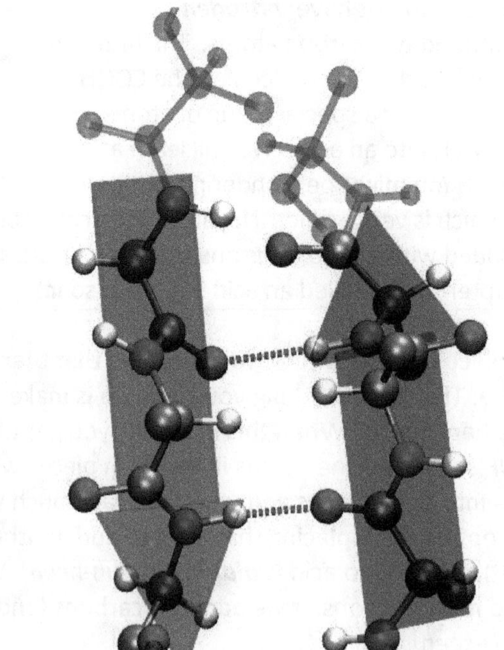

Red= oxygen Blue= nitrogen Large gray= carbon
White/small balls= hydrogen

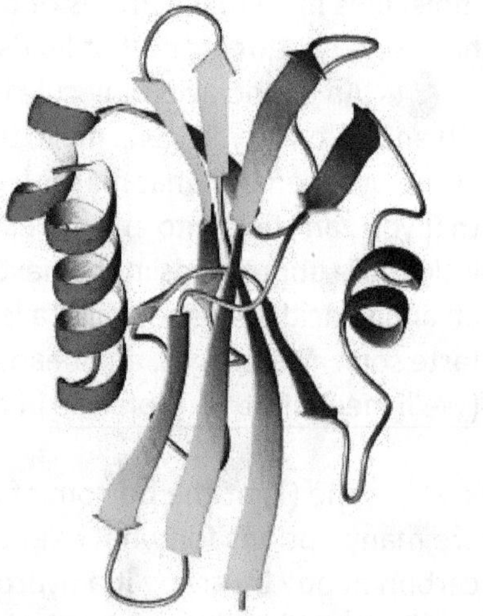

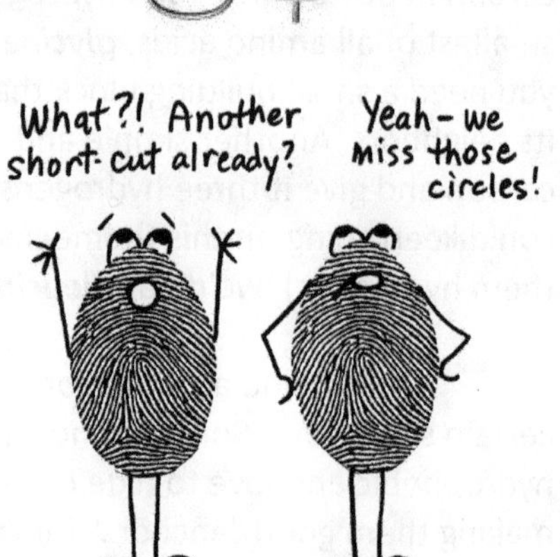

* Beta sheet drawing by Olaf Lenz - The picture was created from the PDB structure 1DX0 using VMD: Humphrey, W., Dalke, A. and Schulten, K., "VMD - Visual Molecular Dynamics", J. Molec. Graphics, 1996, vol. 14, pp. 33-38 by Olaf Lenz, CC BY-SA 3.0, https://commons.wikimedia.org/w/index.php?curid=225980

AMINO ACIDS

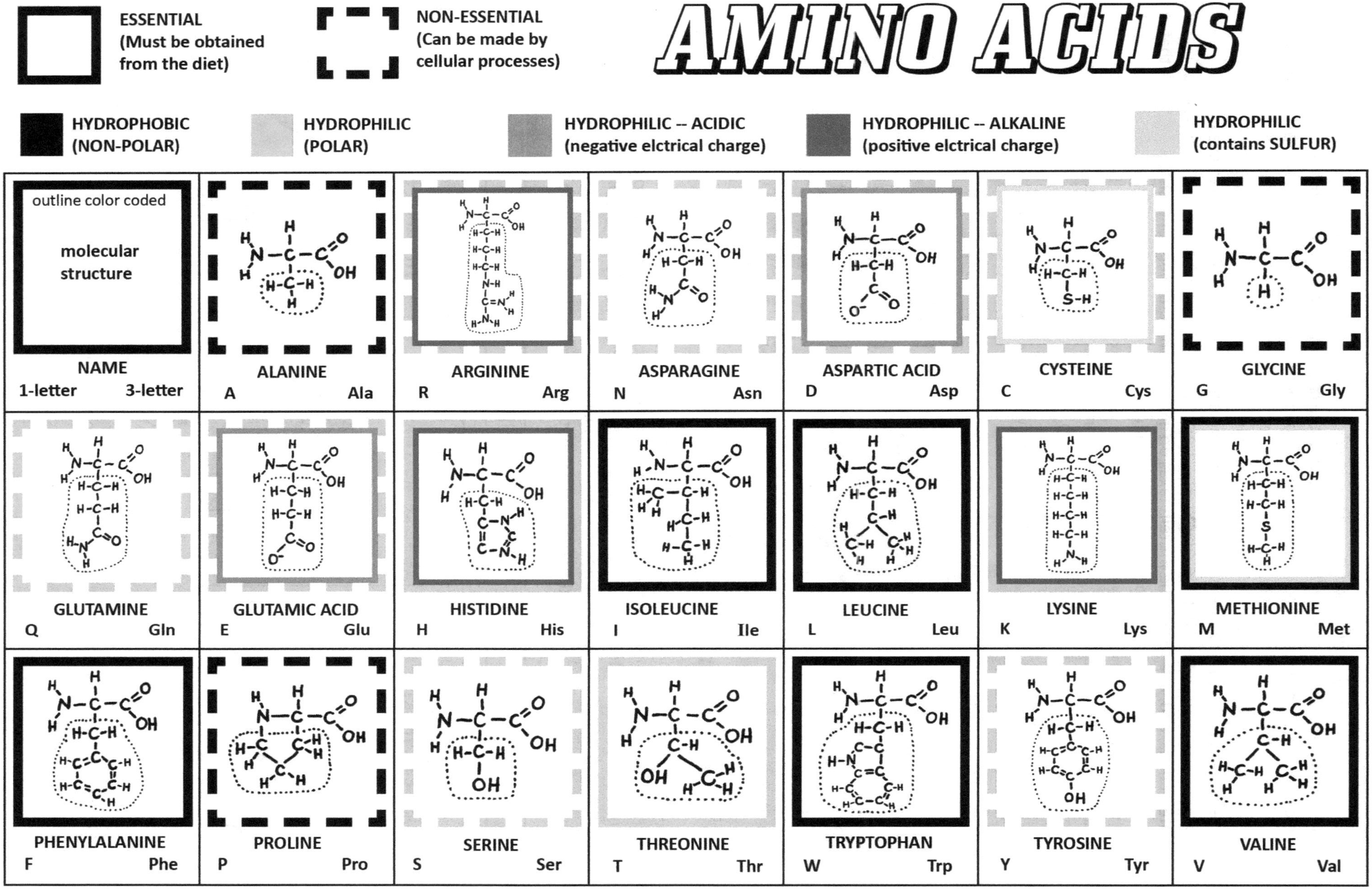

Image created in 2022 by Ellen J. McHenry, who hereby releases it to Public Domain, avialable for any use.

After the amino acid chain makes its coils and folds, a second round of folding occurs. The final shape a protein folds into is determined by how the amino acids in the chain interact with each other. Some of them are hydrophobic or hydrophilic. The water-hating amino acids all try to get together (pulling their part of the strand along with them in the process) to create a special "no water" zone. Water-loving molecules try to move to the outside of the shape (pulling their part of the strand along with them in the process), giving them a better chance of being able to hang out with water molecules. Some amino acids carry a positive or negative electrical charge and will be attracted to aminos of the opposite charge. The amino acid chain is pushed and pulled in various places by all these interactions, and, as a result, the protein takes on a unique three-dimensional shape—a shape that will determine what job it will do. Some proteins end up as pumps in a mitochondrion. A protein called hemoglobin will go to the bloodstream and carry oxygen from the lungs. Some proteins will become ID flags or connectors. Many proteins end up as "messengers," sent as a signal to another cell. There are millions of jobs that need to be done in the body, and there is a specific protein for each job.

HEMOGLOBIN carries oxygen

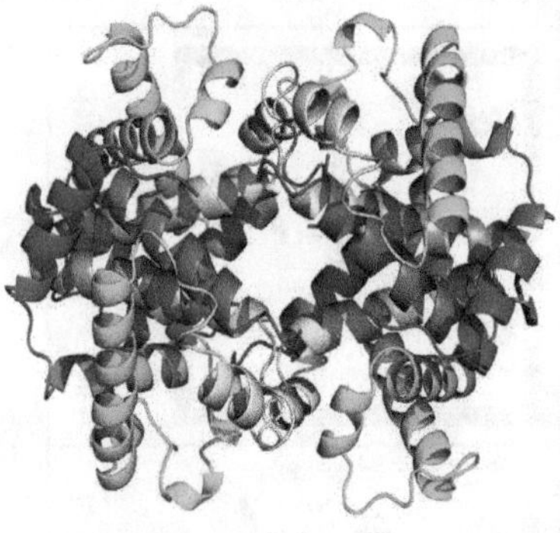

By Emw - Own work, CC BY-SA 3.0, https://commons.wikimedia.org/w/index.php?curid=8814756

AQUAPORIN channel for water

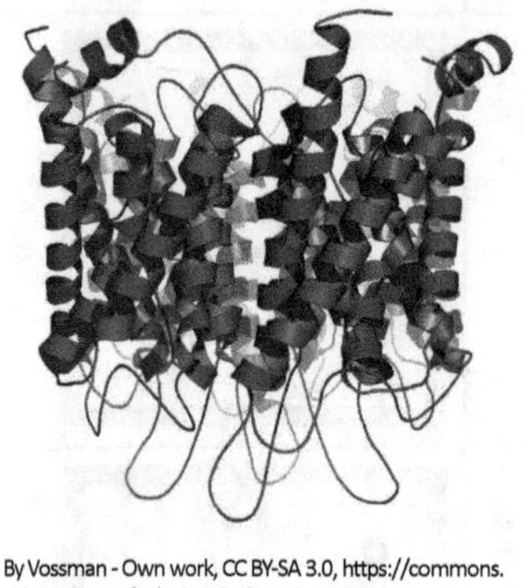

By Vossman - Own work, CC BY-SA 3.0, https://commons.wikimedia.org/w/index.php?curid=1445738

TNFa, a messenger

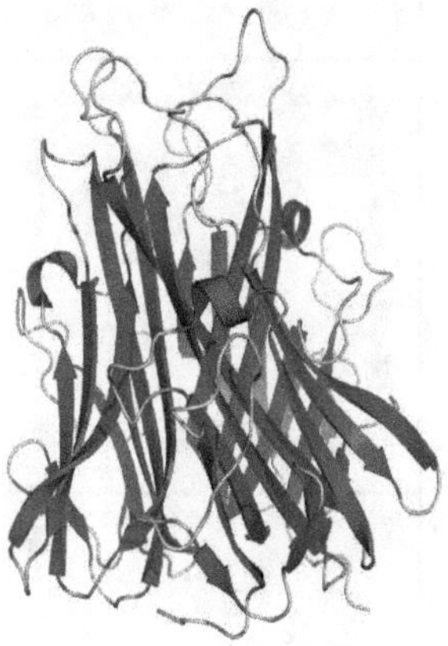

Now let's go back to the scenario we opened with...

There's been some damage to the cell and some pumps must be repaired. Amino acids must be strung together and then folded into the right shape. But a cell doesn't have a brain. It can't figure out what to do. There must be an automatic system that doesn't involve thinking. Sorry, but before we can actually start making proteins for a new pump, we really need to take a few minutes to talk about DNA and the cell's nucleus.

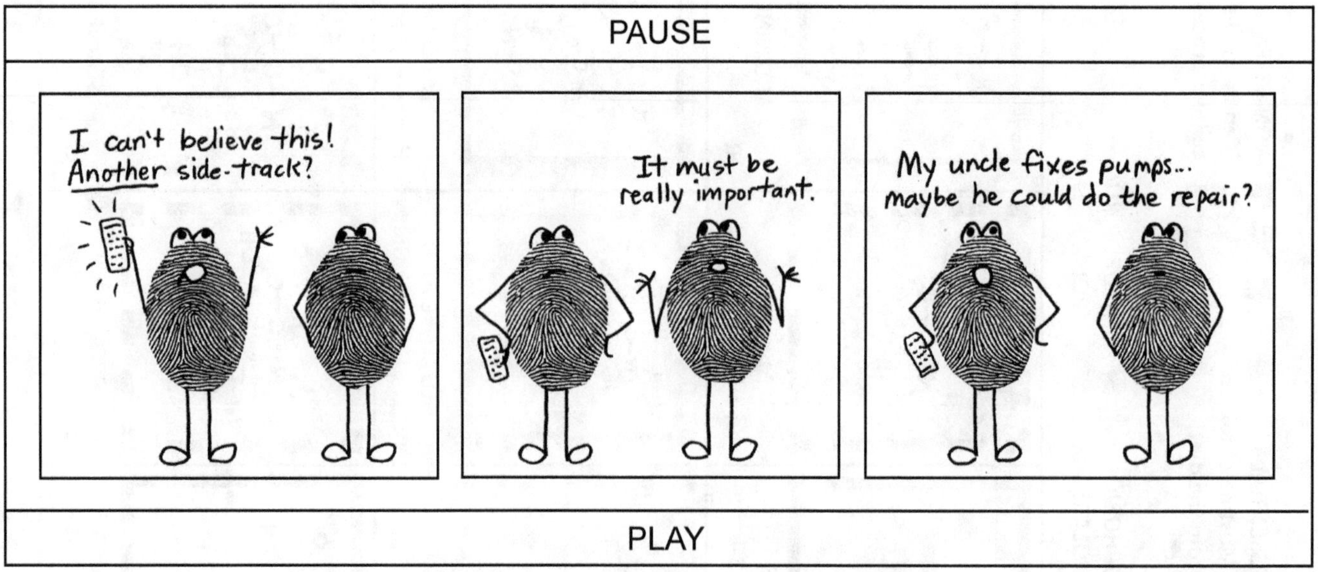

We really will get around to fixing those broken pumps, but it is a complicated process that involves many cell parts. Hang in there!

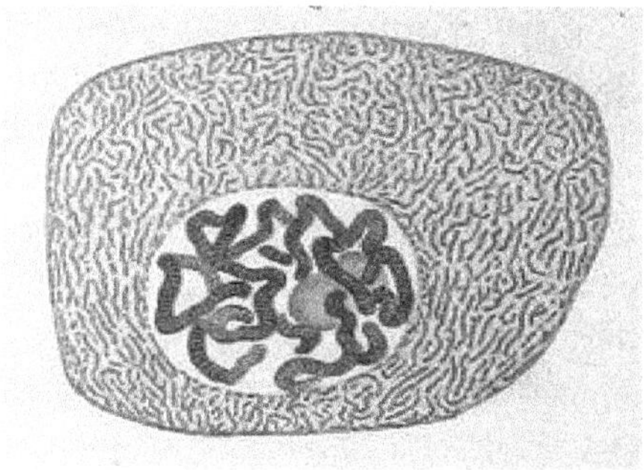

In the middle of the cell there is a very large organelle called the ***nucleus***. You may remember that Robert Brown observed dark dots in plant cells and decided on the name "nucleus." All through the 1700s and 1800s, scientists continued to see these large blobs inside almost every cell they looked at, but they had no way of figuring out what the blobs did. The function of the nucleus remained a mystery until the 1900s, when the electron microscope was invented. Using the electron microscope, scientists were able to see the structures inside the nucleus clearly enough to be able to figure out what role they played in the cell. They determined that the material in the nucleus was critical for the manufacturing of proteins and was probably controlling the process somehow. But this was as far as the electron microscopes could take them. The pictures couldn't tell them exactly how this process worked.

There was disagreement in the scientific community about the source of the cell's information. Many scientists thought it had to be on the proteins themselves. Others believed the information was stored in the nucleus, encoded onto a molecule they had identified as ***deoxyribonucleic acid*** *(dee-OX-ee-RI-bo-new-CLAY-ick)*, or ***DNA*** for short. This isn't an acid like hydrochloric acid or battery acid. The only reason it is called an acid is that it has a certain grouping of atoms attached to it: COOH. Hopefully, you remember COOH from the last chapter, where we saw it on one side of the amino acid. Like the aminos, DNA would not taste sour (acidic) if you ate it.

Most people associate Linus Pauling with his work on vitamin C.

Using x-ray crystallography (crystallizing the molecule, then taking a picture of it using x-rays), scientists had figured out that DNA has three basic parts: ***phosphates***, ***sugars***, and nitrogen-containing ***bases***. They were pretty sure the phosphates and sugars were connected to each other, and a few researchers wondered if they formed some kind of helix shape, like those proteins coils we've been looking at. Three leading researchers who were working on this problem in the 1950s were Linus Pauling, James Watson, and Francis Crick. Linus Pauling was already a famous chemist. He would eventually win four Nobel Prizes. Watson (an American) and Crick (an Englishman) were much younger and just at the start of their careers. Pauling was the first to announce his theory. He proposed a three-sided model. Watson and Crick looked carefully at his model, however, and realized that it disobeyed a fundamental law of chemistry. In his model, Pauling had put together three molecules that would actually repel each other in real life. How could such a famous chemist make such a basic mistake? It seemed almost unbelievable! But when Watson and Crick realized Pauling's mistake, they knew they now had a chance to be the first to solve the DNA mystery.

Watson is on the left, Crick on the right. Their famous model is now in the Science Museum in London.

Working together, and getting some help from x-ray crystallographer Rosalind Franklin, Watson and Crick brainstormed different possibilities for how the molecules could be arranged. One day (Feb. 28, 1953), they hit upon an idea that just clicked. They knew almost immediately that they had solved the mystery. Legend has it that they went into town and walked into a pub (because they were in England) and announced to the crowd, "We have discovered the secret of life!"

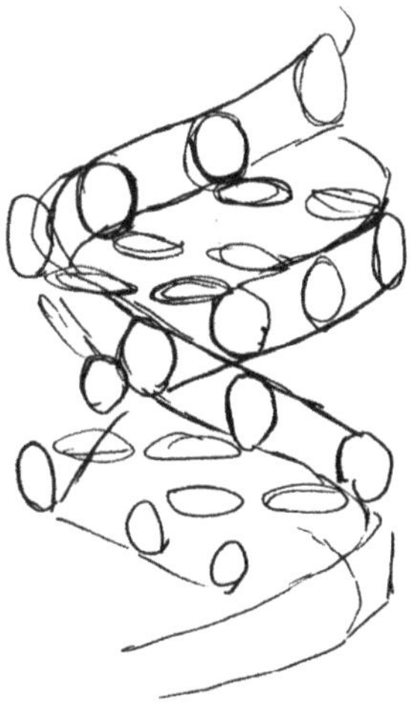

A facsilime (close copy) of Watson's first sketch of his idea about DNA.

Watson, Crick, and Franklin figured out that DNA is made of two helix shapes joined in the middle by "rungs," like a twisted ladder. This shape is often called a ***double helix***. The sides of the ladder are made of an alternating pattern of sugar molecules and phosphate molecules. Sugar, phosphate, sugar, phosphate, sugar, phosphate, etc. It is almost impossible to pick out these sugars and phosphates in this diagram because it is so "cluttered." We'll see them in a simplified diagram in a minute.

Attached to the sugar molecules are molecular structures called ***bases***. Each sugar has one base attached to it. There are four types of bases:

A = adenine T = thymine
C = cytosine G = guanine *(GWA-neen)*

Their names are very well known, but you will most often see them referred to by their letter abbreviations.

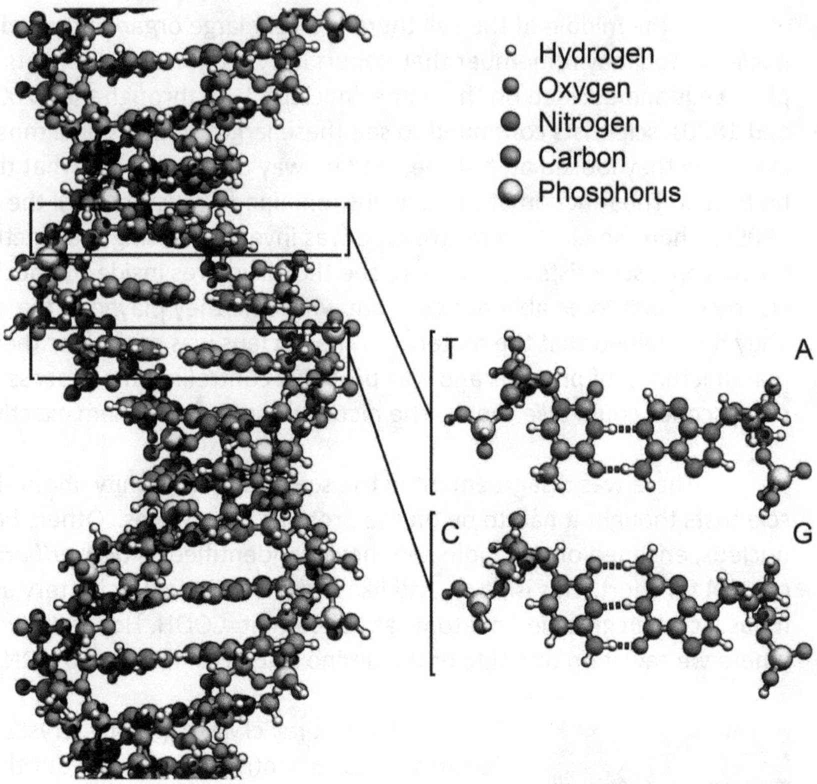

By Zephyris - Own work, CC BY-SA 3.0, https://commons.wikimedia.org/w/index.php?curid=15027555

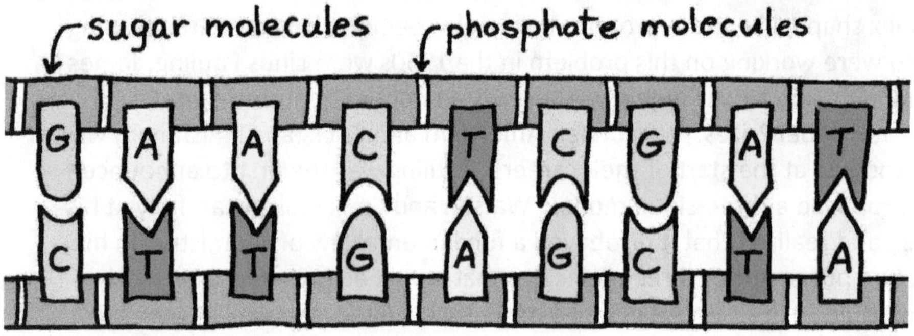

This is an extremely simplified drawing of DNA, with the sugars, phosphates, and bases represented by simple shapes. This makes it very easy to draw a base attached to each sugar molecule. Notice that A always matches with T, and C always matches up with G.

A sugar, a phosphate, and a base fit together to form a ***nucleotide***. A cell makes billions of nucleotides. Nucleotides will be floating everywhere, ready to be grabbed and put into place by the little mechanism that makes copies of the DNA. The diagram with the letters shows the matching bases with complementary shapes, like puzzle pieces. The illustration at the top of the page shows that the bases don't really look like puzzle pieces. What keeps them together is the same type of force that keeps the protein helix and beta sheets together—a weak electrical force called hydrogen bonding.

The arrangement of the bases (A, C, G, T) on the rungs isn't random. The bases come in groups of three, though this isn't obvious by looking at either of our diagrams. A few of these letter combinations actually spell English words, like CAT, GAG and TAG, but mostly they don't look like real words: GGA, CGA, TCG, GGG, etc. These three-letter DNA "words" are called ***codons***. A codon is one "word" in DNA's secret code.

Circle the ten codons between START and STOP:

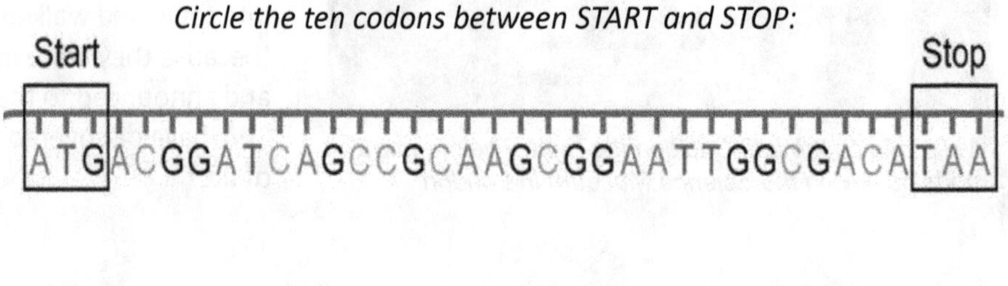

How many three-letter combinations can you make with four letters? There are three possible positions for each of the four letters (first letter, second letter, third letter). If we apply the correct math formula, we get $4^3 = 64$. There are 64 possible three-letter "words" (codons).

PAUSE

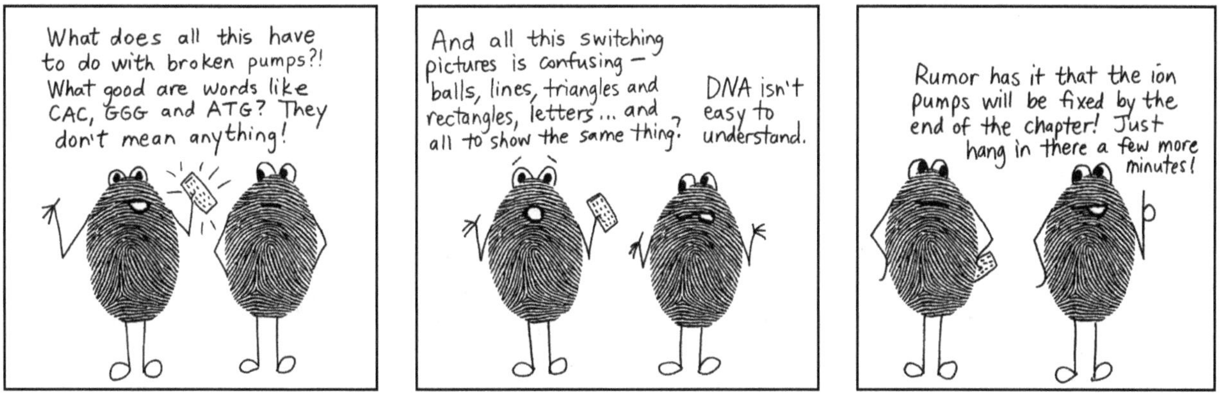

PLAY

We are almost ready to fix that pump! This is where we put what we learned about proteins and protein folding together with the information we are learning about DNA.

What is the cell's pump made of? It's a very long chain of amino acids that is folded up in a certain way. So, all the cell needs to do is make another amino acid chain that is identical to the first one. If it's identical, it will fold up the same way, and presto—a replacement part for the pump!

The DNA in the nucleus has one tiny section that contains the codes for all the amino acids needed to make that new pump part. (This tiny section is called a ***gene***. More about genes in a later chapter.) This section of DNA will have all the codons in the correct order, ready to be "read" and "translated" into a chain of amino acids. It might look something like this (only a lot longer).

ATG AAT GGC GGA GAT TTT TAA GTT AAG AAG GAG CAC TTC TAT ACT GAT CCG TCG TAA

Of course, in the DNA there aren't any spaces between the codons, so it would actually look more like this:

ATGAATGGCGGAGATTTTTAAGTTAAGAAGGAGCACTTCTATACTGATCCGTCGTAA

There's a big problem, though. These instructions can't leave the nucleus. It's not just that they aren't allowed to leave—they physically <u>can't</u> leave. They are too large to fit through the tiny holes in the membrane that surrounds the nucleus. The DNA is stuck in the nucleus. To solve this problem, we have a special messenger molecule called ***messenger RNA***. Messenger RNA is abbreviated like this: ***mRNA***. (When you see "mRNA" you can say either "M-R-N-A" or "messenger RNA.") Messenger RNA will be able to make a copy of the instructions, then exit the nucleus by slithering out through a tiny hole.

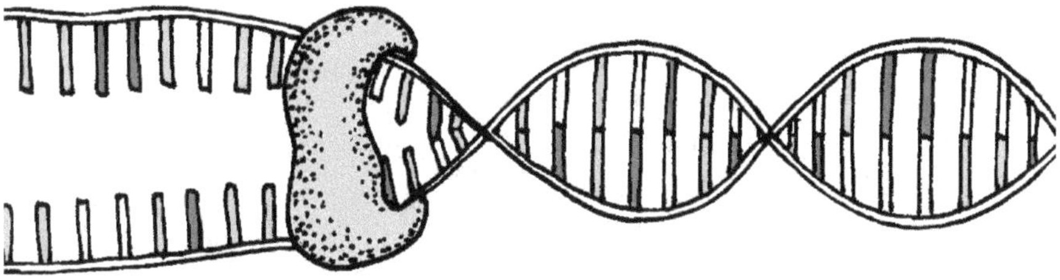

To make mRNA, first a tiny machine (a specialized protein) unwinds and unzips the section of DNA that will be copied. The bonds that connect the A's to the T's and the C's to the G's are fairly weak—it's not hard to pull them apart. Once the process is complete, the DNA zips right back up again.

Next, another specialized protein (that looks and acts a bit like a sled) attaches itself to the side of the DNA that is opposite the side where the coded information is. Scientists used to think that DNA had only one side that contained information and that the other side was simply a mirror image and was "nonsense." Now we know that both sides of the DNA contain information, but the words "sense" and "nonsense" are still used to keep track of which side is being copied by mRNA. The "reader" slides along the "nonsense" side. It knows where to start and stop reading because there are codons that mean "start" and "stop." Fortunately, this reader also happens to be an assembler, and as it slides down the DNA and reads each letter, it grabs the matching letter from the floating supply of nucleotides, and sticks it on the end of the chain it has begun to make. Individual nucleotides go into the reader, and a string of mRNA comes out. This process is called ***transcription***. ("Trans" means "across," and "script" means "write.") The correct name for this "sled" is ***RNA polymerase***, which means "RNA maker."

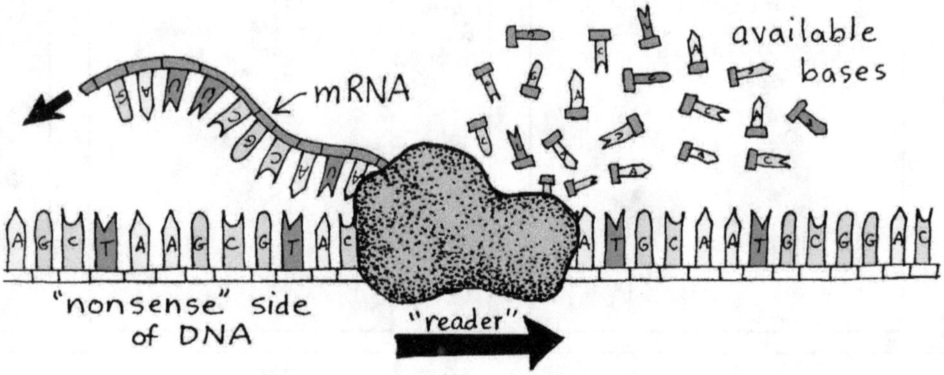

When the mRNA is complete, it looks very similar to DNA, except that it has only one side. It's an exact copy of the "sense" side of the DNA, but with one small difference: **mRNA doesn't have the letter T.** It has A, C and G, but not T. Wherever there should be a T, **there is a U instead**. U stands for ***uracil*** *(YURE-uh-sill)*. The U molecule is very similar to T and works the same way. We just have to get used to using the letter U instead of T. Sorry for having to add a little more confusion to this already long chapter, but that's just how mRNA works.

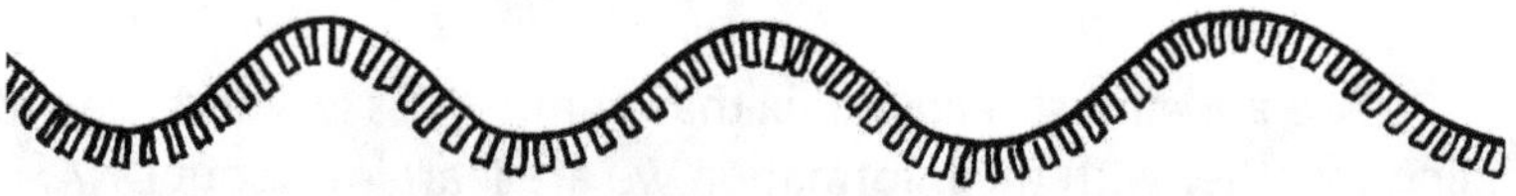

When finished, mRNA looks like one half of DNA.

CLOSE UP VIEW

(Remember, there aren't any letters or geometric shapes in real mRNA. If we could see mRNA, it would look like a long, clumpy string of atoms.)

I'd like to register a complaint about uracil. It's a stupid-sounding word, and it makes this chapter even harder!

When it has finished this process, the mRNA looks exactly like the "sense" side of the DNA (except for having U's instead of T's). Our piece of mRNA is an exact copy of the section of the DNA that has the needed information for making the pump

Then the mRNA snakes out through one of the tiny pores in the membrane surrounding the nucleus. The hole is big enough to let RNA pass through, but small enough that the DNA can't get out. (You could think of it as the nucleus's way of making sure its reference books (instruction manuals) stay in the library. The books are too big to fit through the door!)

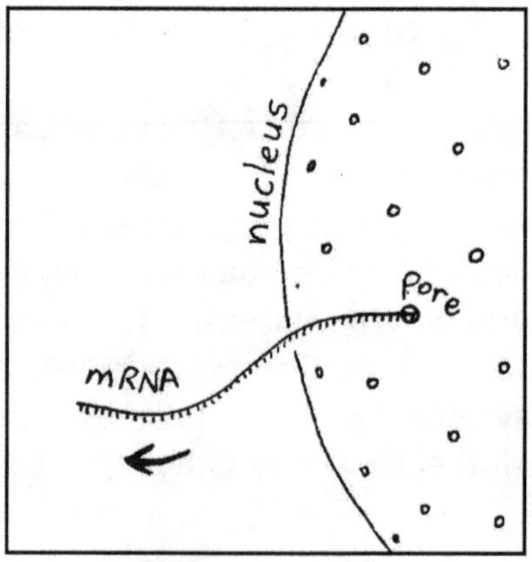

Once outside the nucleus, the mRNA heads for a ***ribosome***. A ribosome is a little "factory" that assembles proteins. It has two halves, one slightly smaller than the other. When the mRNA comes along, the two halves of the ribosome close around it. (This view is sort of a "cut away" view where you can see inside. Normally, you wouldn't be able to see the whole strand of mRNA. You'd just see the ends sticking out.) Ribosomes can float freely in the cytoplasm or they can be attached to a network of tubes near the nucleus.

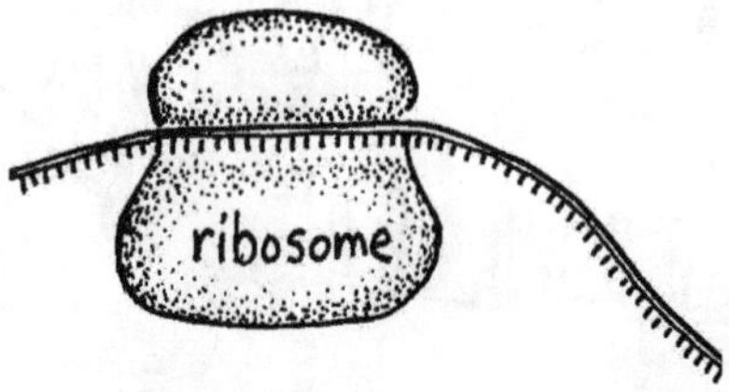

This ribosome looks as though it is one of the free-floating ones. Notice how the ribosome has a long, thin slot that the mRNA fits into. Now we are ready for the next step: ***translation.***

When you translate a book into another language, you replace all the words of one language with words from another language, but without changing the meaning of what is being said. In biological translation, the language of mRNA (written with A, T, C and U) is translated into an actual protein made with amino acids. Translation of languages is done by a translator, a person who understands both languages. Biological translation also has a translator—a molecule called ***transfer RNA or tRNA***. (When you see "tRNA" you can say "T-R-N-A" or "transfer RNA." Either is fine. Books always write "tRNA" because it's shorter and easier, but you can say it either way.)

Like other RNA's, tRNA is a single strand made of a "backbone" (the side of the ladder) consisting of sugars and phosphates, with nucleic acids (C, G, A and U) attached to them. Like mRNA, U replaces T. The two major differences between mRNA and tRNA are size and shape. tRNA is much shorter than mRNA. It doesn't need to be long in order to do its job, as we will see in a minute. Also, tRNA has a definite shape; it's not just a long strand like mRNA is. Artists draw tRNA a number of different ways. In real life, of course, it is a clump of atoms. In textbooks, it is drawn according to what the authors want you to understand about it. All four of these pictures represent tRNA. None of them look just like real tRNA. Each drawing has its strengths and weaknesses.

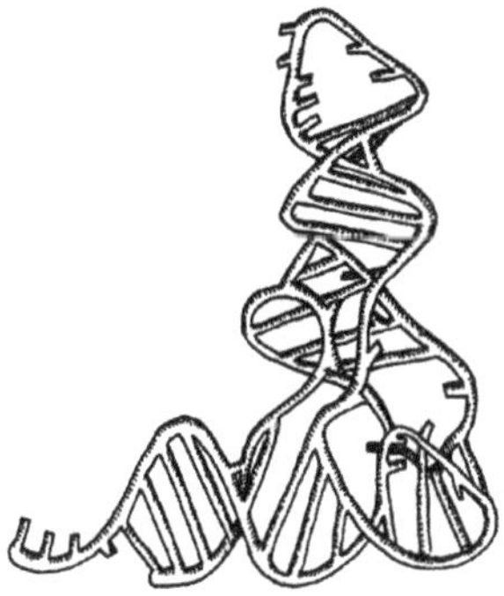

This drawing shows you the helix shape of tRNA and gives you a more accurate sense of its real 3D shape. But it's not helpful for showing how it works.

This drawing is an attempt to simplify the drawing on the left. Like the one on the left, this drawing's main purpose is to show you its 3D shape.

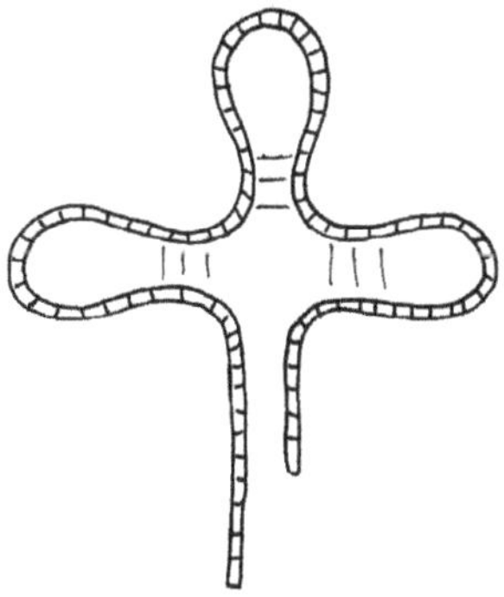

The helix has been flattened out in this drawing. Each loop represents a different part of its twisty helix shape. This shows how tRNA works, but you don't see its real shape.

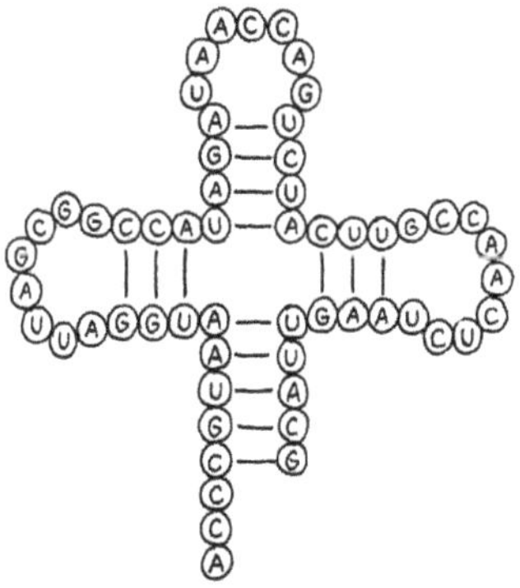

The purpose of this drawing is to show the nucleotide bases (A,C,G,U) that tRNA is made of. This drawing is very good for showing how tRNA does its job.

The first diagram (top left) is trying to show you that tRNA is really a three-dimensional twisted coil of RNA. When the shape is flattened out by artists, it ends up looking something like a clover leaf. At the bottom, where the clover stem would be, is a special "hook" that can carry an amino acid. You can see this "hook" in all four drawings. It's the single strand that sticks off the molecule at the bottom.

PAUSE

PLAY

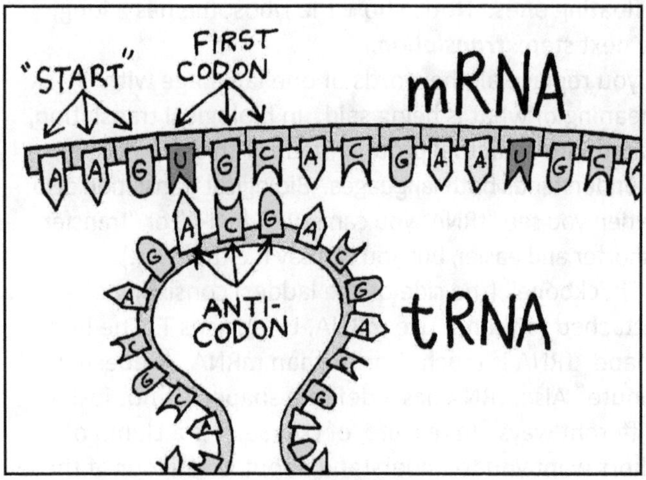

On the top of the tRNA molecule, there are three bases (A, C, and G in this picture) that form a special codon called an ***anti-codon***. The anti-codon matches up with a codon on the mRNA (U, G, C in this picture). The matching of the anti-codon with an mRNA codon is the key to getting the correct amino acids connected in the proper order.

The first codon, AAG, means START. Next, comes the first real codon, UGC, which is the code for the amino acid cysteine. The mRNA code keeps going, with every three letters representing amino acid. Now let's see how these tRNAs do their transferring.

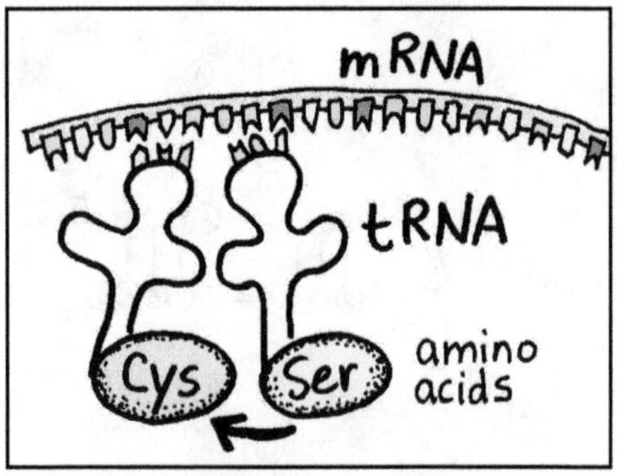

tRNA has a "hook" at the bottom that can carry an amino acid (the one that matches the mRNA code). After the first tRNA is in place, a second one comes along and matches the second mRNA codon. Then the two amino acids that they are carrying are joined together.

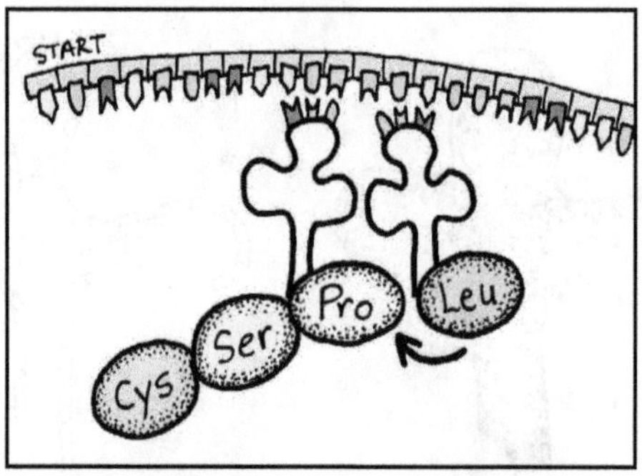

Now something like a relay race begins. The second tRNA must stay in place until a third comes along. After the third tRNA's amino is joined to the second's, it can leave.. The third must stay until a fourth comes along. And so it goes. There must always be two tRNAs on the mRNA, until it reaches the last codon.

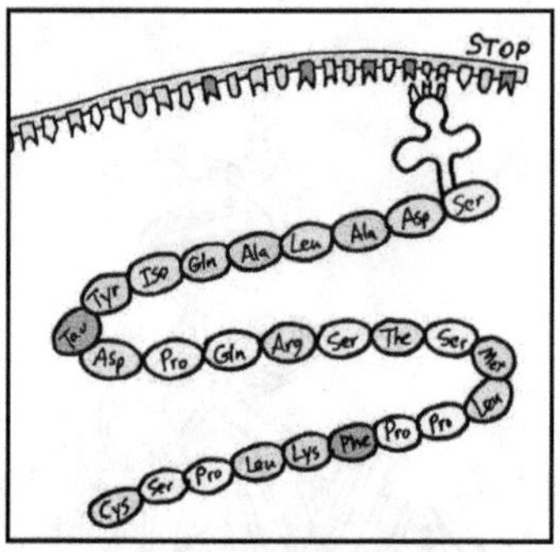

tRNAs "loaded" with amino acids keep coming along and matching up to codons on the mRNA. A very long chain is the result. Finally, the end is reached at a codon that means STOP. Now the protein chain is finished and ready to be delivered.

And now, the protein chain is ready for delivery. A "mailing label" was put on the end of the protein chain during the final steps of the manufacturing process, enabling a motor protein to deliver it to the correct location. The label is a short series of amino acids that is specially made to "dock" at one of the portals (those dots) on the mitochondrion.

Sometimes there are helpers called ***chaperones*** *(SHAP-er-oans)* that travel alongside. Then the long chain will be pulled through the tiny hole. Once inside, the chain will be folded into its final shape, often with the help of chaperone proteins specialized for folding.

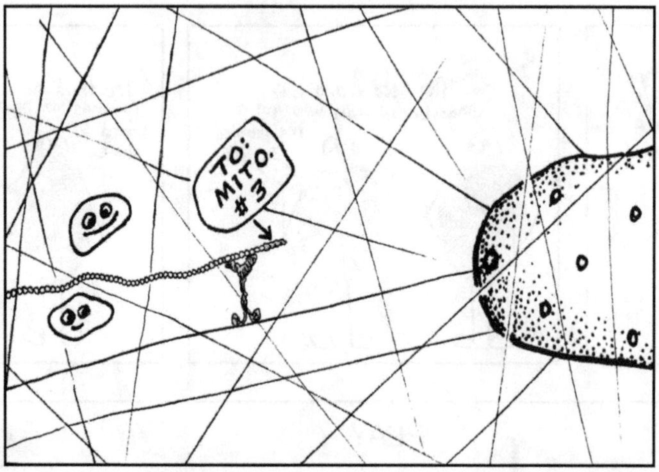

Here is a very silly cartoon of some chaperon proteins helping the amino acid chain to fold into the correct shape. Notice that there is a supervisor who checks for accuracy. If the protein is folded incorrectly, the supervisor will mark the protein as a "fail" and send it to a shredder that will recycle the parts. The process will have to begin again for a second try at getting a properly folded protein. There is also a special “clipper” protein that will remove the delivery tag before the pump is put in place.

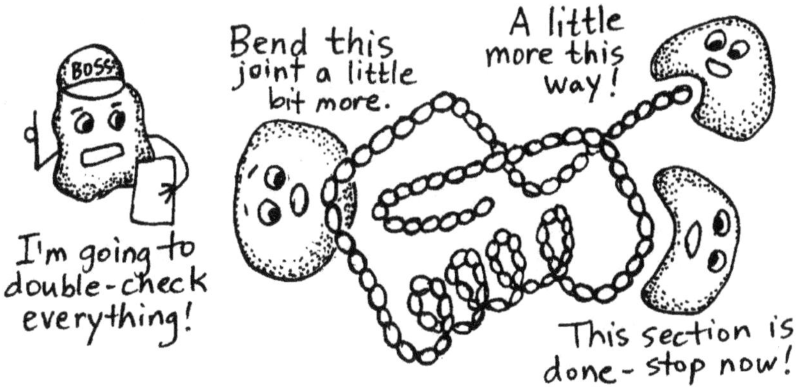

Now our cell has a new problem. How do you get rid of broken parts?

* *

ACTIVITY 5.1 The DNA Song

You can find the music for this song on the YouTube playlist—click on “DNA Song.” If you would like a digital audio file you can find one at www.ellenjmchenry.com. Click on the tab labeled “MUSIC.”

The DNA Song

De-oxy-ribo-nucleic acid,
Watson and Crick worked together with Franklin,*
Learned its shape from x-ray diffraction,**
Double helix, DNA.

Adenine, thymine, cytosine, guanine,
Adenine, thymine, cytosine, guanine,
Adenine, thymine, cytosine, guanine,
Make the rungs of DNA.

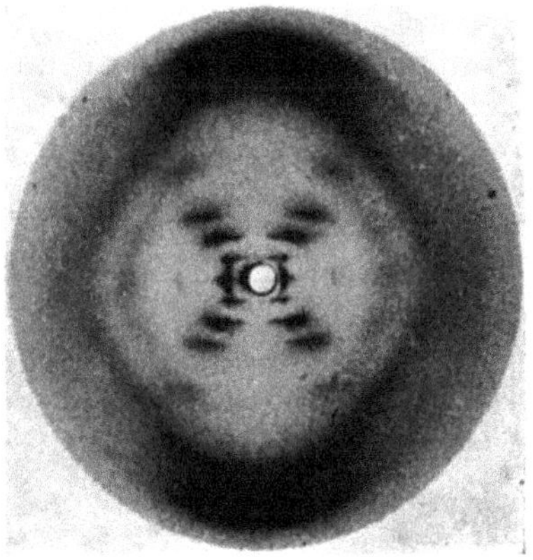

This is the famous image of DNA taken by Rosalind Franklin.

*James Watson was only 25 years old when he moved from America to England to work with Francis Crick. The two of them got help from Rosalind Franklin, an expert in the field of x-ray crystallography.
** Part of the x-ray crystallography process

ACTIVITY 5.2 Watch animations of what we learned in this chapter

The YouTube playlist contains quite a few videos that show animations of the processes described in this chapter. Transcription and translation are much easier to understand when you see them actually happening in an animation. Also, there is a video (animation) of a protein chain being delivered to a mitochondrion, the scenario we followed at the end of this chapter.

Confused about the tRNA molecule? The Wikipedia article on transfer RNA has a 3D animated picture that turns around so you can see it from all angles.

Can you remember what you read?

1) What type of atom is found in proteins but not in lipids or simple sugars?
a) nitrogen b) oxygen c) carbon d) hydrogen

2) TRUE or FALSE? Amino acids taste sour, like other acids.

3) The smallest, simplest amino acid is: a) alanine b) glycine c) leucine d) cysteine

4) How many kinds of amino acids are there? a) 3 b) 4 c) 12 d) 20 e) 64

5) TRUE or FALSE? Fatty acids are able to curl up into a helix shape.

6) TRUE or FALSE? The final shape that a protein takes (after folding) will determine what job it is able to do.

7) Which one of these is NOT a job that a folded protein might do?
a) messenger b) pump c) portal d) flag to identify cell as "self" e) tag for ABO blood types

8) Which one of these does DNA <u>not</u> have? a) phosphates b) sugars c) lipids d) bases

9) Which of these scientists does NOT get credit for helping to discover the shape of the DNA double helix?
a) James Watson b) Francis Crick c) Linus Pauling d) Rosalind Franklin

10) Which of these is NOT in a "nucleotide"? a) sugar b) base c) codon d) phosphate

11) Which base always matches with cytosine? a) adenine b) guanine c) thymine d) uracil

12) Both sides of DNA contain actual information, but when we are talking about just one strip of information (a "gene"), we call that side the "_________" side.

13) Messenger RNA uses "uracil" to replace which base? a) A b) G c) C d) T

14) TRUE or FALSE? DNA can leave the nucleus.

15) How many bases form a codon? a) 2 b) 3 c) 4 d) 20

16) What organelle reads the mRNA and uses the information to assemble a string of amino acids?
a) mitochondria b) ribosomes c) centrosomes d) codons e) amino acids

17) What is transfer RNA designed to carry?
a) an amino acid b) a codon c) mRNA d) information from DNA

18) TRUE or FALSE? Protein chaperones inside the mitochondria prevent the proteins from folding.

19) How does the motor protein know where to take the amino acid chain that will become the new pump?
a) It doesn't, so it is a random process.
b) The mitochondria sends out a chemical message to the motor protein.
c) The end of the amino acid chain (for the new pump) has a sequence that acts as a mailing label.
d) Certain motor proteins always travel to mitochondria, so the ribosome makes sure the amino acid chain gets to the correct motor protein.

20) Where would you find an "anti-codon"? a) on DNA b) on tRNA c) on mRNA d) on a protein

ACTIVITY 5.3 Try an online protein folding game (and help researchers)

If you like to do puzzles and games on your computer, consider going to "https://fold.it" to learn about a computer game that was designed to help researchers figure out how complex proteins are folded. It turns out that human intuition can see solutions that computers can't, even though computers have superior speed and memory. The "Fold It" program didn't start out as a game, but researchers soon realized that solving these protein folding puzzles was actually kind of fun, and that there might be people (even kids and teens) who would be intrigued by this idea and could contribute their time and mental energy. Figuring out the steps in the folding of certain proteins is a key to solving the mystery of many diseases, so the puzzle solvers would also be helping scientists to find cures. The "Fold It" website has everything you need to participate. You need not be an expert in proteins! All you need is an interest in solving visual puzzles and a few spare hours to spend playing the game.

ACTIVITY 5.4 An easy online simulation demo of transcription and translation

This activity can be done in 10-15 minutes. It is a simple demo where you build a protein by clicking on the appropriate codons. https://learn.genetics.utah.edu/content/basics/txtl/.

ACTIVITY 5.5 One last vocabulary word

There was so much new information in this chapter that we tried to keep new words to a minimum. The term "amino acid chain" served us very well during the explanations of transcription and translation. However, **there is a special word for a chain of amino acids,** especially a chain that is fairly long. This is a term that you will undoubtedly run into if you do any further reading about this topic. Find out what this word is by completing the puzzle below. Read the clues fill in the vertical words.

1) Some amino acid chains curl into the _______ helix shape.
2) The organelle that reads mRNA and uses its information to assemble a chain of amino acids.
3) The process described in number (2).
4) The correct name for the reader that reads DNA and makes mRNA: "RNA ______________."
5) The sides of the DNA helix are made of sugars and ________.
6) A ___________ is made of one sugar, one phosphate, and one base.
7) A protein channel that allows water to enter a cell. (Answer on page 40.)
8) This type of atom is found in amino acids but not in lipids or simple sugars.
9) This base replaces T in mRNA.
10) A ______ is made of three bases and is a code for an amino acid.
11) A small section of DNA that has instructions for making a certain protein.

THE ANSWER IS THE WORD FORMED BY THE CIRCLES:

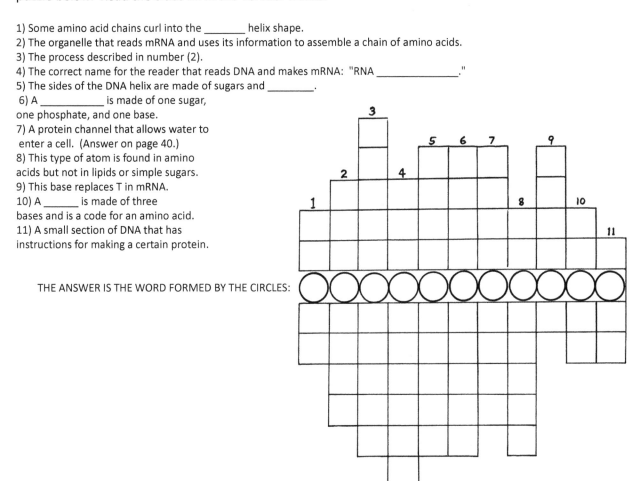

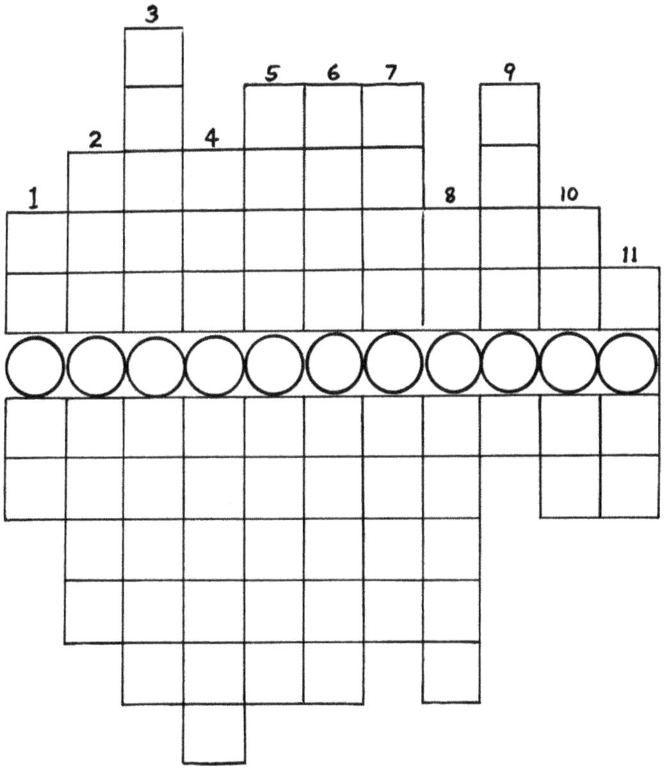

ACTIVITY 5.6 BONUS INFORMATION: Handedness in amino acids and DNA

Did you know that amino acids and DNA can be right or left handed? Amino acids are almost always left handed while DNA is mostly right handed. "Handedness" in molecules is called "chirality." (The Greek word "chiro" means "hand.")

Handedness in molecules is easier to understand if you look at 3D models, not flat drawings. Right and left handed molecules appear to be identical. However, they are different in the same way that a left-handed glove is different from its right-handed mate. You can't put your right hand into a glove designed for the left hand.

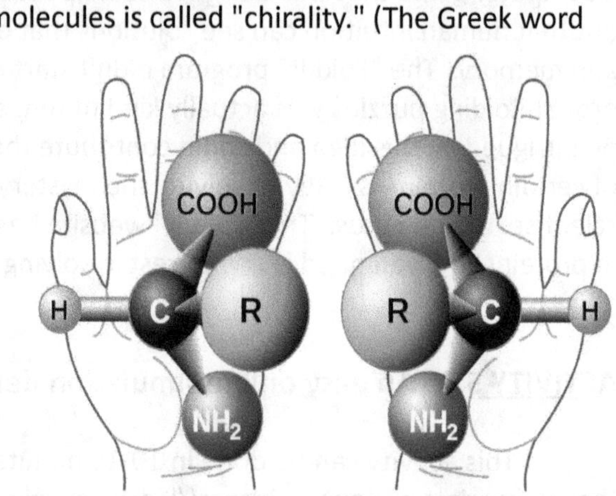

When amino acids are artificially made in a test tube, we get an equal number of right and left handed molecules. But when living cells make them, they almost always make them left handed. (The two exceptions we know about are found in bacteria, and in one messenger molecule in your brain.)

DNA is predominantly right handed. To understand the chirality of DNA, look at the staircase photographs. Imagine walking up the stairs with your hand on the railing. In the staircase marked "R" your right hand will be on the railing. This is a right-handed staircase. When going up the L staircase, your left hand will be on the railing. Now look at the drawings of DNA. Imagine climbing up the DNA rungs. Can you see how they are right and left handed? The left handed DNA looks funny because the amino acids rungs don't match up perfectly. **Another name for left handed DNA is Z DNA because it looks "zig-zaggy."**

L L R R

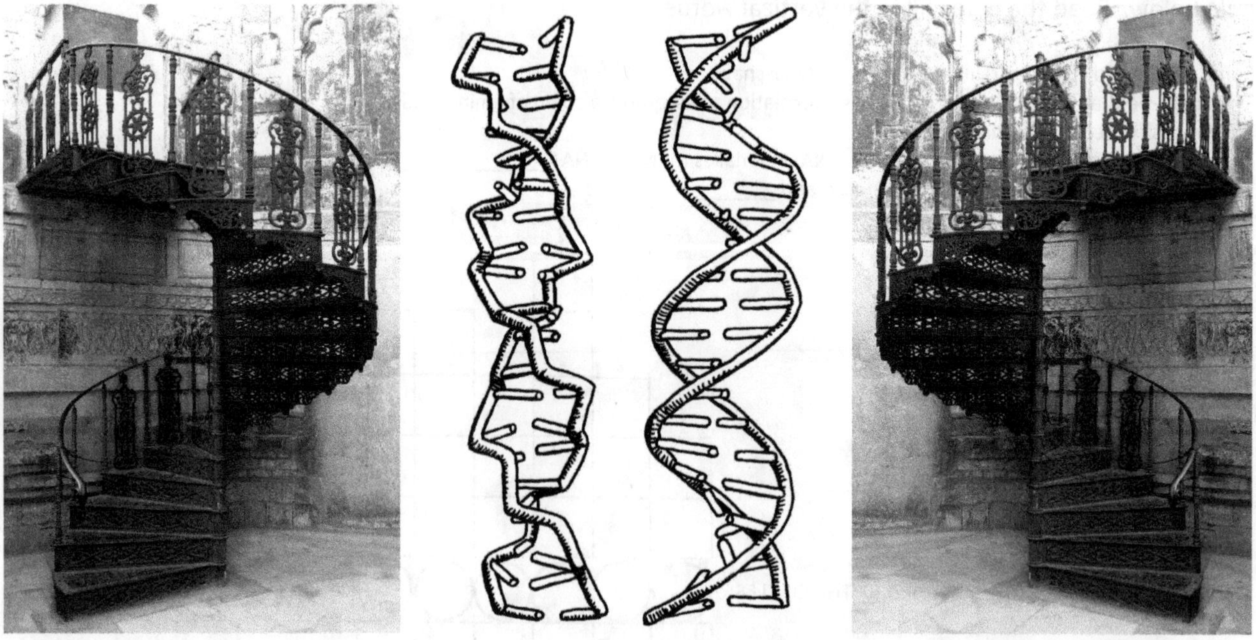

Z DNA was discovered in the 1970s but it wasn't until the 1990s that we had any clues about why it exists and what role it plays in the cell. They usually find Z DNA right behind the RNA polymerase machine that is reading DNA and making mRNA. The DNA helix gets pressed out of shape during this process and can cause the section behind it to temporarily become left handed. In some cases, this turns out to be a good thing, because the mRNA can fold back on itself and stick together, making a double-stranded helix shape similar to DNA. Cells are programmed to freak out over doubled-stranded RNA because many viruses have this type of RNA. We don't want the cell freaking out for no reason, so the Z DNA creates a place where a protective protein can attach. This protein, called ADARI, can counteract this automatic reaction to double-stranded RNA, and stop the cell from beginning to make an anti-viral chemical called interferon. Perhaps in the future we will find out that Z DNA has other functions.

CHAPTER 6: LYSOSOMES, ER AND GOLGI BODIES

Now we have some broken pump sitting around. What does a cell do with garbage? A small percentage of its wastes will leave the cell and eventually be eliminated by the body. However, most "garbage" will be recycled inside the cell. The cell's recycling center is called a ***lysosome** (LIE-so-some)*, from the Greek words "lysis" and "soma." We already know that "soma" means "body." "Lysis" means to loosen, dissolve, or reduce to pieces. So a lysosome is an organelle that loosens, dissolves, and tears things to pieces. If the pieces can then be used for other purposes, that's recycling!

A lysosome is a membrane-bound organelle, meaning that it has an outer membrane made of phospholipids, just like the cell's plasma membrane. The lysosome's membrane differs from the plasma membrane only in the proteins and sugars that are embedded in it. The plasma membrane has portals that let things in and out of the cell, identification tags on the outside, and anchors for the cytoskeleton on the inside. The lysosome doesn't need these structures. It only needs things relevant to doing its particular job—dissolving and recycling proteins, lipids (fats) and sugars (carbohydrates).

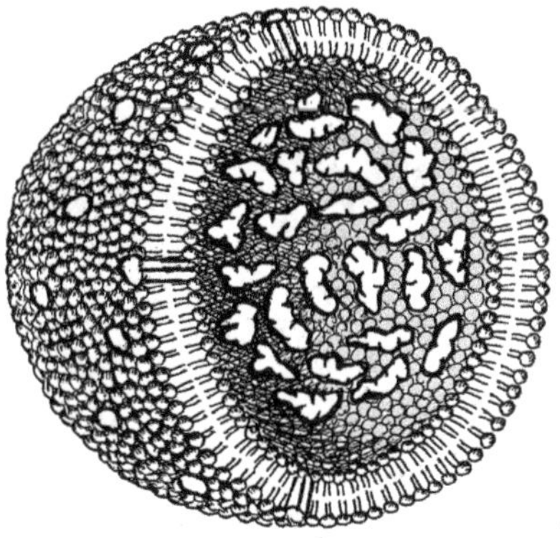

3D CUT-AWAY VIEW

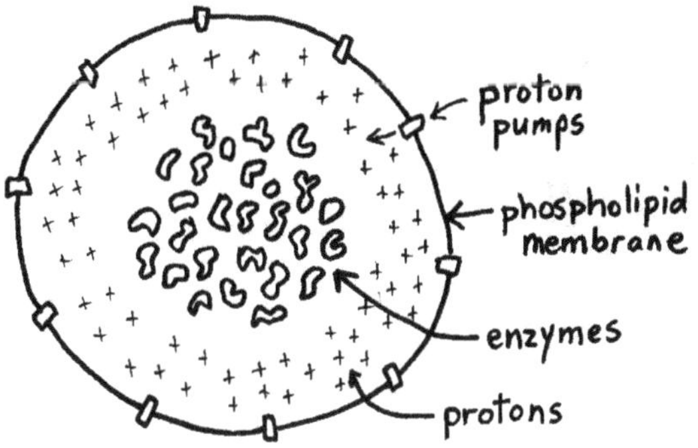

CROSS SECTION DIAGRAM

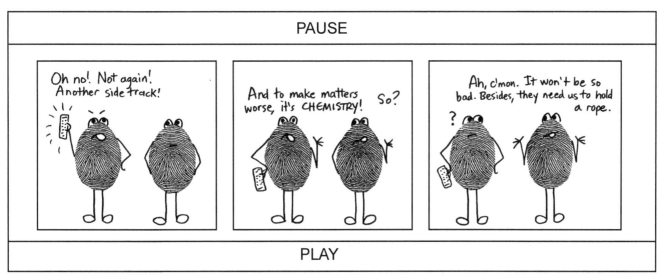

One of the things the lysosome has embedded in its membrane is a particular type of pump called a ***proton pump***. These pumps use ATP energy to bring protons inside the membrane, causing the number of protons inside the lysosome to be much greater than outside. (Left to their own devices, protons would prefer not to be crowded into a lysosome. They have to be forced to come inside, which requires energy.) Why does a lysosome need so many protons inside? The answer requires a very brief chemistry lesson before we go on.

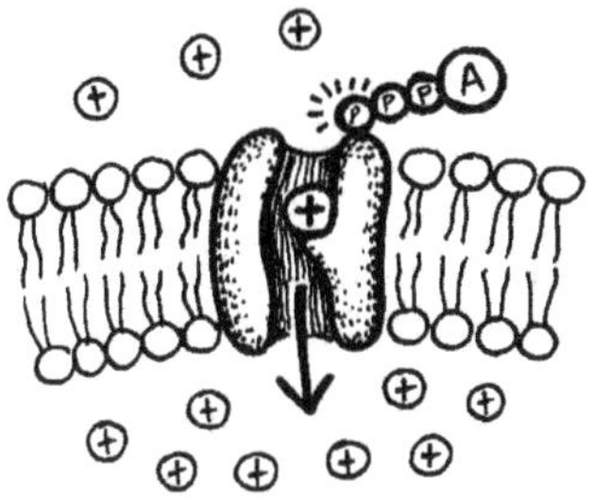

PAUSE

PLAY

A VERY BRIEF CHEMISTRY LESSON

WHAT IS AN ACID? The definition of an acid is: "a substance that accepts (or "wants") electrons." Examples of acids you are familiar with would include lemon juice, orange juice, vinegar, and carbonated beverages. Things that are acidic are tart (sour). Carbonated beverages (except for flavored seltzers) have a lot of sugar added to them to overcome the tartness of the phosphoric acid they contain.

The opposite of an acid is a "base." Bases, which are also known as "alkaline" substances, taste bitter and have often have a soapy texture. Not surprisingly, soap is a good example of an alkaline substance. Baking soda, washing powders and glass cleaners (with ammonia) are also alkaline.

Scientists have a scale for rating how acidic or how alkaline ("basic") something is: the ***pH scale***. The scale goes from 1 to 14. The middle number, 7, is neutral—neither acid nor base. The lower the number, the more acidic the substance is. The higher the number, the more alkaline it is. Plain water that does not contain any dissolved minerals ("distilled" water) is neutral, at 7. Orange juice and vinegar have a pH of about 3. Lemon juice is at about 2.5. The acid in your stomach ranges from 1 to 3. On the other end, baking soda is about 8 and ammonia (often used in cleaning products) is up at 12. (Surprisingly, normal rain water scores a pH of about 5.7. Acid rain is 5.2.)

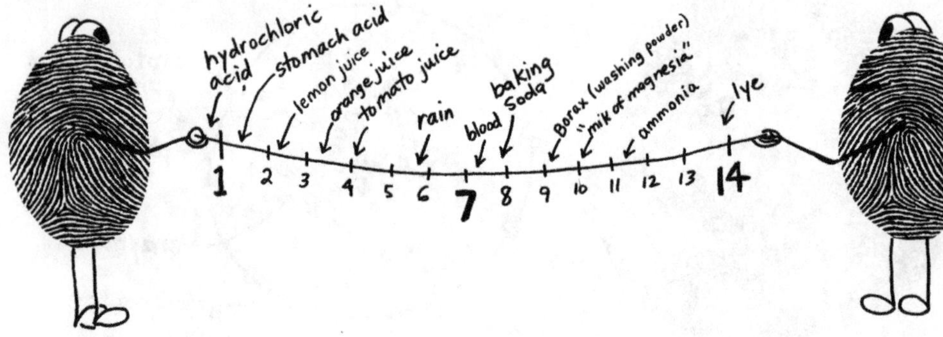

WHAT DOES pH MEAN? It means "***potential Hydrogen***." If you put an electron and a proton together, you get a hydrogen atom. (We saw this at the end of the electron transport chain. Electrons came out of the last pump and were matched with two protons, creating two hydrogen atoms, which were then joined to an oxygen atom to make H_2O, water.) The pH number indicates the degree to which a substance wants to get, or to get rid of, electrons or protons. If it wants to get electrons, it is an acid. The more it wants them, the more acidic it is. If a substance wants to get protons (or the reverse, to get rid of electrons) the more alkaline it is. Confusing? Yeah, a bit. But you don't have to understand it completely to appreciate what a lysosome does.

And now back to the lysosome...

The lysosome is constantly pumping protons inside itself. What happens as a result? Those protons would gladly accept electrons, so, by definition, the interior has become acidic. It has a pH between 4.5 and 5. Why does the lysosome need to make its interior acidic?

Inside the lysosome are proteins called ***enzymes***. An enzyme is a protein molecule that assists in either putting molecules together or tearing them apart. In this case, all of the lysosome's enzymes tear things apart. There are about 40 different kinds of enzymes inside a lysosome. Some of these enzymes break apart proteins, and others break apart lipids (fats) or carbohydrates (sugars) or nucleic acids (DNA and RNA). Each enzyme can only break apart one type of molecule, so many enzymes are needed to break apart the many types of molecules the lysosome must dissolve.

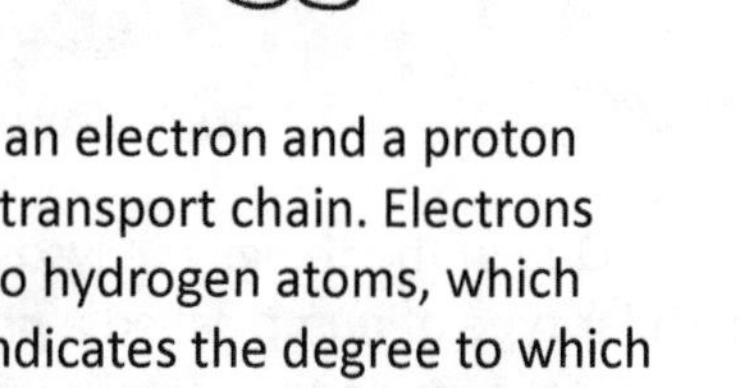

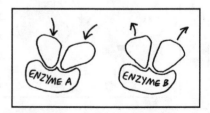

In this picture, enzyme A is joining two molecules. Enzyme B is spltting them.

The enzymes need an acidic environment in order to function properly. It's the same in your stomach. The enzymes in your stomach also need acid in order to do their job, so your stomach makes itself acidic. Pardon the mention of something gross, but if you've ever accidentally burped up stomach juice, you may have felt the acid burn in your esophagus. Your stomach has a protective mucus lining that keeps it from being burned. The lysosome also has ways of protecting itself from its acidic environment and from the action of the enzymes, but scientists are still studying it, so we can't say for sure how it works. However, the fact that the lysosome's enzymes need an acidic environment prevents a major disaster in the cell. What would happen if a lysosome broke open and all the enzymes spilled out? Fortunately, the enzymes would suddenly find themselves in a neutral, not acidic, environment. The (relatively) neutral pH of the cytosol prevents the enzymes from dissolving everything in sight.

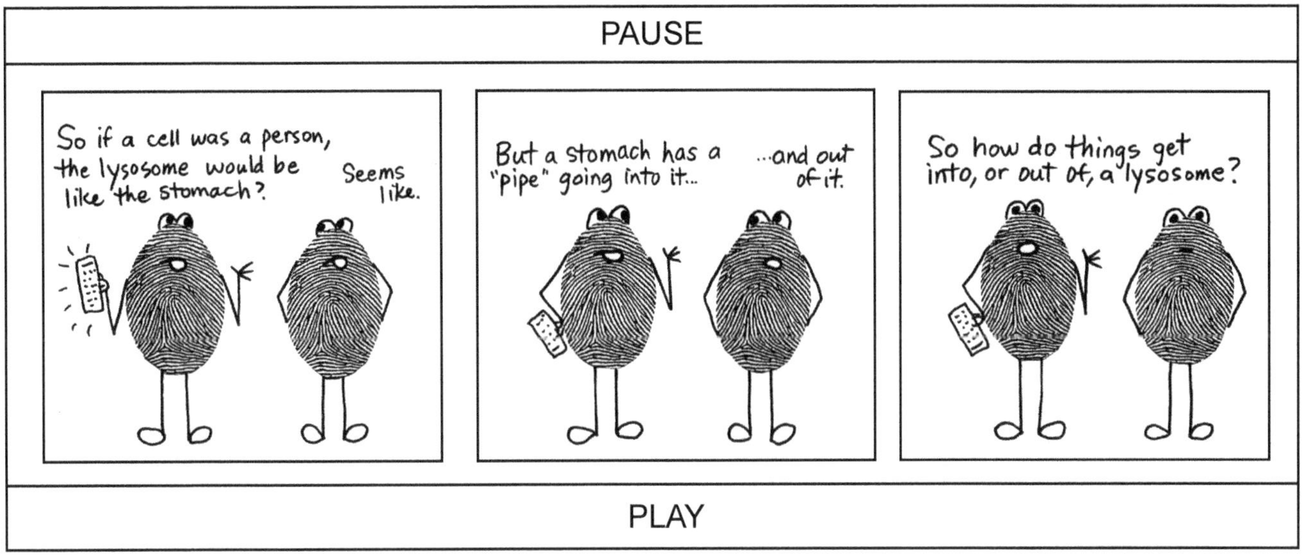

Yes, a lysosome is sort of like a cell's stomach. As you pointed out, it isn't connected to a mouth or intestines. The membrane is continuous all the way around and doesn't have holes. The only way to get into a lysosome is to ***merge*** with it. A good way to see what "merging" means is to observe oil droplets. If you float some oil drops in a bowl of water, you will see that as soon as two droplets touch, they merge and become one larger droplet. Phospholipid membranes work the same way. If the cell can wrap its garbage in a phospholipid membrane, this membrane wrapping will be able to merge with a lysosome. After they merge, the garbage will then be on the inside of the lysosome.

A phospholipid sphere created for the purpose of transporting things (or merging things) is called a ***vesicle***. Vesicles are what cells use instead of cardboard boxes or plastic bags. When they need to ship something or carry something, they put it into a vesicle. Vesicles can be used to hold all sorts of things, including broken pumps. We'll see lots of vesicles in future chapters.

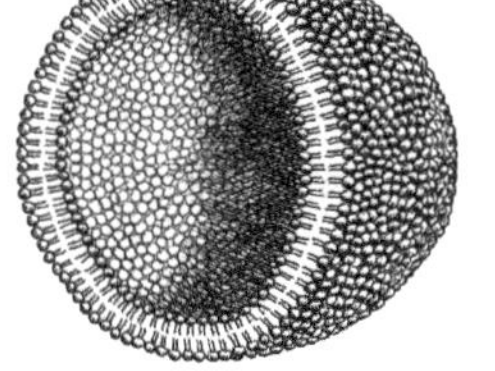

Imagine putting the broken pump into a vesicle and then giving it a push towards a lysosome. It drifts off and gradually approaches the lysosome... it touches the lysosome... and then, presto—it's gone! As soon as it touched the lysosome, the vesicle seemed to disappear. Of course, it didn't really disappear, it simply merged into the lysosome. What used to be the membrane of the vesicle is now part of the membrane of the lysosome. The broken pump now finds itself inside the lysosome. In a matter of seconds, the enzymes reduce the old pump to a pile of amino acids. Then the lysosome will put those amino acids back into the cytoplasm, expelling them through special portals in its membrane. The pieces can either be used by the cell to make more proteins, or they can be exported out of the cell where they will be picked up by lymph vessels.

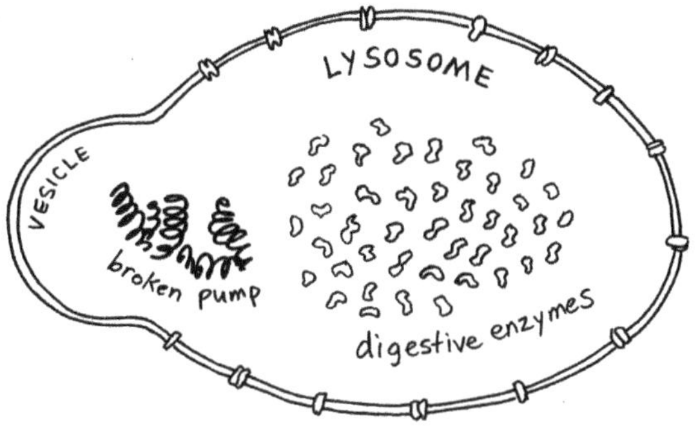

A lysosome can also digest things that come into a cell from outside. Some cells take in food particles by creating a "dent" in the plasma membrane that will eventually be pinched off and turn into a vesicle. Even bacteria can be gotten rid of in this way.

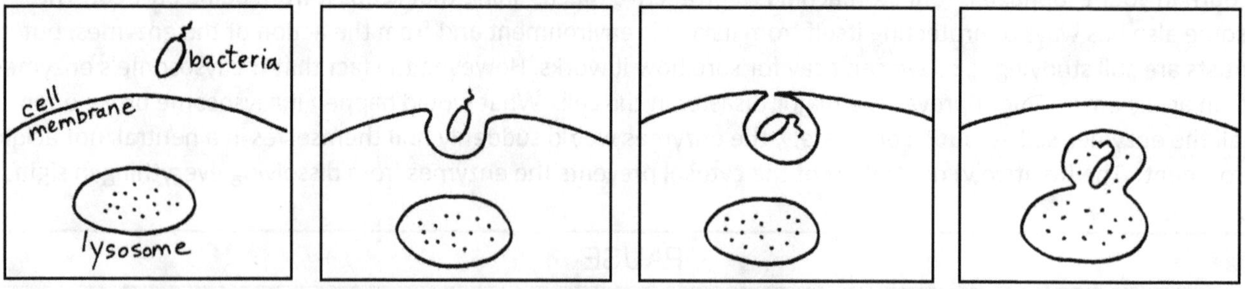

A certain type of white blood called a macrophage (meaning "big eater") has lots of lysosomes. If you watched the video of a white cell chasing down a bacterium, you've seen this cell in action. When the macrophage catches a foreign invader, it will engulf it and create a vesicle to bring it inside the cell. Then the vesicle will merge with a lysosome and "digest" the invader, breaking it down into amino acids, fats, and sugars. These raw materials will then be released into the cell and used to make things the cell needs.

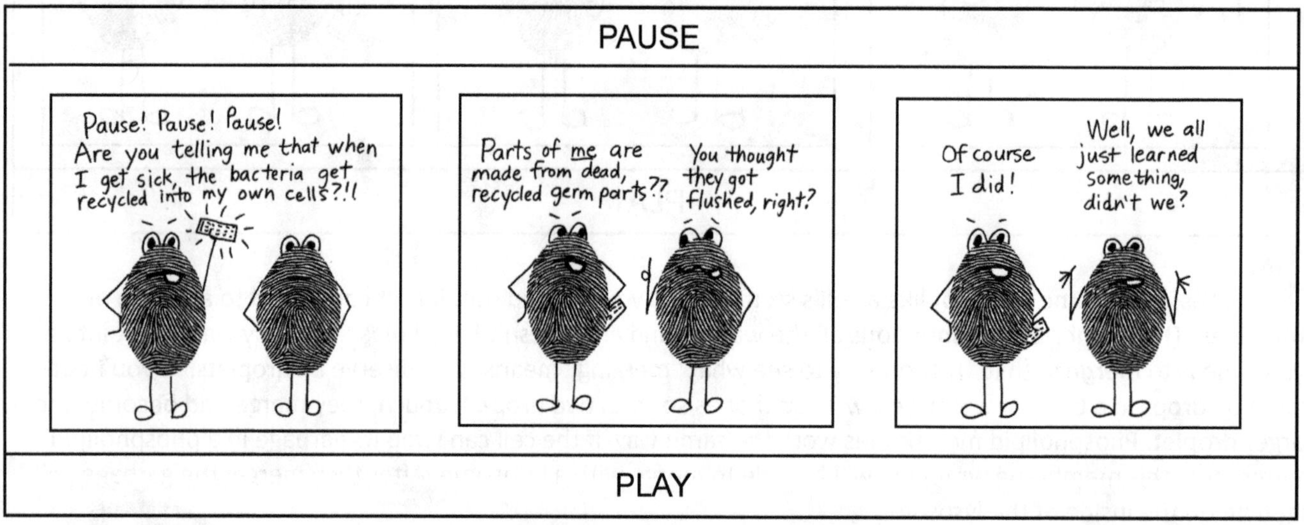

Yes, your body is partly made of recycled germs. Amino acids, fats, and sugars are exactly the same, no matter where they are found. It's how they are arranged (into a virus or bacteria, or into cell parts) that makes us call them "good" or "bad."

Let's learn even more about lysosomes. How does the cell make them? The first step is to make the enzymes that will be inside it. An enzyme is a specialized protein. In the last chapter we saw how ribosomes take mRNA (that came from the nucleus) and work together with tRNA to hook amino acids together into a chain called a ***polypeptide***. The ribosome that made our polypeptide was working all by itself as it floated in the cytoplasm. The polypeptide chain was transported to the mitochondrion where it was folded and put into place. This process won't work for making a lysosome. You can't make a phospholipid ball, then stuff it with enzymes. The ball must be created around the enzymes. This process is fascinating but a little complicated. Buckle your cellular seat belts and hold on for a rather long explanation, but hopefully at the end you'll say, "Wow, cells are amazing!"

The process starts with the assembly of the digestive enzymes. A ribosome begins to read the mRNA instructions for making one of the enzymes. The first 50 or so "rungs" on the mRNA have a special meaning. This first part of the code means, "Take me to the ER!" In this case, the ER isn't the Emergency Room. It's something called the ***endoplasmic reticulum***. *(EN-do-PLAZ-mik reh-TICK-yu-lum)*

The name of this organelle sounds more difficult than it actually is. "Endo" just means "in" or "within." "Plasmic" refers to the cell's cytoplasm (the fluid, or "cell gel"). So "endoplasmic" means "in the cytoplasm of the cell." Not too hard. The word "reticulum" comes from the Latin word "reticulatus," meaning "net-like" or "network." So this organelle isn't round like a mitochondrion or a lysosome. In fact, it's a complicated network of long tubes that are connected at various places. Even scientists find these long words a bit of a bother and they hardly ever use them. They almost always refer to the endoplasmic reticulum by its initials, ***ER***. From now on, we'll do as the cell biologists do, and we will write "ER". When you see "ER" you can say "E-R' or you can say "endoplasmic reticulum," whichever you prefer.

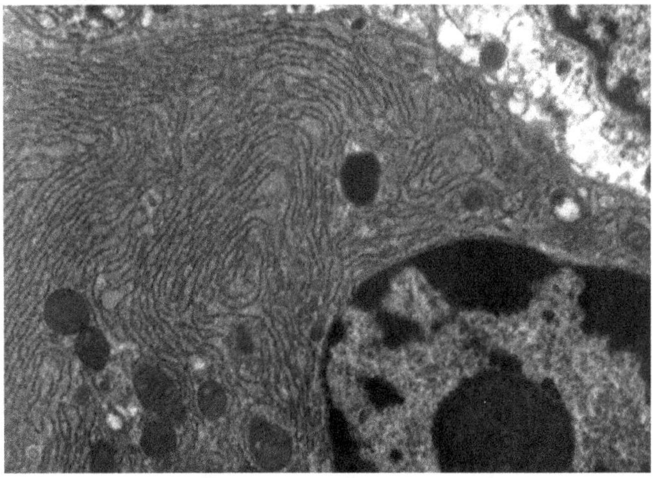

The ER is the squiggly stuff. The nucleus is the large (not smaller) oval area in the lower right corner. The spots in the ER are mitochondria. Sometimes mitochondria and ER work together.

Now back to our story. When the ribosome begins reading this particular piece of mRNA (the instructions for how to made a digestive enzyme), the first part of the code tells the ribosome to take it to the ER. Why? Because the ER is the organelle that can put phospholipid membranes around things. The enzymes must be inside a membrane for reasons we shall see shortly. Follow along with these pictures as we find out what happens next:

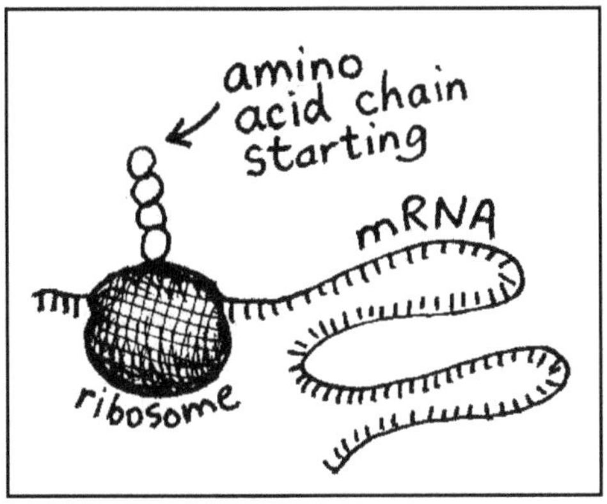

1

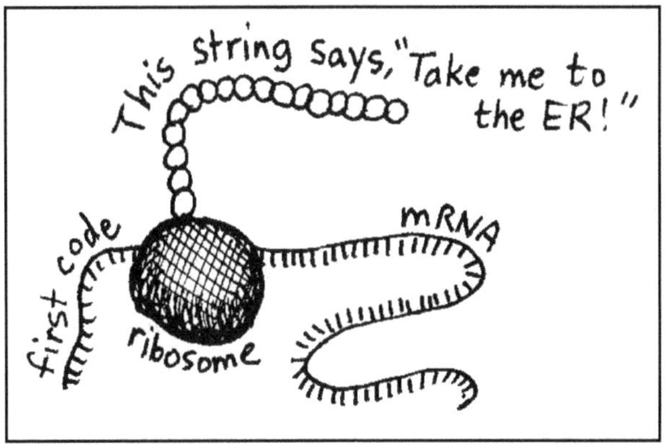

2

1) The ribosome is just starting to make the amino chain.

2) The first part of the code is complete. It says, "Go to the ER." The ribosome cannot make any more of the chain until it has anchored itself to the ER. The mRNA just waits.

3) There are little "docking ports" on the outside of the ER, where a ribosome can come and attach itself, sort of like a boat coming in to a dock. In fact, there are even little "tug boat" proteins that go out and help guide the ribosome into the dock! There are two parts to the little dock. One part is the place where the ribosome is secured so it doesn't drift off (like the "cleat" on a real dock). The other part, right next to it, is a portal (a hole). The ribosome sticks the end of the protein through the portal so that it goes inside the ER.

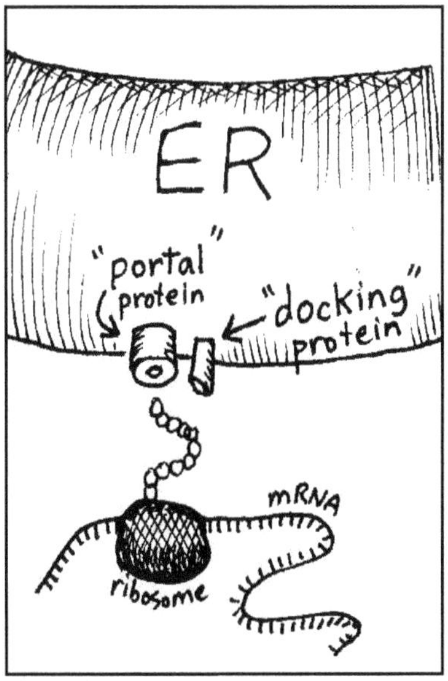

3

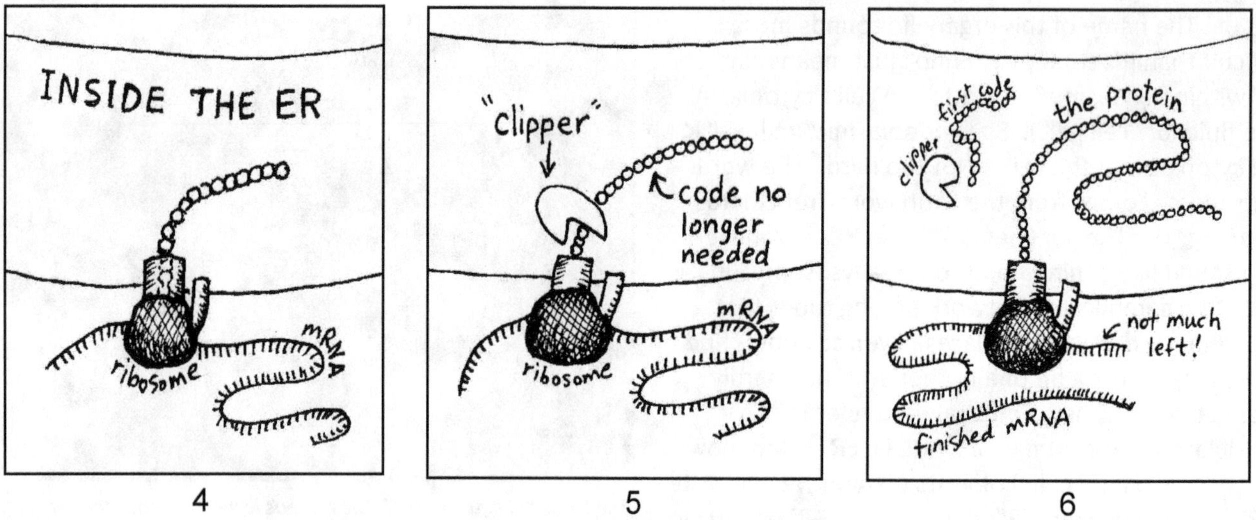

4) In this picture, the ribosome has successfully docked and has inserted the first part of the chain.

5) Little "clippers" come over and cut off the end of the protein because that part of the message isn't needed anymore. (The message said to go to the ER, and that has now been accomplished.)

6) Once that out-dated message is clipped off, the ribosome starts reading the mRNA again. (This picture doesn't show all the tRNAs that are bringing their amino acids over as the ribosome needs them.) As the protein chain grows, the ribosome feeds it through that portal and into the ER.

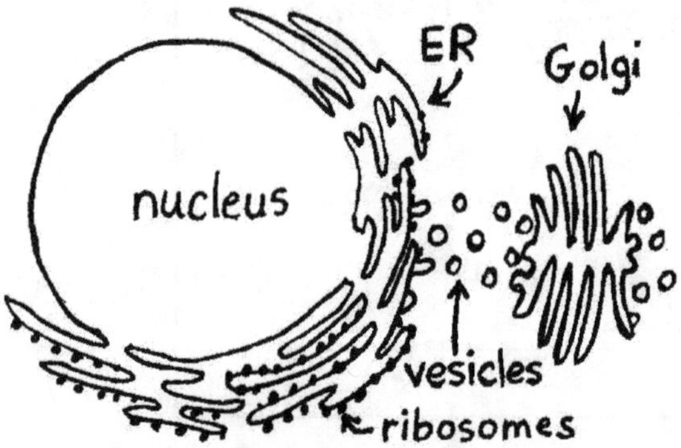

When the ribosome finishes, it detaches from the ER and goes off to find a new piece of mRNA and begins to build other proteins. This ribosome might build another one of these digestive enzymes or it might go off and build something else. Ribosomes are not specific to particular jobs. They can read any piece of mRNA that comes their way.

Meanwhile, inside the ER, the newly formed protein isn't alone. Many other digestive enzymes are being formed. The ER now moves all these enzyme proteins to one of its tubes that is very close to another organelle: the ***Golgi body***. (*Goal-jee*) The Golgi body (or ***Golgi apparatus***) is named after Italian scientist Camillo Golgi, who discovered it in 1897. Golgi bodies stay close to the ER. They don't touch it, but they stay close.

The ER is made of phospholipid membrane, just like mitochondria, lysosomes, and vesicles. The ER now pushes the enzymes outward so they make a bulge in the membrane. Then it pinches off the bulge, and voila! It becomes a vesicle. The vesicle will attract a motor protein and it will carry the vehicle over to a Golgi body.

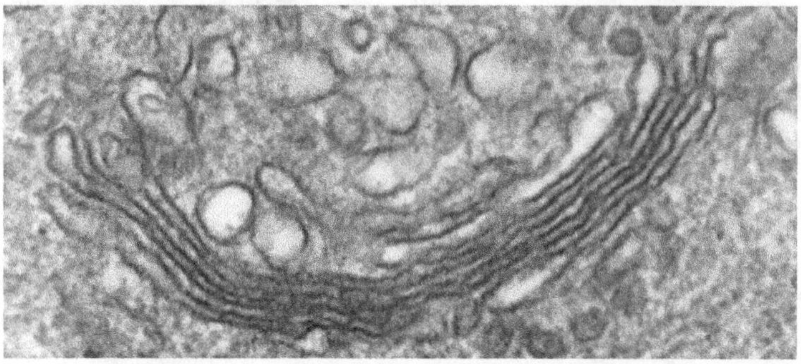

The long, flat shapes are the Golgi body. The round things are vesicles.

Golgi bodies are surrounded by phospholipid membrane, so all a vesicle has to do is make contact with a Golgi, and it instantly merges with the Golgi's membrane. Whatever the vesicle was carrying is then inside the Golgi.

Now for the weird part—the Golgi body itself. It looks like a stack of pancakes with bumps stuck to the first and last cakes. But that's not the weirdest part. The Golgi body is in constant flux. The overall shape stays pretty much the same, but the parts are always changing. Watching an animation is the best way to see this, but here is a quick description.

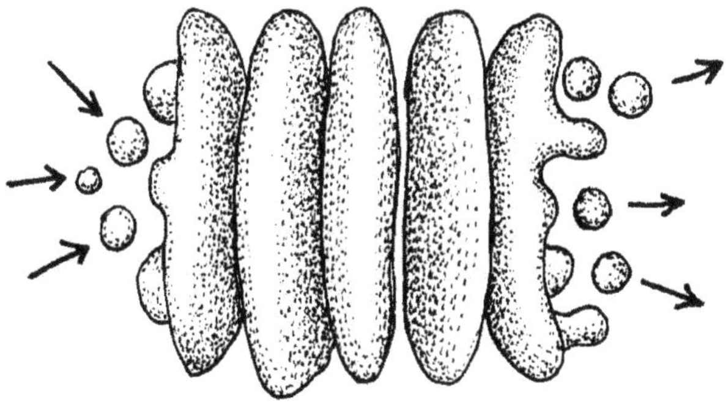

The first pancake receives all the vesicles that are coming from the ER. They merge with the first pancake. Then there are about three pancakes in the middle, then a final pancake. That last pancake is constantly breaking apart into vesicles. Those bumps on the last pancake are vesicles budding off. Eventually the last pancake completely disintegrates and the pancake right behind it moves in and takes its place. Then the next pancake down the line has to move up and take the place of the pancake that just moved into last position. And so on it goes, back to the first pancake. As the first pancake moves up to take the place of the second, new vesicles move in and start forming a new first pancake. Meanwhile, there are little protein "machines" inside the middle pancakes that somehow manage to stay in those middle cakes despite all the moving around.

(SIDE NOTE: This theory of how a Golgi body's pancakes constantly shift (called the "maturation theory") is generally accepted by most scientists, but not everyone is convinced. If you browse around on the web, you might find a site that doesn't agree with this theory.)

All scientists agree that despite any shifting of Golgi pancakes, the little protein "machines" inside the Golgi body stay in place. These proteins are similar to the chaperone proteins we met in the last chapter. The Golgi's chaperone proteins act like post office workers packaging, sorting, and labeling the proteins and enzymes that come through. In the case of our lysosomal enzymes, the Golgi chaperones fold them into the right shape, tag them with (sugar) labels that identify them as lysosomal enzymes, gather all of them into one place, then package them into vesicles.

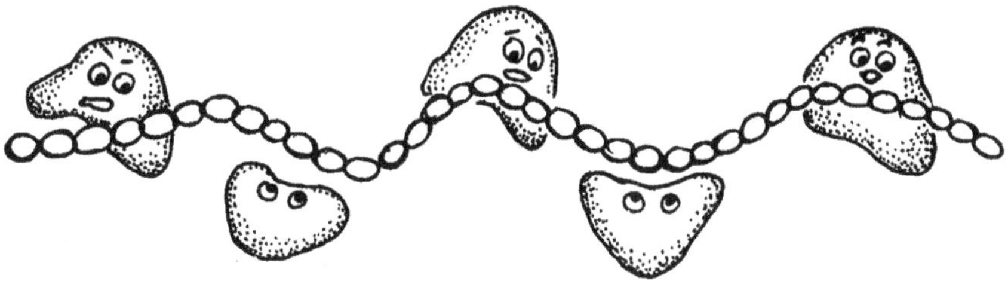

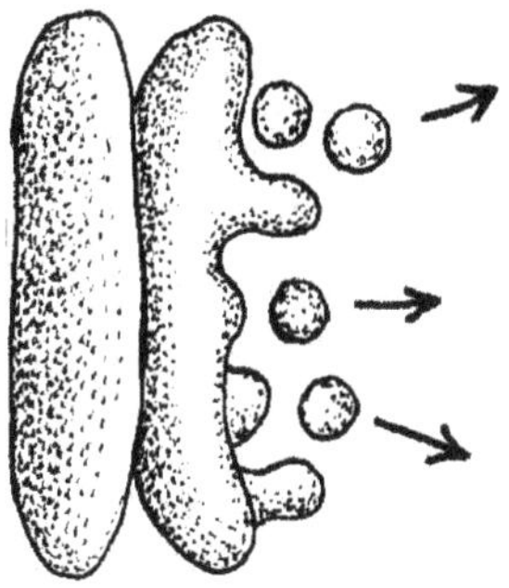

Some of the vesicles budding off the last pancake on this Golgi are "baby" lysosomes. The Golgi body made sure that the future lysosomes have two special types of proteins in their membranes: proton pumps and "portal" proteins that allow the digested particles to exit the lysosome and go back into the cytoplasm. The pumps and portals were manufactured in the ER in much the same way that the enzymes were. These baby lysosomes will drift off into the cytoplasm and either merge with a larger lysosome or will grow into a new lysosome.

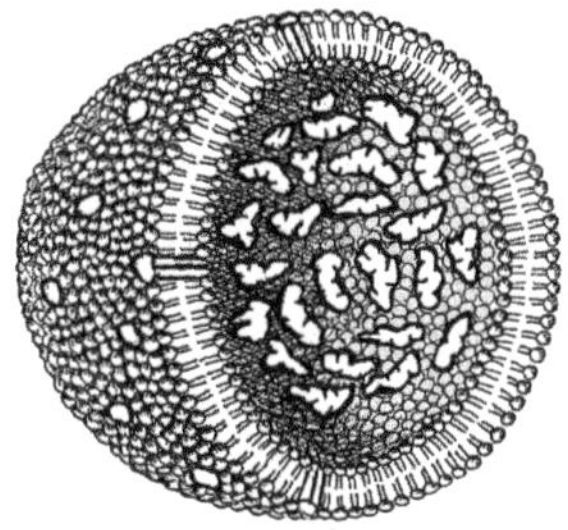

Those little dots on the membrane are portals and proton pumps. The clumps on the inside are digestive enzymes.

Golgi bodies process other things, too, not just lysosomal enzymes. Cells make all kinds of things. Sometimes they even make things that need to be delivered to other cells, even cells that are in another part of the body. The Golgi's job is to put the correct label on each protein that comes through so that it will be delivered to the right place. If the items in question must be delivered outside the cell, then the Golgi puts them into a vesicle that will drift all the way out to the plasma membrane and merge with it. The result of merging with the outer membrane is that the contents of the vesicle end up outside the cell. The Golgi is sort of like a packing and processing factory and a post office all in one. It packs proteins and gets them to where they need to be.

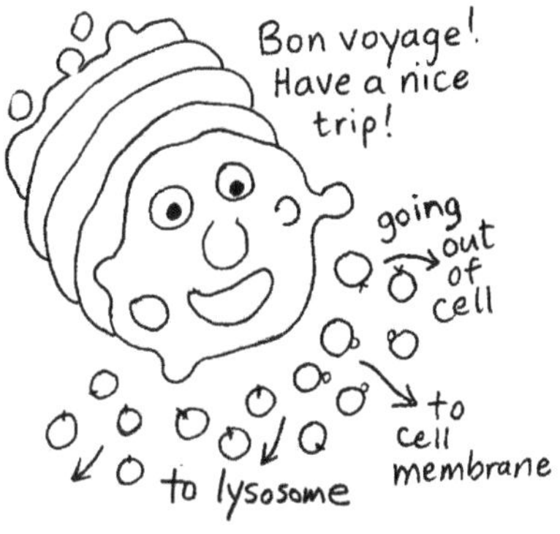

PAUSE

PLAY

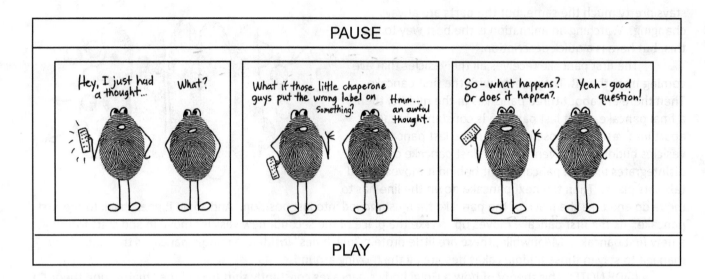

In the case of lysosomal enzymes, if they don't get labeled properly, they don't end up inside a lysosome. You might get "lost" enzymes wandering around the cell unable to do their job, and you have empty lysosomes with no enzymes inside. What a disaster! You'd have cellular garbage piling up and no way to get rid of it. The cells would eventually be so full of garbage that they could not function anymore. When something like this happens, it is called ***lysosomal disease.*** There are many kinds of lysosomal diseases because there are many different kinds of lysosomal enzymes. Babies born with a lysosomal disease have very severe health problems and rarely live beyond the age of six or eight. The most famous lysosomal disease is Tay-Sachs disease. Most adults have heard of this disease but they have no idea what causes it. It is a genetic disease, meaning the root cause is a mistake in the coding of the person's DNA. Since DNA is inherited by children from their parents, one way to try to eliminate these diseases is by testing young people who will become parents and making sure that two people with the same DNA mistakes don't marry each other. Usually, if one of the parents has normal DNA, it won't matter if the other parent has the mistake; the children will be born without the disease. Meanwhile, researchers are also trying very hard to find ways to make faulty Golgi bodies start tagging the enzymes correctly. (Some of the protein puzzles from the protein folding game mentioned at the end of the last chapter could be a part of lysosomal disease research.)

* *

<u>ACTIVITY 1</u> Notes about one of the supplemental videos

You can access the video from the playlist (look for the title "Protein Modification; Gogli"), or you can download it from its website addres at: **http://vcell.ndsu.edu/animations/proteinmodification/movie-flash.htm** Don't let the vocabulary in this video scare you off. (Turn off the sound if it does!) This video will show you the process we explained in this chapter. You will see vesicles going from the ER to the Golgi, proteins being processed within the Golgi, then being tagged and sent to a lysosome.

Here is a little dictionary of terms, so you don't get confused while watching:

- hydrolase = digestive enzyme (they are gray or green in this video)
- oligosaccharide = a sugar tag (Sugars are shown as little colored hexagons. "Oligo" means "few.")
- glycosylation = adding a sugar
- M6P = the "mailing label" that means "deliver this to a lysosome" Mannose-6-Phosphate
- endosome = baby lysosome
- cis cisterna = first pancake (where vesicles are merging)
- trans cisterna = last pancake (where vesicles bud off)
- M6P receptor proteins = the docking sites on the Golgi membrane that anchor only enzymes with the M6P tag on them (sort of like "seats" on the vesicle "shuttle buses")

Can you remember what you read?

1) Which of these describes something a lysosome does?
   a) Makes proteins out of amino acids. b) Breaks down protein chains into amino acids.
   c) Makes amino acids from nucleic acids. d) Helps proteins fold properly.

2) The environment inside a lysosome is very: a) acidic b) basic c) warm d) salty

3) Lysosomes have a special pump in their membranes that brings in lots of _____.
   a) ATP b) proteins c) electrons d) protons

4) How does the lysosome get its enzymes?
   a) It makes its own enzymes. b) They come in from the bloodstream.
   c) A vesicle brings the enzymes from the ER and Golgi body.
   d) The enzymes are brought into the lysosome by pumps in its membrane.

5) TRUE or FALSE? If a lysosome breaks open, the cell is in big trouble because the enzymes will digest cell parts.

6) TRUE or FALSE? The Golgi body processes only enzymes destined for lysosomes, not for any other organelle.

7) Why do ribosomes bind to the docking ports on the ER?
   a) So they don't get lost in the cytosol. b) Because the ribosome would fall apart otherwise.
   c) Because the first part of an mRNA strand told them to go there.
   d) Because the ER is the only place you can find a supply of amino acids for tRNA.

8) Which of these organelles is NOT a membrane-bound organelle?
   a) ribosome b) lysosome c) mitochondria d) endoplasmic reticulum

9) TRUE or FALSE? Lysosomes can only digest proteins.

10) TRUE or FALSE? Golgi bodies usually stay close to the endoplasmic reticulum.

11) What part of the body is a lysosome most like? a) heart b) stomach c) liver d) mouth

12) How do the Golgi's own "worker" proteins stay inside the Golgi body even though the Golgi body is constantly shifting and changing around them?
   a) They anchor themselves to the proteins in the membrane. b) They stay inside vesicles.
   c) They tag themselves so they don't get lost. d) We don't know.

13) Which is the correct order for the process of making an enzyme for a lysosome?
   a) Golgi body, ER, vesicle, ribosome b) vesicle, ribosome, ER, Golgi body
   c) ribosome, ER, vesicle, Golgi body d) ER, vesicle, Golgi body, ribosome

14) TRUE or FALSE? Some enzymes put things together, others take things apart.

15) TRUE or FALSE? Vesicles are made of phospholipids.

16) TRUE or FALSE? Ultimately, the source of information for how to make a lysosome comes from the DNA.

17) The word root "pep" is always associated with: a) proteins b) sugars c) lipids d) DNA e) RNA

18) Which one of these is NOT true about Golgi bodies? a) It contains several dozen kinds of enzymes.
   b) It uses sugar tags as mailing labels. c) It forms vesicles. d) It assembles proteins from amino acids.

19) TRUE or FALSE? Some of the vesicles that leave a Golgi body might contain materials that need to be "dumped" outside the cell.

20) TRUE or FALSE? You can die if your lysosomes don't work right.

ACTIVITY 6.2

TIME TO REVIEW!

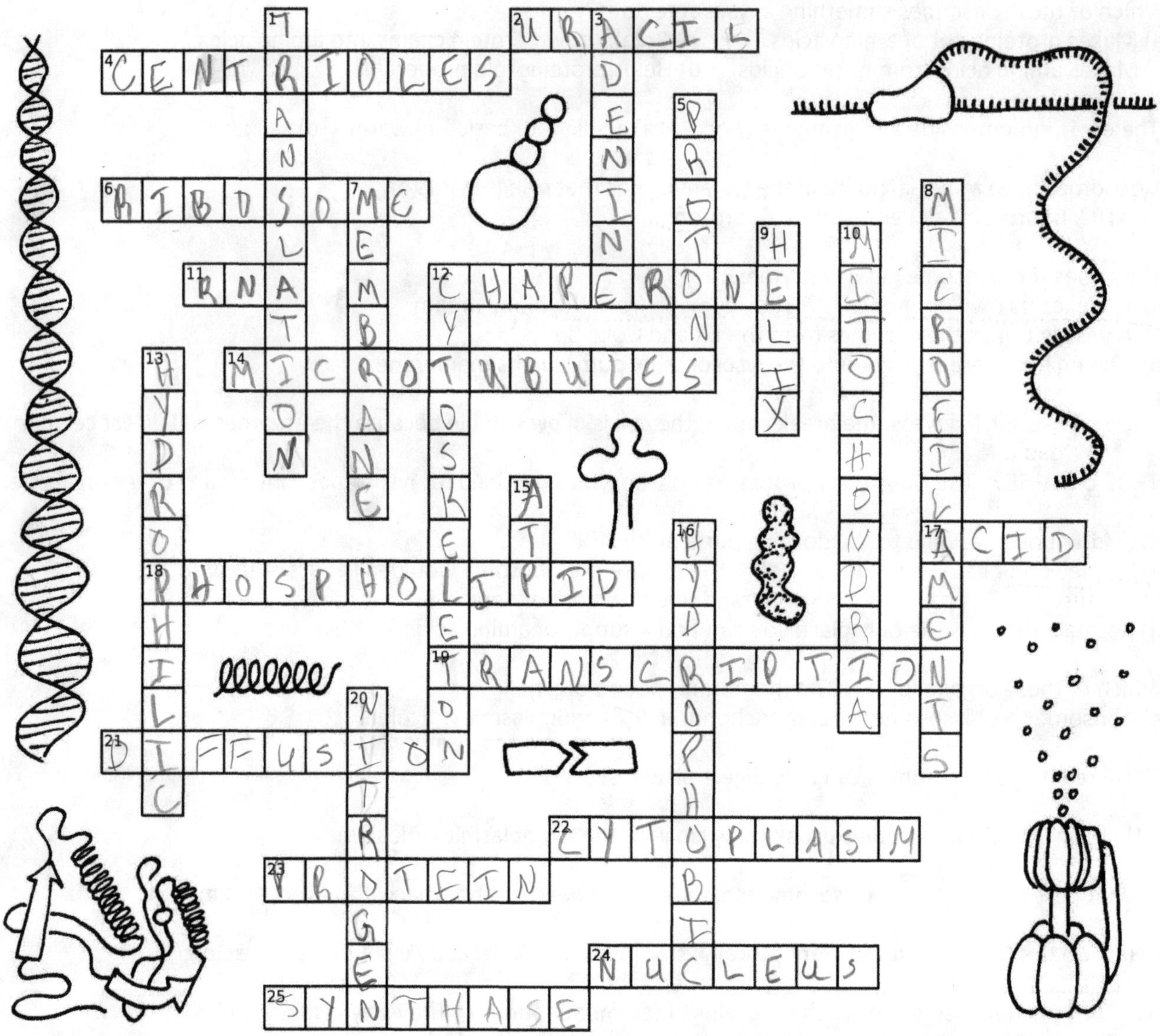

ACROSS

2) This replaces thymine in mRNA and tRNA
4) This is the central organizing area for the cytoskeleton
6) This organelle is the cell's protein manufacturing unit.
11) This contains the information on how to make every protein in your body.
12) This type of protein supervises protein folding.
14) These are part of the cytoskeleton and act as "high-ways" for motor proteins to travel on.
17) A molecule with this group of atoms (COOH) will always be an _____.
18) Membranes are made of this molecule.
19) The process of "reading" DNA to make mRNA
21) This is when a lot of something goes to an area where there is less of it.
22) This is the fluid inside a cell.
23) This is made of a very long chain of amino acids, folded up into a 3D shape.
24) DNA is found inside the _______.
25) The machine that pops the P back onto ADP is called ATP ________.

DOWN

1) The process where tRNA turns a code into a real protein.
3) This nucleic acid is abbreviated as "A."
5) These go down through the ATP synthase machine.
7) This is what the outer layer of a cell is called.
8) These are the smallest fibers in the cytoskeleton.
9) This common folding shape looks like a coil.
10) This is the organelle that turns ADP back into ATP.
12) This is the network of fibers that gives the cell its shape as well as acting as a system of "roads."
13) This means "water-loving."
15) This is the basic energy unit that cells use.
16) This means "water-fearing."
20) This element (type of atom) marks an organic molecule as a protein.

CHAPTER 7: THE NUCLEUS, AND HOW RIBOSOMES ARE MADE

In the last chapter we were introduced to the endoplasmic reticulum and saw how it can create vesicles. Some of its vesicles become lysosomes. In this chapter, our goal will be to find out how ribosomes are made, but in the process we will learn more about the ER and about the nucleus to which it is attached.

We took a quick look at the nucleus in chapter 5 when we learned how DNA is copied by mRNA in order to make proteins. We learned that DNA never leaves the nucleus, making it necessary to send mRNA copies of the DNA out into the cytoplasm. We saw the single-strand mRNA leave the nucleus through a pore in its membrane. In order to understand ribosomes, we need to go back inside the nucleus and find out more about it. When we get to the center of the nucleus, we'll see ribosomes being made. Let's start from the outside and work our way in.

The membrane surrounding the nucleus is twice as thick as the cell's plasma membrane. There are two phospholipid bilayers instead of one. This double-thick layer is called the ***nuclear envelope***. One reason for having a double layer is so that the nucleus can be connected to the ER without actually being part of it. The outer layer of the nuclear envelope is "continuous with" (is part of) the outer membrane of the ER.

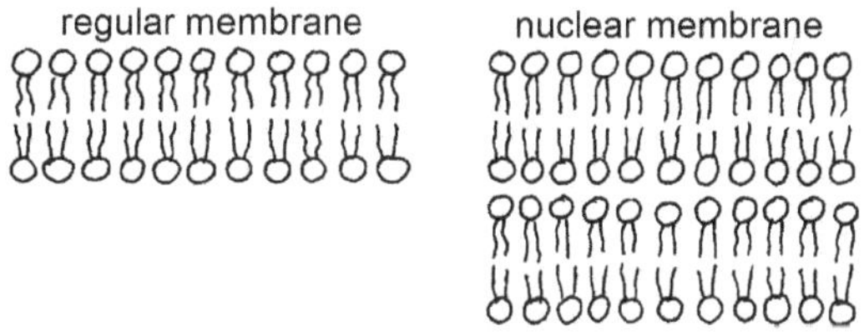

In this picture you can see that the outer membrane of the nucleus becomes the ER membrane. The inner layer is not connected to the ER. The inner layer forms sort of an interior sac around the contents of the nucleus. We also drew a few ribosomes sticking to the ER, inserting their proteins, as we learned in the last chapter. ER that has ribosomes stuck to it is called ***rough ER***. When scientists first saw it under their microscopes, it looked like it had a rough texture, so they called it rough ER. You can also see nuclear pores in this picture. We'll take a closer look at these pores on the next page.

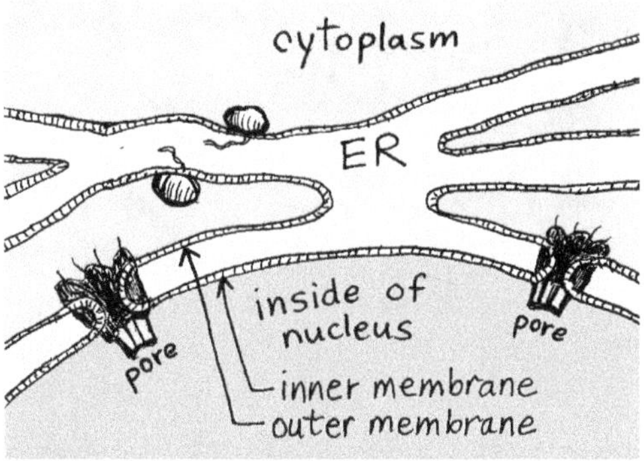

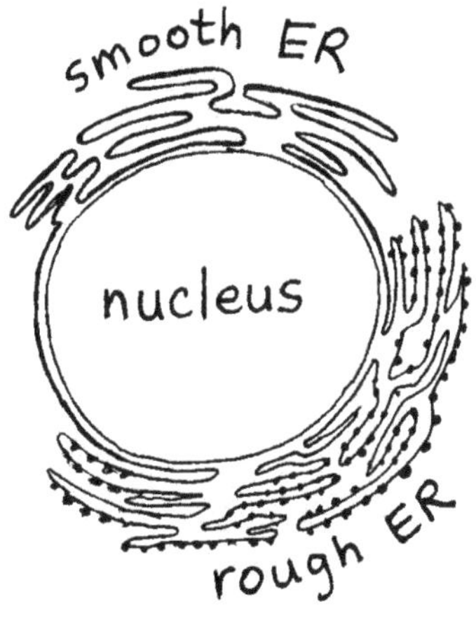

So if there is *rough* ER, might there be *smooth* ER, too? Yes. ER that doesn't have any ribosomes stuck to it is called ***smooth ER***. It has the same basic structure as rough ER, but it does things that the rough ER does not do. The exact jobs the smooth ER does depends on what kind of cell it is in. If it is in a muscle cell, one of its main jobs is to store calcium atoms that will be used in muscle contraction. (Where does smooth ER get its calcium? Only from what you eat and drink. Make sure you eat foods with plenty of minerals and vitamins!)

In all cells, the smooth ER helps to manufacture lipids (fats), especially a certain type of lipid called ***steroids***. Steroids are a whole family of molecules that control body processes such as growth of muscle and bone, preventing inflammation, breaking down carbohydrates, and developing and maintaining characteristics that make us look and act male or female. (That's quite a diverse list. These things don't seem related, do they?)

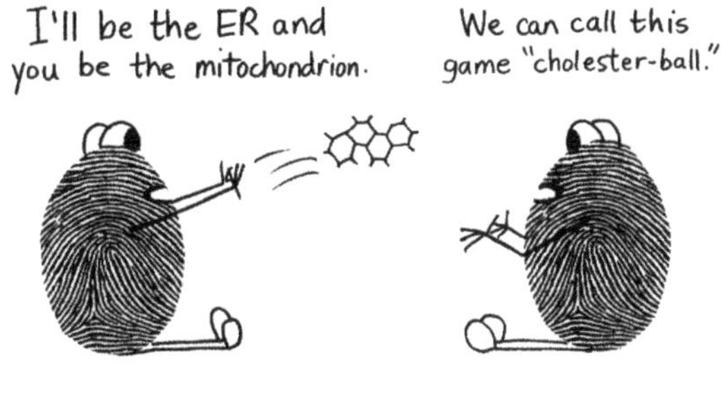

In some cells, the smooth ER and the mitochondria work together to make steroids. They pass the molecules back and forth, each one adding something or clipping something off until the steroid molecule is finished. (The mitochondria have a few other jobs besides making ATP.) After the steroid molecule is finished, proteins inside the ER will tag it so that it gets delivered to the right place.

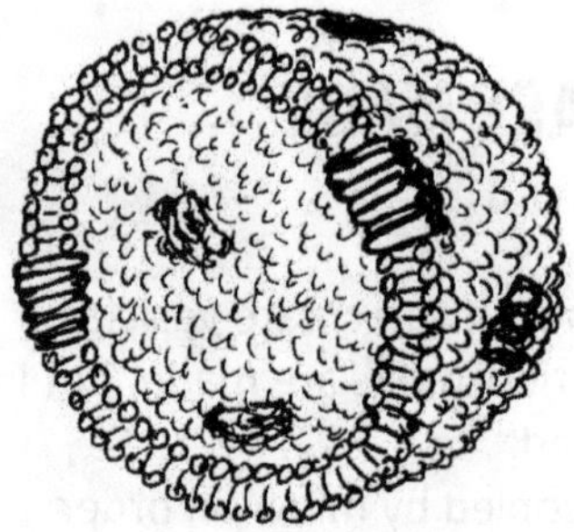

Lipid rafts embedded in a vesicle, ready for delivery.

The smooth ER is also the place where both phospholipids and cholesterol rafts are manufactured. The rafts are delivered to the plasma membrane by using a vesicle. The ER creates a bulge by manufacturing a lot of new phospholipids. (There are specialized protein gadgets in the ER that know how to assemble phospholipid molecules.) The bulge expands and grows larger. The bubble is then pinched off and becomes a free-floating vesicle. The rafts are embedded in the vesicle's membrane. The vesicle will be towed out to the edge of the cell by a kinesin motor protein. Then all the vesicle has to do is touch the plasma membrane and it will immediately merge with it and become part of it. The lipid rafts suddenly find themselves part of the plasma membrane.

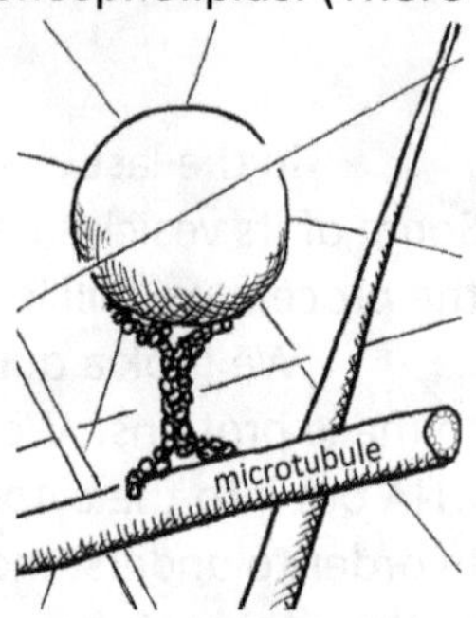

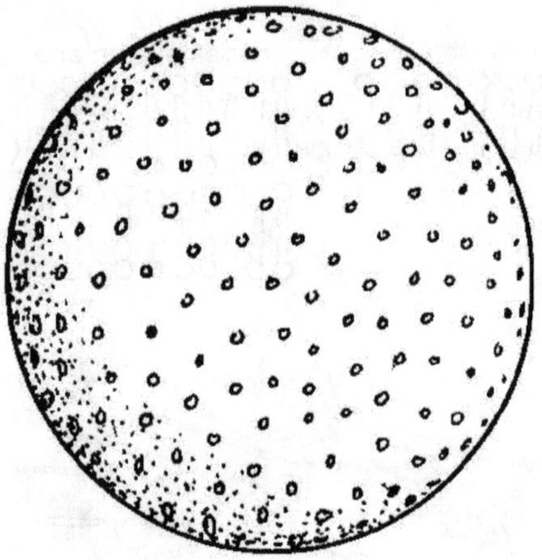

Now let's begin to explore the nucleus. The nucleus's double-layer membrane contains 3,000-4,000 ***pores*** that let things in and out. The picture here on the left shows pores, but not nearly enough of them. The problem with illustrations is that by the time you get the pores small enough to be accurate, you'd no longer be able to see them! So here we have out-of-scale pores that are too large, but at least you can see them.

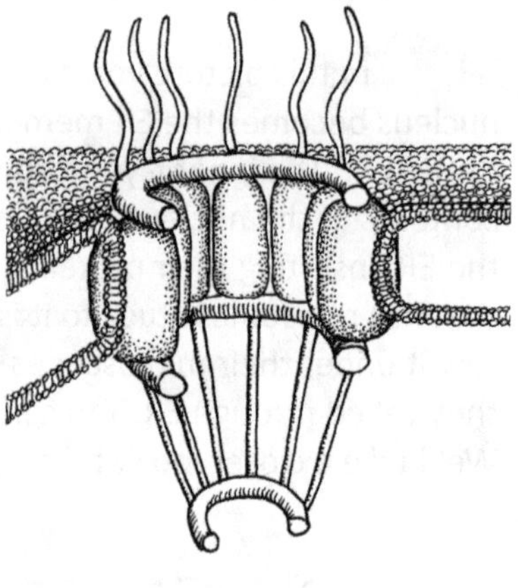

A close-up view of these pores shows them to be very strange-looking indeed—sort of a combination of a basketball hoop and a collar. In this drawing, the pore has been sliced in half. Notice the phospholipid bilayer. It's been sliced, too. The double bilayer is actually a continuous sheet all the way around the pore. The basket sticks down into the nucleus. The ring at the bottom can open and close to some degree. Chaperone proteins collect around the opening, checking on the "traffic" coming in and going out. Those filaments (strings) on the top are probably for anchoring the nucleus to the cytoskeleton, but they could have other jobs, too. This is an on-going area of research right now, so we can't say for sure what the filaments do or don't do.

The nucleus is filled with fluid similar to the cytosol found in the rest of the cell. However, the fluid in the nucleus has its own special mix of proteins and minerals floating around in it (sort of like its own soup recipe), so cell biologists felt a need to create a special name for it. They decided to call it ***nucleoplasm***. So if we say something is found in the nucleoplasm, that means it is inside the nucleus.

Now we are ready to examine the contents of the nucleus. Floating in the nucleoplasm is something we've already looked at: DNA. You'll remember that DNA is critically important to the cell because it contains all the information that the cell needs in order to manufacture new cell parts. How much DNA is there in the nucleus? If you could take all the strands of DNA and lay them end to end, they would measure about 6 feet (2 meters). That probably doesn't impress you. Six feet. Big deal. But you have to remember how incredibly thin DNA is. If you could enlarge this six-foot strand of DNA until it was as thick as a piece of thread, its length would measure over 500 miles (800 kilometers)! That's a bit more impressive. If we enlarged the nucleus by the same amount that we enlarged the DNA, the nucleus would be about the size of an average bedroom. So imagine storing 500 miles (800 km) of thread in your room. How would you do it?

The first thing you would probably do is wind the thread onto spools. You would need over 2,000 spools (and endless hours of winding!), but it could be done. Then, after all your thread was wound onto spools, you could organize the spools into boxes or onto shelves. Would all 2,000 spools fit in your room? Amazingly enough, yes, they would all fit onto shelves on just one wall.

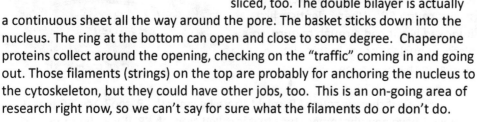

You'd have most of the room still open. Spools are great for organizing things that are very thin and very long, so it's not too surprising that cells use spools, too. The cell's spools are called ***nucleosomes***, and are made of protein "balls" called ***histones***. Eight histones join together to make one nucleosome spool. One nucleosome spool can wind a section of DNA that is about 146 rungs long. You can see from the picture that the DNA loops around the nucleosome twice. A ninth histone protein (called H1) acts like a piece of tape and keeps the two loops of DNA from falling off the spool.

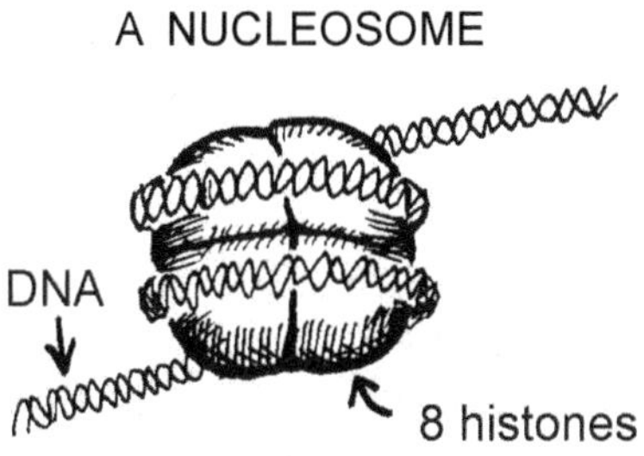

DNA researchers are pretty sure that the binding histones (the pieces of tape) are partially responsible for controlling which parts of the DNA get copied. When the DNA is tightly wound onto a spool, it is harder (or even impossible) for the mRNA to copy it. There are also protein "switches" on the nucleosomes that control the winding and unwinding of the spools. Additionally, there is another set of proteins whose job it is to turn these molecular switches on and off. (The DNA certainly is well guarded!) When the switches are flipped off, the spools relax and let the DNA unwind so that it can be unzipped by the RNA polymerase "sled." When the switches are flipped back on, the spools tighten again and the DNA is no longer accessible.

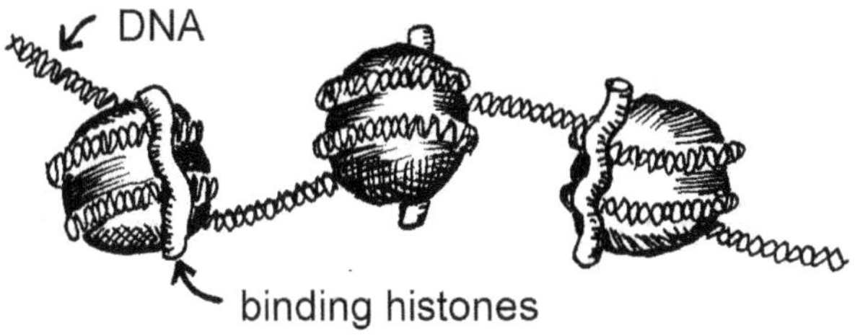

Since a cell doesn't have shelves or boxes, it must find another way to organize its spools. This drawing shows you what it does; it organizes the nucleosomes into larger clusters. This string of clusters is called ***chromatin***. (The Greek word "chroma" means "color," so "chromatin" means "colored stuff." When early cell scientists stained cells to look at them under their microscopes, the chromatin of DNA stained easily and showed up well on their slides. They didn't know what else to call it because they didn't know what it was or what it did.) Strands of chromatin can be very long, but the cell has efficient ways of organizing them even further. However... we're going to leave you in suspense for a while because that part of the story fits in better with chapter 9.

Here is another illustration of DNA wound around the histone spools. The blue stuff is the protein of the histones. The thin, coiled orange thing is the DNA. This lets you appreciate how simplified most illustrations are. However, simplification is often helpful. The drawings on the previous page were much easier to understand than this one. Even this illustration is simplified. We can't see all the coiled shapes of the proteins.

Pictures of DNA very rarely show it with its associated proteins. Not only is it coiled around proteins, it is also surrounded by chaperones.

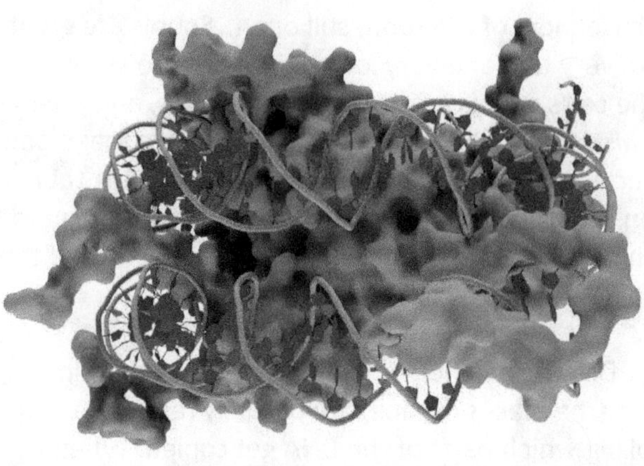

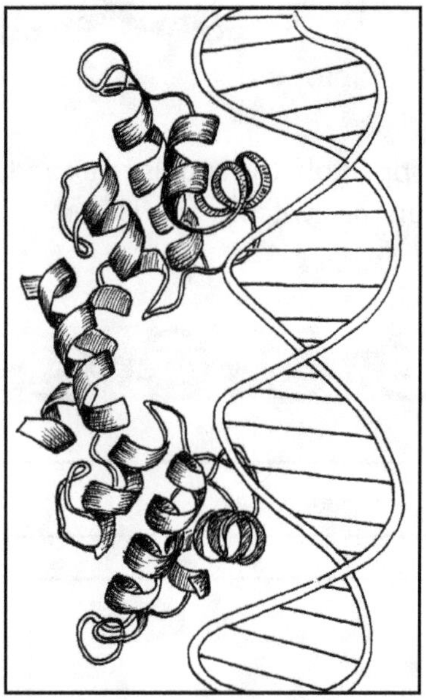

Here we see just one protein next to DNA. This protein is completely curled into helices (plural of helix).

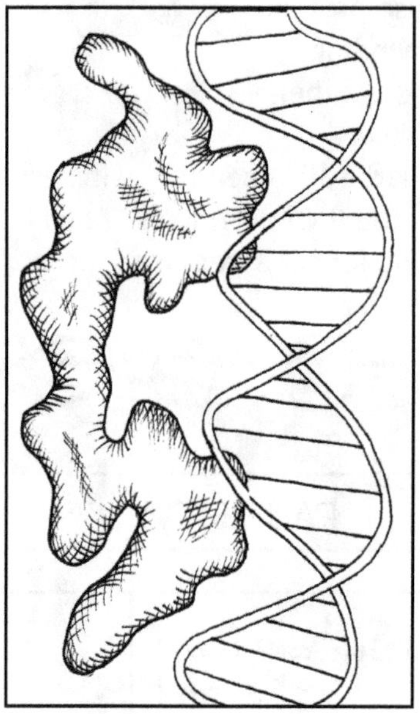

Here the same protein has been simplified and is drawn as a blob, instead of as helices.

The chaperone-type proteins control the winding and unwinding, and the zipping and unzipping, of the DNA. They "know" when to allow transcription and when to prevent it. They "know" which sections of DNA the cell will need and when it will need them. They "spell check" and edit the DNA when necessary. They watch over the DNA and try to keep it safe. If it gets damaged, they repair it. They are the unsung heroes of the cell! These proteins, like all proteins, are primarily made of helices and beta sheets, but they also include some parts with cool names like "Zinc Fingers" and "Leucine *(loo-seen)* Zippers." You can see why artists leave them out when they draw DNA. They make the drawing much more complicated. If you are just studying DNA itself, the proteins add an unnecessary layer of complexity.

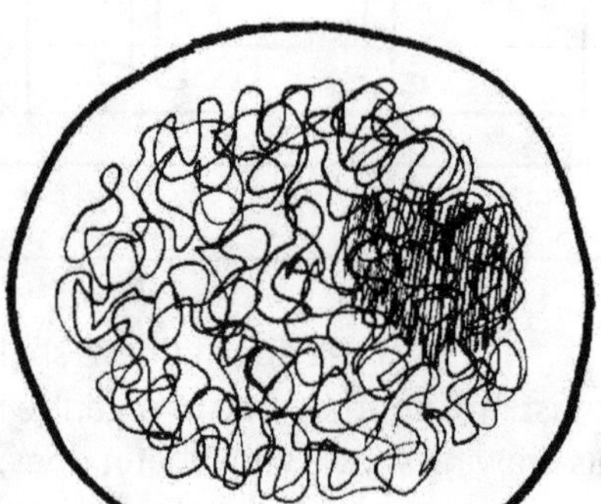

Now we're ready to move on to ribosomes. Let's zoom out a little bit and look at the whole nucleus. It looks sort of like a plate of spaghetti. The spaghetti, of course, is the DNA. At this distance, you can't see the helix shape of the DNA, nor can you see the proteins we just talked about. It just looks like a tangle of strings. See that dark patch that looks like a meatball? It's a very dense part of the DNA and it contains the instructions for ribosomes. (DNA that has instructions for ribosomes is called "ribosomal DNA" or "rDNA.") There's not just one set of instructions—there are thousands of them. The code for ribosomes is repeated over and over again. It's like when a library has multiple copies of a popular book that everyone wants to read. In the cell, so many ribosomes have to be made (hundreds per minute) that if there was only one copy of the instructions, the cell wouldn't be able to keep up with the demand. There's a special name for this dark area: ***the nucleolus***.

Now for the part that will make you shake your head for a moment while you process it. Ribosomes are made mostly of RNA. Yes, they are made of the same stuff they "read." To avoid confusion, the RNA that crumples up to form a ribosome is called rRNA, where the "r" stands for "ribosomal." So now we have <u>mRNA</u> that leaves the nucleus and attaches to a ribosome, and we have <u>rRNA</u> that folds up and <u>becomes</u> a ribosome.

When new ribosomes are needed, one of those ribosomal code areas on the DNA (rDNA) opens up and that little "sled" slides along and "reads" it, just like the transcription we saw in chapter 5. The nucleoplasm has lots of nucleic acids floating around in it, waiting to be assembled into either rRNA or mRNA. Usually that little sled (RNA polymerase) produces mRNA that will be used to make proteins. In this case, the RNA produced by the sled isn't a code, it's the finished product! That piece of RNA is rRNA because it will fold up and become part of a ribosome.

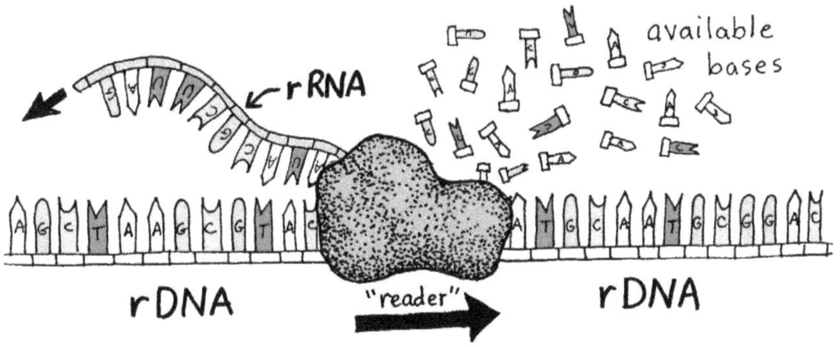

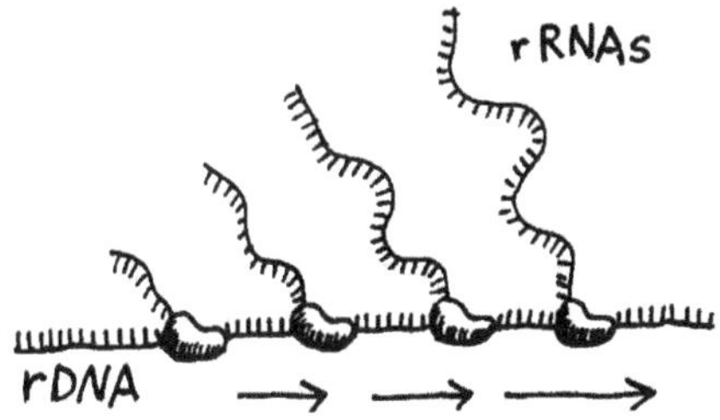

Many sleds can be on the same rDNA. They often follow each other closely. They can even be on both sides.

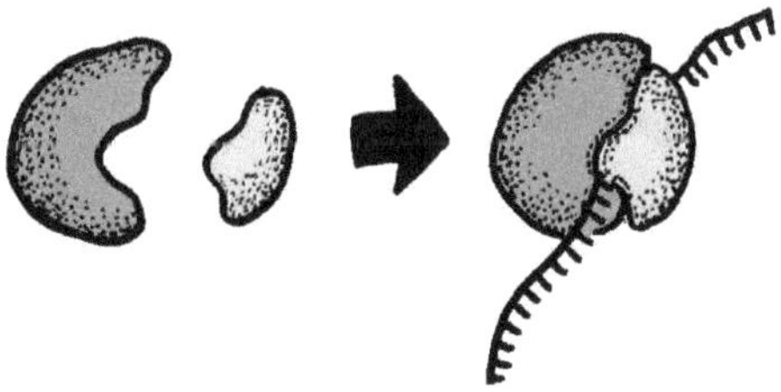

Ribosomes are made of two halves called ***subunits.*** There is a large subunit and a small subunit. Each subunit floats around in the cytoplasm by itself. Through a complex signaling process, the two halves know to come together around a piece of mRNA. When they are done making the protein, the subunits separate and float around by themselves again.

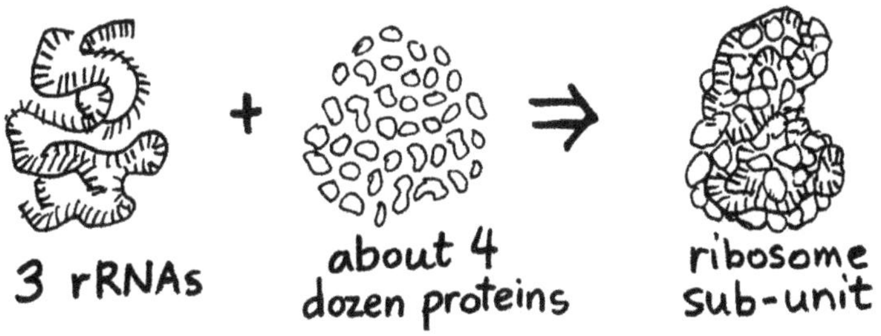

The large subunit is made of three different stands of rRNA. The small subunit is made of only one strand. In addition to rRNA, ribosomes have some protein molecules, too. (About two-thirds of a ribosome is rRNA and about one-third is protein.) The proteins probably help the rRNA to keep its shape after it is folded. Oddly enough, these ribosomal proteins come from outside the nucleus. Ribosomes floating in the cytoplasm produce them. The protein chains are brought into the nucleus through the pores and are then shuttled over to the nucleolus. The proteins will stick to the rRNA in various places, but the final assembly won't happen until it is exported out of the nucleus. (Remember, the pores are not very large. Fully assembled ribosomes would not fit through them.) One of the proteins contains a "pass code" that will be read by the chaperone proteins that control the pores. This pass code will cause the chaperone proteins to push the protein out into the cytoplasm. Once in the cytoplasm there is plenty of room for the proteins to fold up into their final shape, becoming mature ribosomal units.

These new ribosomes can translate any piece of mRNA that comes their way. They might end up sticking to the ER and making proteins for lysosomes. Or perhaps they will synthesize enzymes that will be secreted out of the cell and sent to another part of the body. Some will end up making more proteins for more ribosomes. A few will unknowingly read mRNA that was inserted into a cell by a virus and will unwittingly help the virus to make more copies of itself.

In the next chapter, we'll meet the last of the major organelles and we'll see how they all work together to keep the cell alive. We'll see how cells eat and drink, and we'll also discover that they make more than one kind of toxic waste that must be dealt with. Combustion makes pollution no matter where it happens!

Can you remember what you read?

1) ER is the abbreviation for ______________________________.

2) Why is rough ER rough?
   a) It has a scratchy surface. b) It has ribosomes stuck to it.
   c) It has ribosomes inside of it. d) It has many pores in its membrane.

3) What is the ER connected to?
   a) the inner membrane of the nucleus b) the Golgi bodies
   c) the outer membrane of the nucleus d) nothing

4) Which of these jobs would the smooth ER never do?
   a) store calcium ions b) make lipids c) make ATP d) make steroids for hormones

5) Where do lipid rafts come from? a) smooth ER b) ribosomes c) the nucleus d) Golgi bodies

6) If you unwound all the DNA in one of your cells, how long would it be?
   a) 500 miles (880 km) b) from here to the moon c) a few inches (a few centimeters) d) 6 feet (2 meters)

7) TRUE or FALSE? The membrane of the nucleus is the same thickness as the cell's outer membrane.

8) What are nucleosomes most like? a) thread b) shelves c) switches d) spools

9) TRUE or FALSE? The nucleolus is where rDNA is located.

10) TRUE or FALSE? The nuclear DNA has only one set of instructions for how to make ribosomes.

11) These get embedded into the walls of a vesicle so they can become part of a cell's outer membrane.
    a) lipid rafts b) enzymes c) steroids d) chaperone proteins

12) How many histones are there in each nucleosome spool? _____

13) Ribosomes are made of how many subunits? ____

14) TRUE or FALSE? Multiple "sleds" (RNA polymerases) can "read" the DNA at the same time.

15) Which is the correct "recipe" for a large ribosomal subunit?
    a) 2/3 DNA and 1/3 protein b) 2/3 protein and 1/3 DNA
    c) 2/3 RNA and 1/3 protein d) 2/3 protein and 1/3 RNA

16) Chromatin is made of: a) nucleosomes b) histones c) ribosomes d) DNA

17) TRUE or FALSE? Each ribosome specializes in making one particular type of protein.

18) TRUE or FALSE? Ribosomes make DNA.

19) TRUE or FALSE? Ribosomes make RNA.

20) TRUE or FALSE? Ribosomes are completely assembled before they leave the nucleus.

ACTIVITY 7.1 **Use a decoder to find the answers to these clues**

To find out the answers to the following questions, you will the decoder strips at the bottom. Cut off the two strips (they are side by side to make less cutting work for you) and then glue them end to end at the marked glue tab. Then, find a pencil or pen around which you can wind this very long strip so that the the lines match up perfectly. (HINT: I used a standard-size USA pencil.) If you have wound correctly, the short lines will all line up to make a long line one either side of the pencils. Then you will be able to see matches between number and letters. **The letter to the right of the number** is the one it matches with. *(HINT: If everything is matching up correctly, the number 10 should line up with the letter Z.) NOTE: If you are afraid of making a cutting mistake, copy this page before you cut, so you have a back-up.*

AMAZING CELL TRIVIA

1) The kinesin motor protein was first discovered in a nerve cell in a ___ ___ ___ ___ ___ ___ ___ ___ ___ ___
5 22 19 12 17 15 24 8 22 6

2) The Golgi body was discovered by Camillo Golgi while examing a brain cell from an ___ ___ ___
1 11 7

3) This type of cell does not have any mitochondria: ___ ___ ___ ___ ___ ___ ___ ___ ___ ___ ___ ___.
4 23 6 3 7 1 1 6 14 23 7 7

4) Cells found in this organ can have as many as 2,000 mitochondria: ___ ___ ___ ___ ___
7 22 9 23 4

5) This organelle is the only one outside the nucleus that contains DNA: ___ ___ ___ ___ ___ ___ ___ ___ ___ ___ ___ ___
18 22 17 1 14 13 1 12 6 4 22 19

6) This type of cell is often used for research because its organellese are so similar to human cells: ___ ___ ___ ___ ___
21 23 19 15 17

7) How many "rungs" are there on one copy of your DNA? 3 ___ ___ ___ ___ ___ ___ ___
3 22 7 7 22 1 12

8) If you took the DNA out of all of you cells and carfully laid all the strands end to end, they would stretch from the

___ ___ ___ ___ ___ to the ___ ___ ___ and back 600 times.
23 19 4 17 13 15 8 12

9) Almost half of of the information in human DNA can also be found in a ___ ___ ___ ___ ___ ___ ___
14 19 3 3 19 5 23

10) Body heat is produced by special mitochondria in the cells of a tissue called ___ ___ ___ ___ ___ ___ ___ ___
3 4 1 11 12 16 19 17

CORRECT NAMES

11) The correct name for the "folds" of the mitochonrial matrix: ___ ___ ___ ___ ___ ___ ___
14 4 22 15 17 19 23

12) The more complicated name for the Electron Transport Chain: ___ ___ ___ ___ ___ ___ ___ ___ ___
1 20 22 6 19 17 22 9 23

___ ___ ___ ___ ___ ___ ___ ___ ___ ___ ___ ___ ___ ___ ___
2 13 1 15 2 13 1 4 21 7 19 17 22 1 12

13) The correct name for the end of the Golgi body that faces the ER: ___ ___ ___
14 22 15

14) The correct name for the end of the Golgi body that doesn't face the ER: ___ ___ ___ ___ ___
17 4 19 12 15

15) The inside area of lysosomes and the ER is called the ___ ___ ___ ___ ___ (this is a common biological term)
7 8 18 23 12

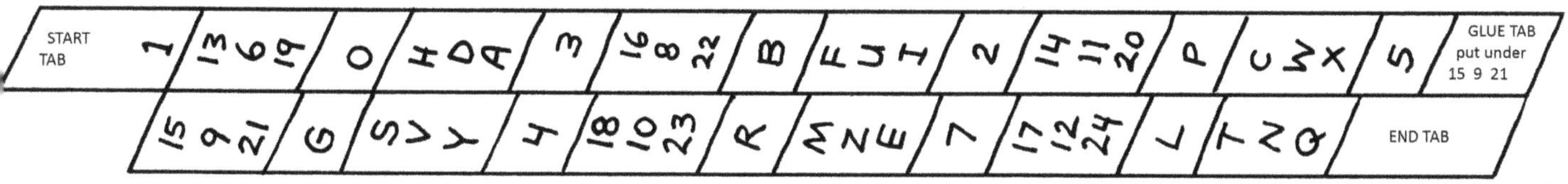

ACTIVITY 7.2 Look at a 3D rotating virtual ribosome

Look up "ribosome" on Wikipedia.com. Don't worry, you don't have to read the article. Just scroll down a bit and take a look at the pictures to the right of the article. Two of the images are animated. The RNA is shown in an orange-ish color, and proteins are in bluish-purple. (The areas that look olive green are just areas that are in shadow.) Scroll down to the one that is labeled "50S subunit." This is the large subunit. Watch carefully as the ribosome turns around. For a split second you will be able to see a bright green dot inside. This represents the "active site" where the ribosome brings the mRNA together with the tRNAs and hooks the amino acids together.

ACTIVITY 7.3 Virtual labs

1) Use this website to learn how DNA can be extracted from cells:
   https://learn.genetics.utah.edu/content/labs/extraction/
2) Learn about gel electrophoresis, a technique for sorting strands of DNA that are different lengths:
   https://learn.genetics.utah.edu/content/labs/gel/

ACTIVITY 7.4 What do the lysosome's proton pumps look like?

This illustration shows the proton pump used by lysosomes. It isn't an accident that it looks a lot like ATP synthase. It is essentially the same machine, except that it runs in reverse, using ATP energy to move protons (instead of moving protons to make ATPs).

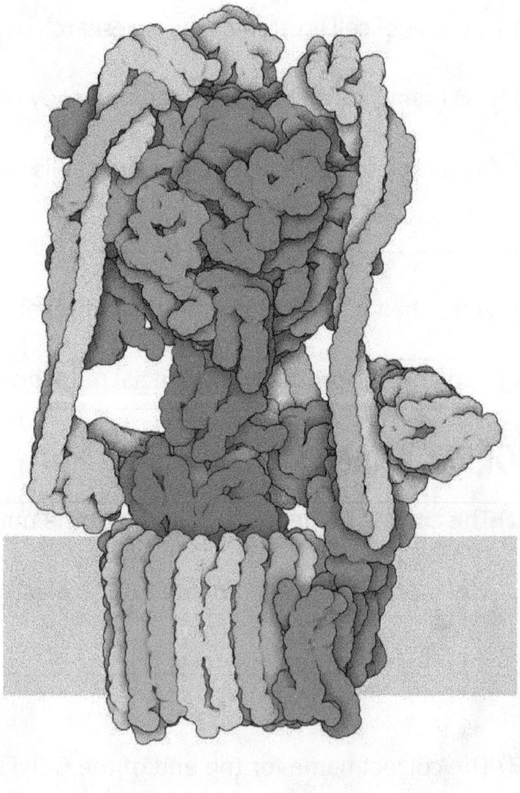

This image is from the Protein Data Bank
https://pdb101.rcsb.org/motm/219

CHAPTER 8: CELL METABOLISM AND PEROXISOMES

The activities that go on in a cell involve either breaking things down or assembling things. We've seen how the lysosome breaks things apart. In this chapter we'll learn that mitochondria do some disassembly, too, and we'll also meet a new organelle that takes things apart. The assembly jobs are divided up between the ribosomes and the ER, with the Golgi bodies helping out. This constant cycle of breaking things down, then using the parts to make other things, is called ***metabolism***. (This word comes from the Greek word "metabolikos," which means "changeable.") There are special words for the two parts of the cycle. The breaking-down part is called ***catabolism*** ("cata" means "down") and the building-up part is called ***anabolism*** ("ana" means "up"). There is variation in how these words are pronounced; however you pronounce them will be fine. Don't worry if you find it hard to remember which is which. At least the words will sound familiar when you meet them again later.

Catabolism starts when you take a bite of food and begin to chew it. Your saliva contains enzymes that begin breaking down starches right away. You can observe this happening if you put a cracker into your mouth and hold it there a while, letting it dissolve. The cracker mush will start to taste sweeter. This is the action of an enzyme called ***amylase***. It breaks apart the starch molecules and turns them into simple sugars. It's not as sweet as a mouthful of white cane sugar, but the cracker will be noticeably sweeter.

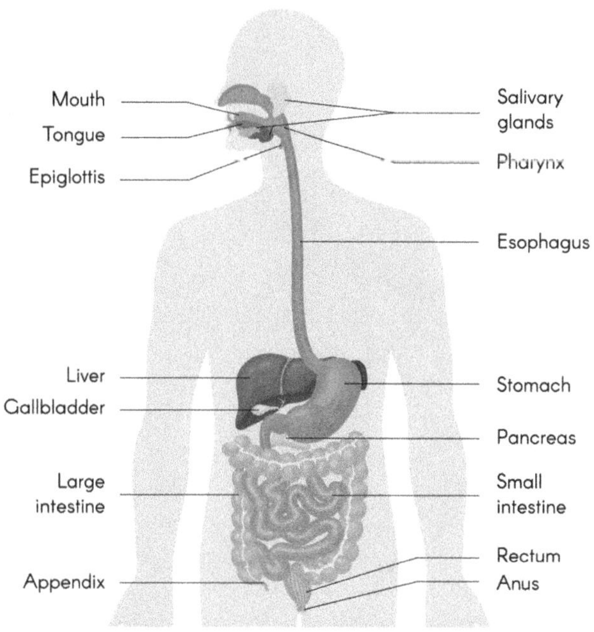

When your meal gets to your stomach, more digestive enzymes go to work breaking down the various types of food you've eaten. For example, there's an enzyme called ***pepsin*** that begins the process of breaking down proteins. The enzymes in the stomach function at their best when in an acidic environment (sound familiar?), so your stomach also secretes ***hydrochloric acid***. To protect itself from its own acid, the cells lining the stomach secrete a protective mucus layer. The stomach secretes a total of about six different enzymes. The enzymes are manufactured by ribosomes in specialized stomach cells.

From the stomach, the food goes into the very first part of the intestines, the duodenum (*du-ODD-en-um*), and meets enzymes that have come from the pancreas. These enzymes start breaking down fats, and continue to break down starches, sugars and proteins. Also, the liver makes a contribution at this point—a dark-colored substance called ***bile***. Bile works a bit like dish soap does on greasy dishes. It helps the enzymes to break down fats.

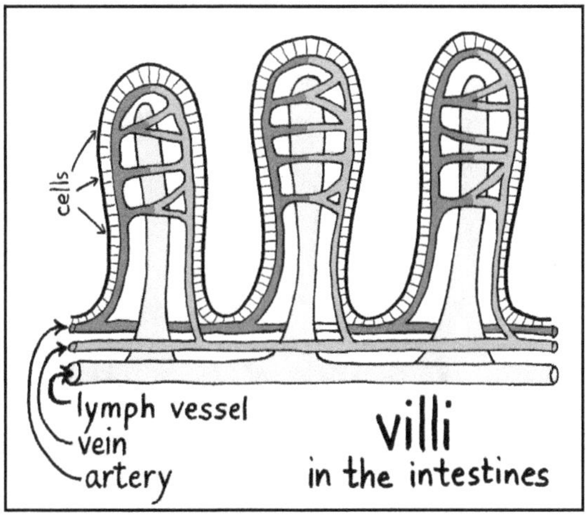

The walls of your intestines are lined with microscopic finger-like things called ***villi***. The villi contain microscopic blood vessels and lymph vessels that come very close to the surface. As the wet, mushy food mix (which no longer resembles what you ate) is pushed along through the intestines, it comes into close contact with the villi. Sugars, amino acids, water, salt, and water-soluble vitamins (such as B and C) pass through the cells on the outside of the villi (a layer that is just one cell thick) and then into the ***capillaries***. The tiny capillaries inside the villi are attached to slightly larger vessels, and those vessels turn into even larger vessels. Eventually, all the blood vessels end up as part of a large vein that goes to the liver. (We will take a closer look at villi cells in chapter 10.)

PAUSE

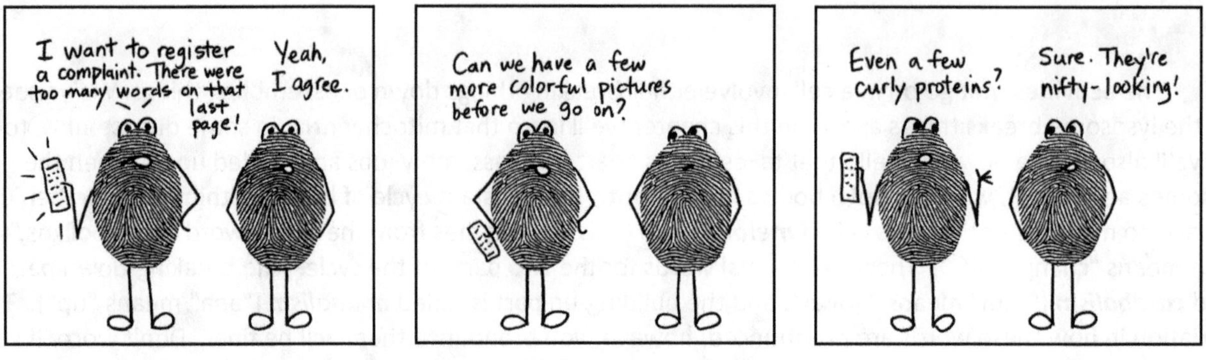

PLAY

Ok, no problem. Let's look at some of these digestive enzymes. They are made of protein (chains of amino acids). Each enzyme is designed to break apart one particular type of chemcal bond.

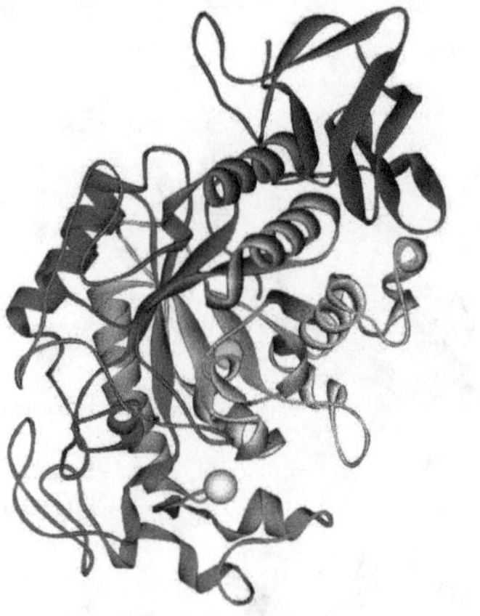

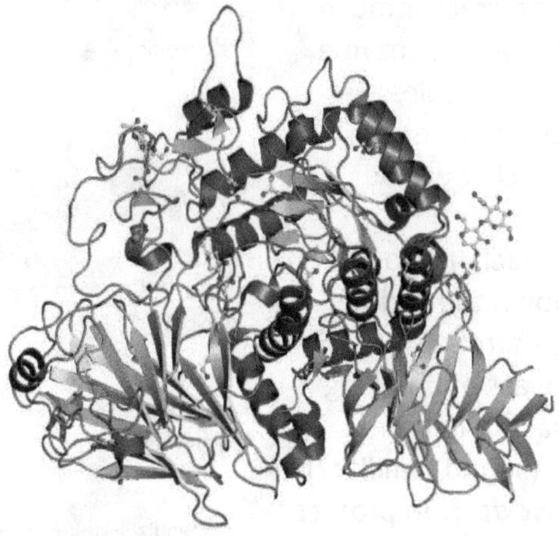

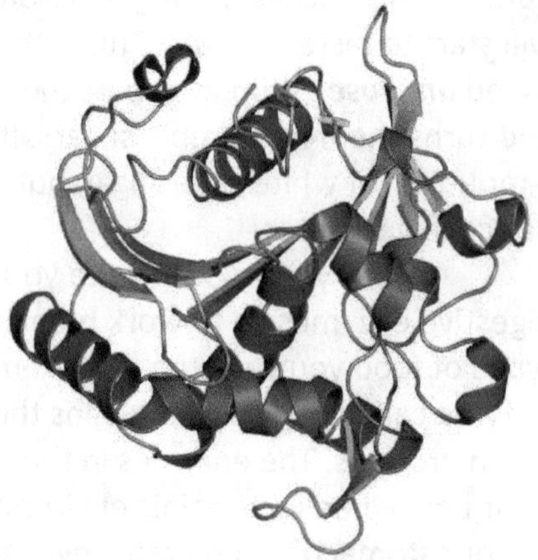

AMYLASE *(AM-ill-ase)* breaks apart the long chains of glucose molecules found in starches. The green dot is an atom of chlorine.

MALTASE breaks apart maltose, a molecule made of two glucose sugar molecules. Maltase can't break apart long chains; it can only act on the maltose molecule.

LIPASE breaks apart lipids (fats). The pancreas makes lipase, which is released into the small intestine. Lysosomes also make and store lipase.

This mostly-green protein (below) is called chymotrypsin *(KIE-mo-TRIPS-in)*. It breaks apart amino acids. The process of separating two amino acids requires a water molecule, so it is called ***hydrolysis*** *(hi-DROL-uh-sis)*, from the Greek "hydro," meaning "water," and "lysis," meaning "breaking apart."

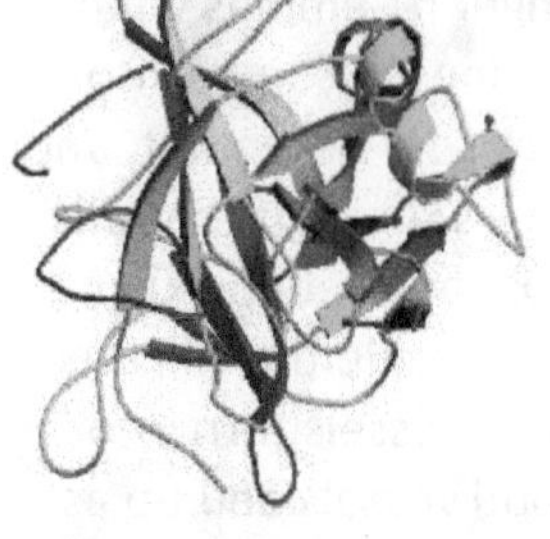

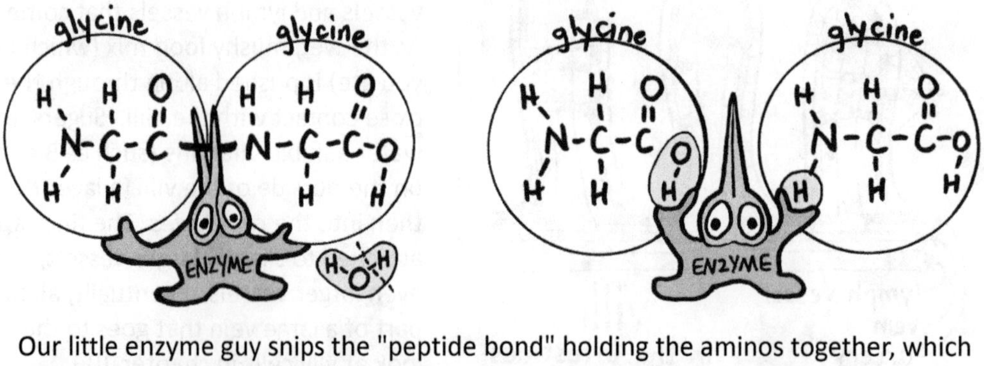

Our little enzyme guy snips the "peptide bond" holding the aminos together, which creates a ragged end on each end. One gets patched with OH and the other with H.

Back to the story of digestion; the tiny nutrient molecules had entered the blood stream...

Now the liver gets a chance to make final changes to the nutrient molecules and to strain out things that shouldn't go to the cells. If there are extra nutrients, the liver can store some of them for later use. Then the nutrients leave the liver (by way of the bloodstream) and are carried throughout the body.

Fat molecules, and vitamins A and D, follow a different route. The cells of the villi put the fats into little things that resemble vesicles. Remember, fat molecules are hydrophobic and hate being in water. Blood is mostly water. The fat molecules will not float along happily in the bloodstream as sugars will. So the fat molecules are packaged into spheres called ***chylomicrons*** (*KY-lo-MY-krons)*. The chylomicrons look like the micelles we learned about in chapter 2. Notice all the hydrophobic tails pointing to the inside. The tiny lipid (fat) molecules enjoy this environment inside the chylomicron where they are surrounded by hydrophobic tails. (The round things in the membrane are specialized proteins; we don't need details about them right now.)

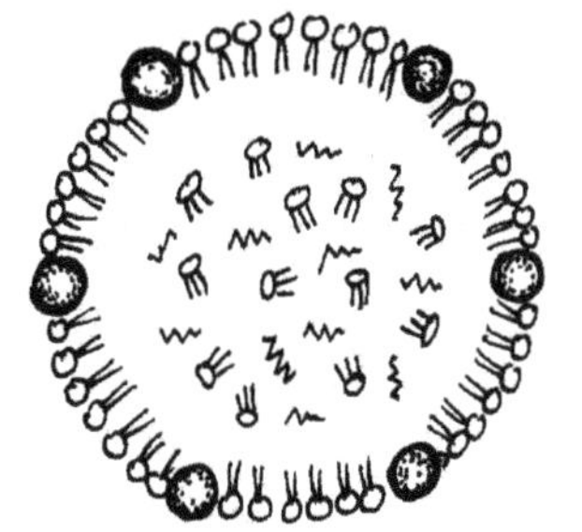

Fat molecules in a chylomicron

The chylomicrons are then absorbed into the lymph vessels in the villi. (Look at the picture of the villi again. Notice the lymph vessel in the middle.) The chylomicrons travel up through the lymph system, through special vessels designed for carrying just lymph fluid, and are dumped into your bloodstream at a location just under your collar bone. The lipids in the chylomicrons will make several trips through the liver where they will be modified even more. Digestion sure is complicated!

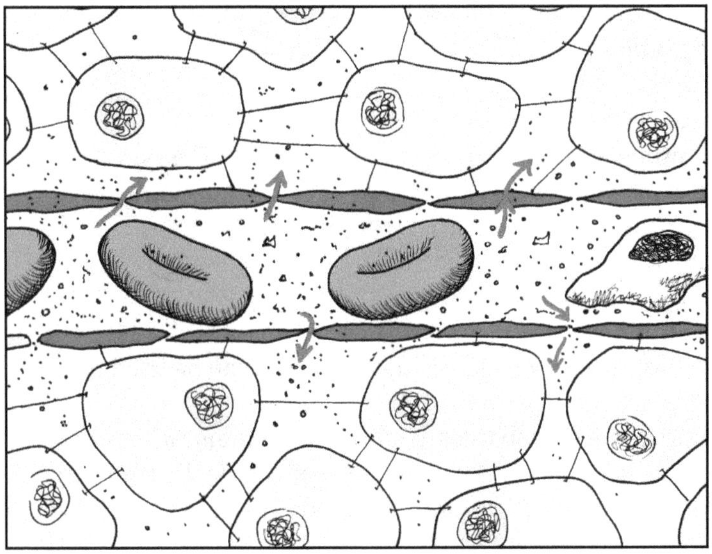

Now that we've got all these nutrients floating around in the blood, let's see how the cells are able to use them. To get nutrients, a cell must be very close to a blood vessel. This is why your blood vessels branch off and get smaller and smaller. They have to get so small that they can go in and among all your billions of cells. This picture shows a capillary (a microscopic vessel) running along between some skin cells. The cells of the capillary (in red) are shown as a cross section view, which is why they look long and thin. Their real shape is similar to a curved pancake. The donut-shaped things are red blood cells carrying oxygen. The fried egg thing is a white blood cell (the kind that eats bacteria).

Those little lines between the cells are pieces of protein called ***desmosomes***. They keep the cells linked together, but in a flexible way. In this picture, the desmosomes look like they are attached to the cells' membranes, but they are actually connected to the cytoskeleton.

The little dots coming out of the capillary represent sugars, amino acids, fatty acids and nucleotides. Water, oxygen and minerals are also present in the fluid, but are even smaller than these nutrients. The pressure of your blood pushing through the capillary will force many of these tiny nutrients out through the cracks and into the space between the cells. This space between the cells is called the ***interstitial*** space *(in-ter-STISH-ul).*

Once the nutrients are in the interstitial space, the cells can take them in by using one of these methods:

1) DIFFUSION: Oxygen and minerals can go right through the membrane.

2) PORTAL PROTEINS: Some small or medium-sized molecules can be taken inside the cell by membrane-bound portal proteins.

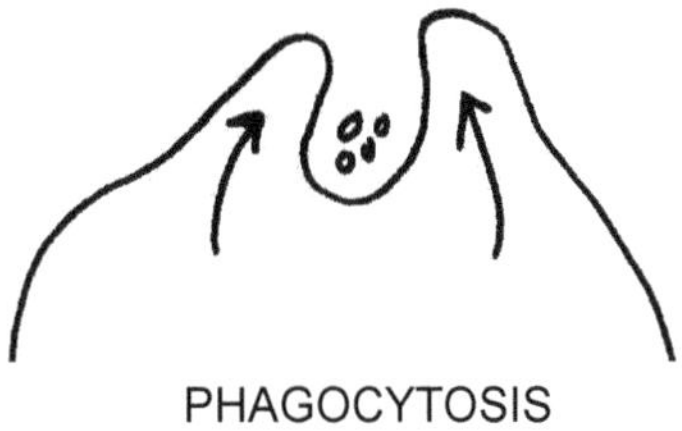

PHAGOCYTOSIS

3) PHAGOCYTOSIS: Large molecules are taken in by something called ***phagocytosis*** *(FA-go-cy-TOE-sis)*. "Phago" is Greek for "eat" and "cyto" (as we know) means "cell." Phagocytosis is how a cell "eats." Little pseudopods (pushed out by rapidly growing microfilaments) reach out and surround a food particle. The pseudopods close in around the particle and a vesicle is formed. If the particle needs to be digested, this vesicle can be sent to merge with a lysosome.

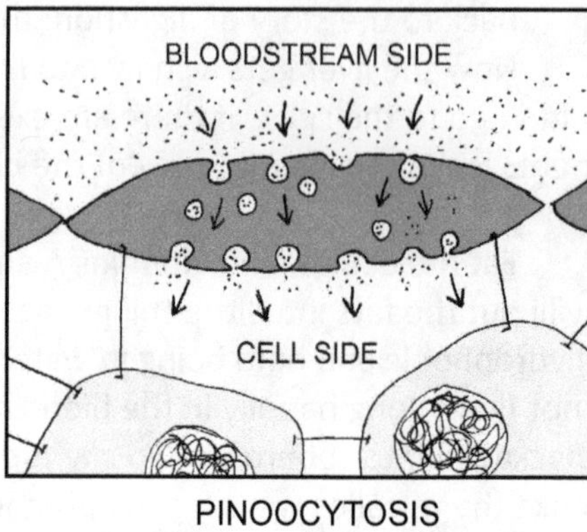

PINOOCYTOSIS

4) PINOCYTOSIS: When a cell takes in tiny amounts of fluid from its environment, it's called ***pinocytosis*** *(PIN-o-cy-TOE-sis)*. Some biology books define pinocytosis as "drinking" and phagocytosis as "eating." But this is a little misleading, because a cell doesn't have to do pinocytosis to get water. Water normally goes in through a portal called aquaporin. Pinocytosis is basically the same as phagocytosis, just on a smaller scale. Most cells are constantly taking in little "sips" of the fluid that is around them. The flat cells that form the walls of the capillaries do a lot of pinocytosis. The sides of the cells that are on the inside of the capillaries take fluid in, then the cells "spit out" fluid on the other side—the side that faces those skin cells. Almost all cells do some pinocytosis, but some do more than others, depending on what type of cell it is. These capillary cells do a lot of pinocytosis because it's part of their job.

So that's how cells take things in. Cells can also get rid of things using basically the same methods as above, just in reverse. Carbon dioxide can diffuse out of a cell, go through the interstitial space, and into the blood in a capillary. The blood will take the carbon dioxide to the lungs where it will be exhaled.

When a cell does the reverse of pinocytosis, it's called ***exocytosis***. (You can figure out what "exo" means because you know the word "exit.") Sometimes a vesicle that is used for exocytosis is called a ***vacuole***. Vacuoles are just big, empty vesicles sitting around the cell. In plant cells, vacuoles play an important role in allowing the cell to expand or shrink while still maintaining its shape. The vacuole can getter bigger or smaller quickly and easily, according to how much extra water is in the cell. Plant vacuoles can be very large, sometimes filling half of the space inside the cell. But we digress... this book is about animal cells, not plant cells. Animal vacuoles are basically very large vesicles, and they can be used to store things or to export (dump) things outside the cell.

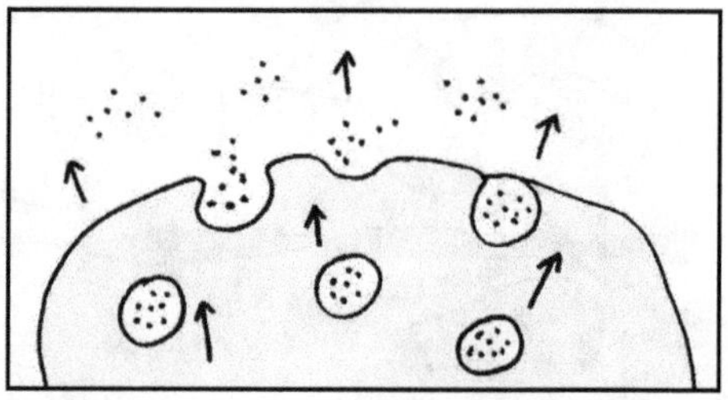
EXOCYTOSIS

After the cell dumps stuff into the interstitial space, the fluid in the interstitial space drains into lymph vessels. The lympth fluid travels through the lymph vessels that go up through your chest and finally empty into your blood stream at a place under your collar bone (as we mentioned before). The blood then goes to the liver and kidneys for final clean up.

Now that we've got all your food digested and the nutrients delivered to your cells, let's see what your cells do with them. But first, let's categorize the nutrients. They can be put into four basic categories: ***amino acids*** (from proteins), ***nucleotides*** (the individual units of nucleic acids), ***fatty acids*** and ***sugars***.

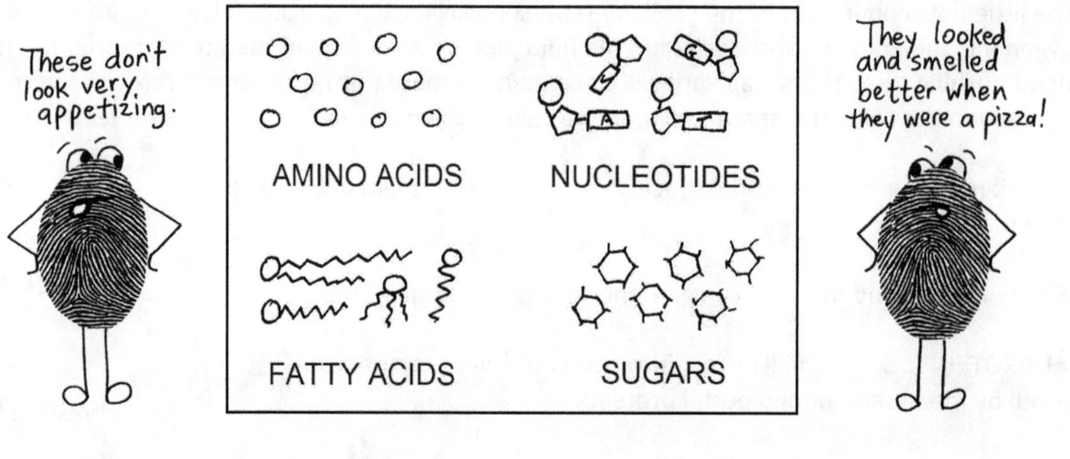

AMINO ACIDS

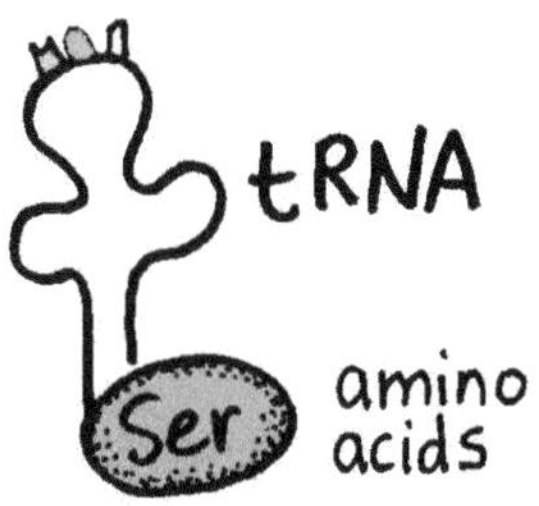

When individual amino acids (that came from digested proteins) are taken into the cell, the tRNAs can come and pick them up and take them over to the ribosomes, where they will be used to make proteins that the cell needs. Inside the cell, these could be cytoskeleton fibers, membrane portals and pumps, enzymes, chaperone proteins, or messenger molecules. Some proteins end up outside the cell, too, forming the framework on which bones and tendons are built. Nine amino acids are "essential" and must come from your diet (the ones in the dashed line circles on the chart on page 39). Your cells can assemble the non-essential aminos, though they will also gladly accept them ready-made, from your digested food.

NUCLEOTIDES

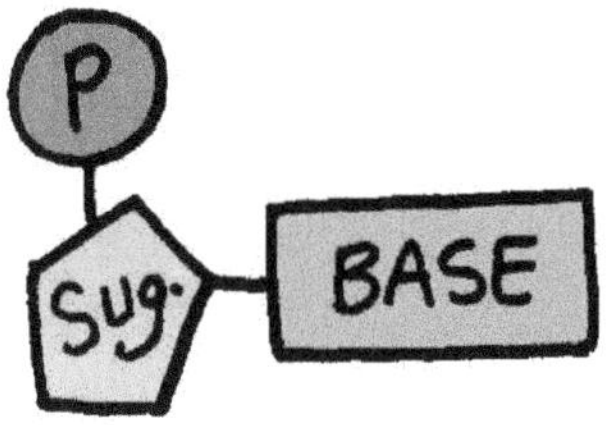

The single unit of a nucleic acid is called a ***nucleotide***. We met these in chapter 5 when we first learned about DNA. A nucleotide has three parts: **a sugar, a phosphate and a base**. There are five bases: A (adenine), T (thymine), C (cytosine) and G (guanine), and U (uracil). Some nucleotides may come from things you eat, (even plant cells have DNA in their nuceli), but many of the nucleotides floating inside your cells have been recycled, and used to be part of the DNA or RNA of other cells. Your cells are programmed to live only a certain amount of time, then they "die" and their parts are recycled. Also, we learned that bacteria and viruses can also be recycled. The nucleotides in the DNA of a bateria or virus are identical to your own. If a cell finds that it doesn't have enough nucleotides, it can also manufacture them "from scratch" using phosphates, ribose sugars and other molecules that are floating around the cytoplasm.

FATTY ACIDS

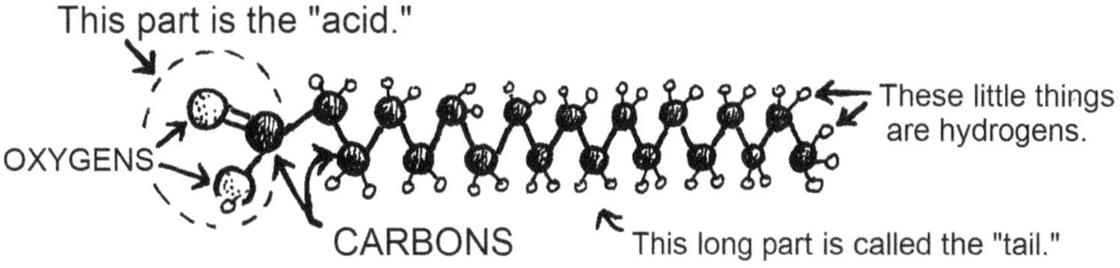

The fats and oils that you eat get digested (with help from lipase enzymes) down to microscopic fatty acid molecules. Even though they are microscopic, many of them are still pretty long—up to several dozen carbon atoms in the chain. Remember, the ultimate goal of of digesting fats is to break the bonds between the carbon atoms. When these bonds are broken, energy is released. But also remember that this process has to occur in many small steps.

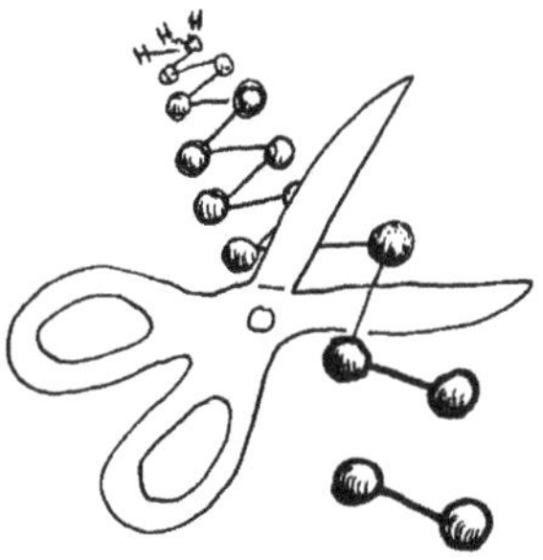

The next step is called ***oxidation*** and occurs in the mitochondria. The mitochondria take the fats and start chopping off pairs of carbon atoms using a very complicated chemical process that we'll just call "enzyme scissors." As we've discussed previously, cells use enzymes like we use scissors and staplers. Some enzymes cut things apart, others put things together. All you need to know is that in this oxidation process, the mitochondria snip up the carbon chains into pairs of carbon atoms. Then they stick a molecule called "CoenzymeA" onto the end, using what we will call an "enzyme stapler." This produces an extremely important and useful molecule called ***acetyl-CoA*** *(ah-SEE-till co-AY)*. Acetyl-CoA can be used in construction or be "burned" for energy. You might want to think of acetyl-CoA as a piece of lumber. Like acetyl-CoA, a piece of lumber used to be part of something larger (a tree). And like acetyl-CoA, a piece of lumber can be used for more than one purpose—you can use it to make a useful object such as a piece of furniture or the walls of a house, or, if you were cold and needed some heat, you could chop up the lumber and use it to make a wood fire. The cell can use acetyl-CoA as a raw material to make some cell parts, such as the "switches" on the histones that tighten and untighten the coiled DNA in the nucleus. Or, the cell can "decide" to use the acetyl-CoAs as fuel. To "burn" acetyl-CoAs, mitochondria have a chemical process that we are going to call the "Krebs factory."

Acetyl-CoA is "burned" using a process called the ***Krebs cycle*** (named after scientist Hans Krebs). Just to really confuse you and make science harder than it has to be, this cycle goes by several different names: the "Citric Acid Cycle," or the "Tricarboxylic Acid Cycle," or... you'll love this one... the "Szent-Györgyi-Krebs cycle." So many to choose from! We just decided that "Krebs" sounds pretty cool and is a word you'll be likely to remember, so we'll go with Krebs.

The Krebs cycle is like a factory that takes in acetyl-CoA's and puts them on an assembly line where workers (enzymes, of course) process them (a 10-step process) and harvest the energy from the carbon bonds. Two things come out of the factory: some ATPs (technically they are "GTPs" but they can function as ATPs) and some high-energy electrons that are taken by a "shuttle" molecule, NADH, to a nearby electron transport chain. Like many factories, the Krebs cycle produces carbon dioxide, CO_2 as a waste product. The CO_2 will be taken away to the lungs to be exhaled. It's strange to think that you can exhale the carbon atoms in the greasy snack you just ate! Perhaps a plant will take in the carbon dioxide you exhaled and use it in photosynthesis to make new starches for you to eat.

SUGARS

All the starches and carbohydrates you eat (pasta, bread, potatoes, cookies, etc.) are digested down to these simple sugar molecules: glucose, fructose, and galactose. The cells take in these simple sugar molecules and use them for energy, or for building things like "mailing labels" in Golgi bodies or those sugar strings that coat the outer membrane.

Sugars used for energy are sent to the Krebs factory where they can be burned for fuel, but first they need to be turned into acetyl-CoA's because the Krebs factory can't deal with anything other than acetyl-CoA's.

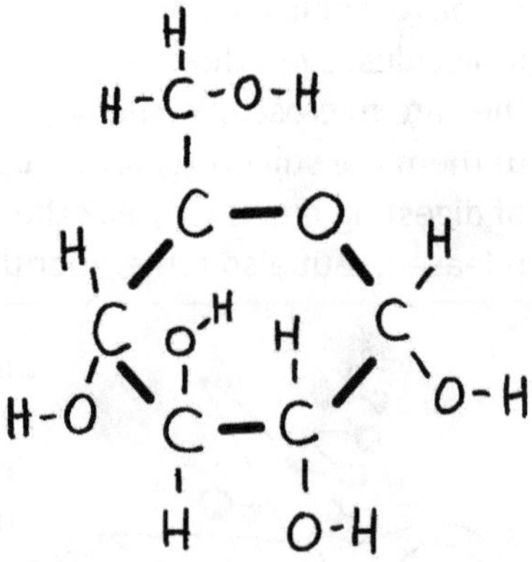

GLUCOSE

The process of preparing glucose for the Krebs cycle is called ***glycolysis***. ("Glyco" means "sugar," and you already know what "lysis" means.) Glycolysis doesn't require a special organelle. It can take place anywhere in the cytoplasm. There are 10 steps in glycolysis, just like there are 10 steps in the Krebs cycle. Many of these steps involve popping phosphates on and off. (It turns out that phosphates are very important in many cellular processes, not just for ATPs.) The workers who do the popping on and off of the phosphates are enzymes, of course.

Both the Krebs cycle and glycolysis are very complicated processes. In order to really understand them, you need to know a lot of college-level chemistry. In case you are curious about what these cycles would look like in a college textbook (or perhaps an AP high school biology book), we'll show you. (For those who are not curious, try not to look, and skip ahead to the text as fast as you can!)

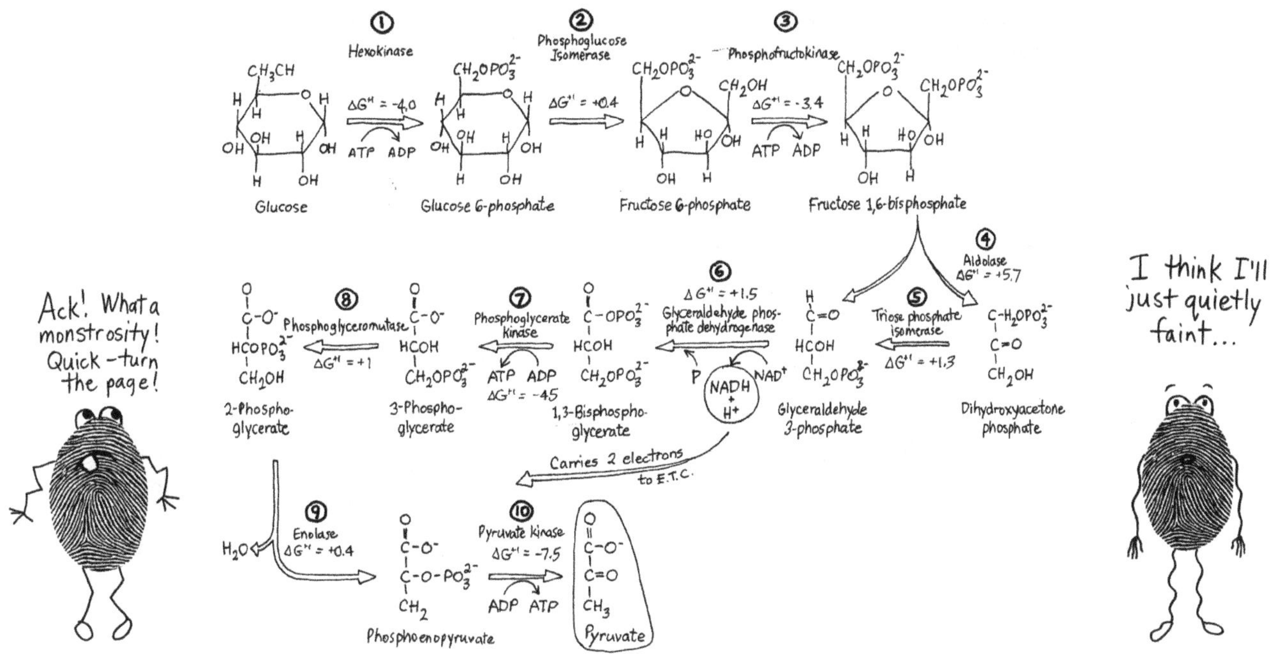

The worst part is that after all of this, the necessary molecule, acetyl-CoA, still hasn't been produced! You end up with something called ***pyruvate***, which is a 3-carbon molecule. The Krebs factory can only process 2-carbon molecules. But the good news is that at least some energy was produced: 2 ATPs.

Now the cell has two options of what to do with pyruvate, depending on whether oxygen is available. Somehow, the cell knows whether there are oxygen molecules waiting at the end of the electron transport chain. If there isn't any oxygen there, the cell has no choice but to use a chemical process whereby pyruvate is turned into lactic acid. Energy is released, but lactic acid has a nasty side effect. It's what makes your muscles burn when you lift weights or run too fast. Not the best option.

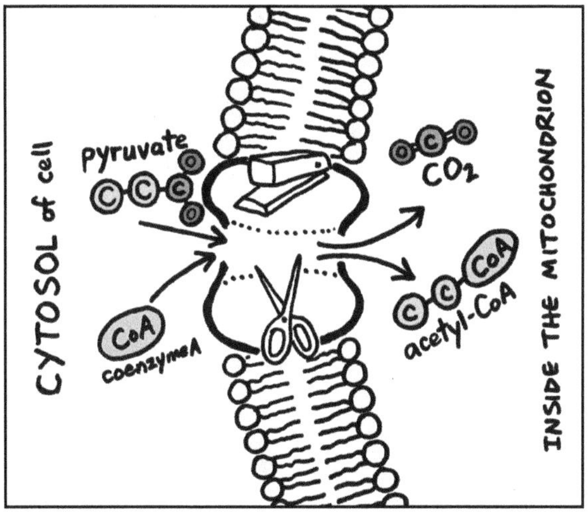

Normally, oxygen is present, so most of the time, the cell is able to convert pyruvate into acetyl-CoA by pushing it through a special portal enzyme located in the membranes of the mitochondria. This diagram shows you that the enzyme does both scissor and stapler jobs. It lops a carbon atom off the pyruvate and releases it as CO_2. Fortunately, that particular carbon has two oxygens attached, so one little snip and a CO_2 goes floating off. The other carbons have an oxygen or some hydrogens attached to them, but these atoms are not shown in order to keep the pyruvate looking as non-threating as possible. The CoA gets stapled onto the remaining two carbons of pyruvate, turning it into acetyl-CoA. (Finally!)

You already know what will happen to the acetyl-CoA. It will most likely be sent to the Krebs factory where the bond between the carbon atoms will be broken and the energy will be released as either ATPs or as electrons carried by NADH (or a very similar molecule called $FADH_2$). Or, as you undoubtedly remember, your cells also have the option of using acetyl-CoA to build things.

If everything goes perfectly and all parts function as they should, your body can make 36 ATPs from one glucose molecule. However, current research suggests that often little things go wrong here and there and not every glucose ends up generating the full 36 ATPs. They estimate that your cells get an average of about 30 ATPs per glucose. But if you are ever asked how many ATPs can be derived from one glucose molecule, say 36.

This whole process, starting with glycolysis and ending with ATPs coming out of the ATP synthase machine, is called ***cellular respiration***. The goal of cellular respiration is to provide ATPs for all the activities that go on in the cell. Your cells process billions upon billions of glucose molecules and fatty acid molecules every day. All cells, even plant cells, use cellular respiration to produce energy.

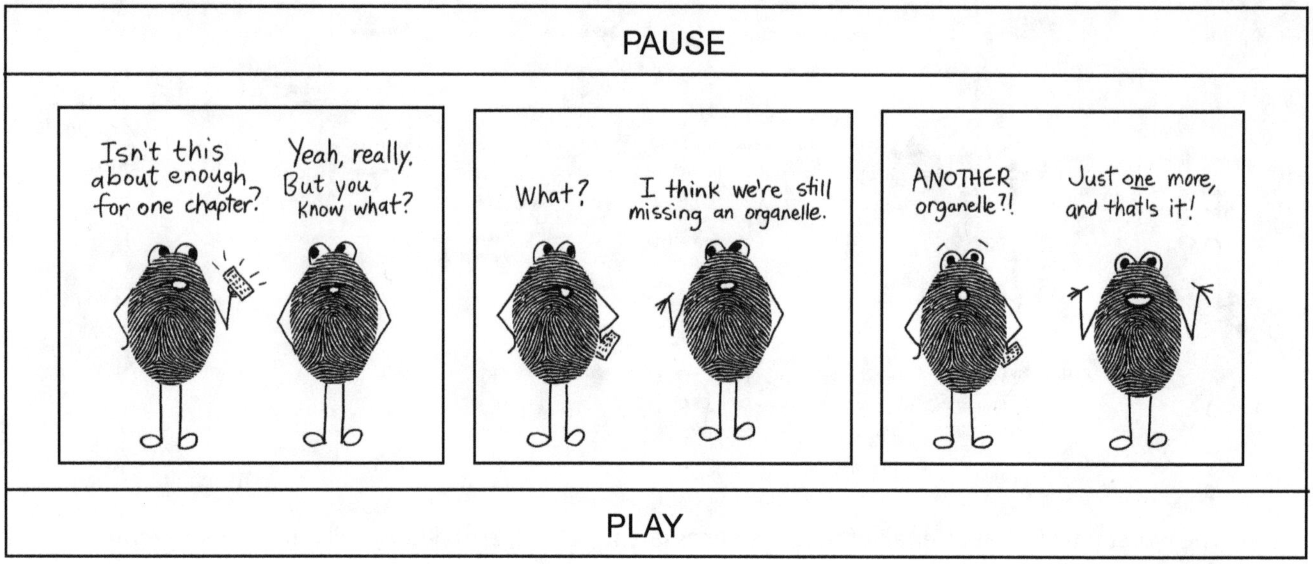

The cell has a problem we haven't mentioned yet. Some of the fatty acid chains that come into the cells are too long for the mitochondria to deal with. These extra-long fatty acid chains are sent over to an organelle called a ***peroxisome***. The peroxisome looks a lot like a lysosome, but it doesn't have any proton pumps in its membrane so it doesn't have an acidic interior. Its interior is called a "crystalline core."

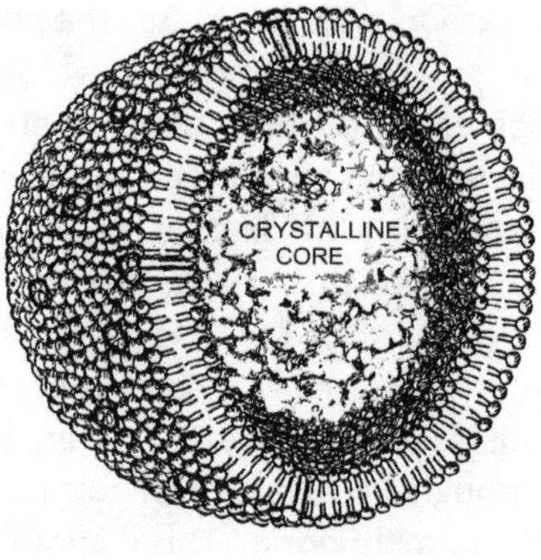

Inside a peroxisome

The peroxisome uses a specialized pair of "enzyme scissors" to break apart long carbon chains into shorter lengths that can be sent over to the mitochondria for further processing.

The peroxisome can do some other things, as well. It is able to break apart toxins such as alcohol molecules. It can rearrange the structure of some amino acids and turn one amino acid into another. It can build lipid molecules that are critical to the proper functioning of your brain and nerves. It can even make cholesterol (necessary for rafts and for other things).

Peroxisomes are amazing little organelles! You wouldn't be able to live without them. But now for the bad news: during the process of breaking down the fatty acids, a ***by-product*** is produced. A by-product is an extra chemical that is produced along with the main chemical you want to produce. When the fatty acids are oxidized, a chemical called ***hydrogen peroxide*** is produced. You may be familiar with this chemical. Many people use hydrogen peroxide to sanitize cuts and wounds. It's great for killing germs. Unfortunately, it can also kill cells. The peroxisome must get rid of the hydrogen peroxide before it harms the cell. To do this, it uses an enzyme called ***catalase***.

The chemical formula for hydrogen peroxide is $\mathbf{H_2O_2}$. You will notice how similar this is to the formula for water, $\mathbf{H_2O}$. If you could remove an oxygen atom from hydrogen peroxide, you would get a harmless water molecule plus an atom of oxygen. This is what the enzyme catalase does. It splits hydrogen peroxide molecules into water and oxygen. Water and oxygen are very useful to the cell, so this story ends "happily ever after."

$$H_2O_2 \xrightarrow{\text{catalase}} H_2O + O_2$$

hydrogen peroxide — water — oxygen

Your liver cells contain a lot of peroxisomes. The cells can adjust to the work load. If there are many toxins to process, the cells will signal their peroxisomes to split in half, doubling their number in a very short time.

Can you remember what you read?

1) Reducing nutrients down to their individual components is called:
a) metabolism b) catabolism c) anabolism d) oxidation

2) What does pepsin break down? a) proteins b) nucleic acids c) fats d) carbohydrates

3) What enzyme is produced by the salivary glands in your mouth?
a) catalase b) amylase c) dehydrogenase d) hydrochloric acid

4) What tiny structures line your intestines? a) capillaries b) lymph vessels c) villi d) chylomicrons

5) Where would you find hydrochloric acid?
a) in your stomach b) in your intestines c) in your duodenum d) in your liver

6) What is inside a chylomicron? a) glucose b) phospholipid membranes c) amino acids d) fatty acids

7) What do you call the protein fibers that form connections between cells?
a) ligaments b) desmosomes c) chromosomes d) tendons

8) Which one of these is NOT a way that something could get into a cell?
a) diffusion b) portal proteins c) exocytosis d) pinocytosis e) phagocytosis

9) A very large storage vesicle in a cell is called: a) peroxisome b) chylomicron c) vacuole d) lysosome

10) What is in the space between cells? a) nothing b) blood c) lymph fluid d) water

11) Where do long fatty acid chains get oxidized (cut apart with "enzyme scissors")?
a) in the cytosol b) in mitochondria c) in lysosomes d) in the ER

12) Where does glycolysis (mainly) occur?
a) in the interstitial space b) in the peroxisomes c) in the nucleus d) in the cytosol

13) Which of these is burned as fuel in the "Krebs factory"?
a) glucose b) acetyl-CoA c) long fatty acid chains d) amino acids

14) Which of the following is NOT a job that peroxisomes do?
a) break down alcohols b) break down fats c) make amino acids d) get ATPs from glucose

15) How many ATPs can be produced from one glucose molecule? a) 2 b) 10 c) 32 d) 36

16) What is the end product of glycolysis? a) acetyl-CoA b) glucose c) citric acid d) pyruvate

17) What does the Krebs "factory" produce as a waste by-product?
a) oxygen b) carbon dioxide c) pyruvate d) electrons

18) What toxic molecule does catalase break down into oxygen and water?
a) peroxide b) pyruvate c) hydrogen peroxide d) alcohol

19) What molecule is needed in the hydrolysis (breaking apart) of two amino acids?
a) water b) oxygen c) glucose d) tRNA

20) Which of these can break apart fat molecules? a) amylase b) maltase c) lipase d) pepsin

ACTIVITY 8.1 **Amylase at work**

If you've never tried the experiment mentioned in the second paragraph of this chapter, give it a try. You will need a piece of cracker or bread that no, or very little, added sugar. Hold the piece of cracker or bread in your mouth without chewing for about a minute. If you think it has not gotten sweeter, put in a fresh piece for an instant comparison.

ACTIVITY 8.2 **Online video games about cells**

If you are a fan of video games, check out these free online games where you control a little spaceship thing and shoot organelles and invaders. https://biomanbio.com/HTML5GamesandLabs/Cellgames/Cells.html

ACTIVITY 8.3 **Who am I?**

Here are the possible answers:

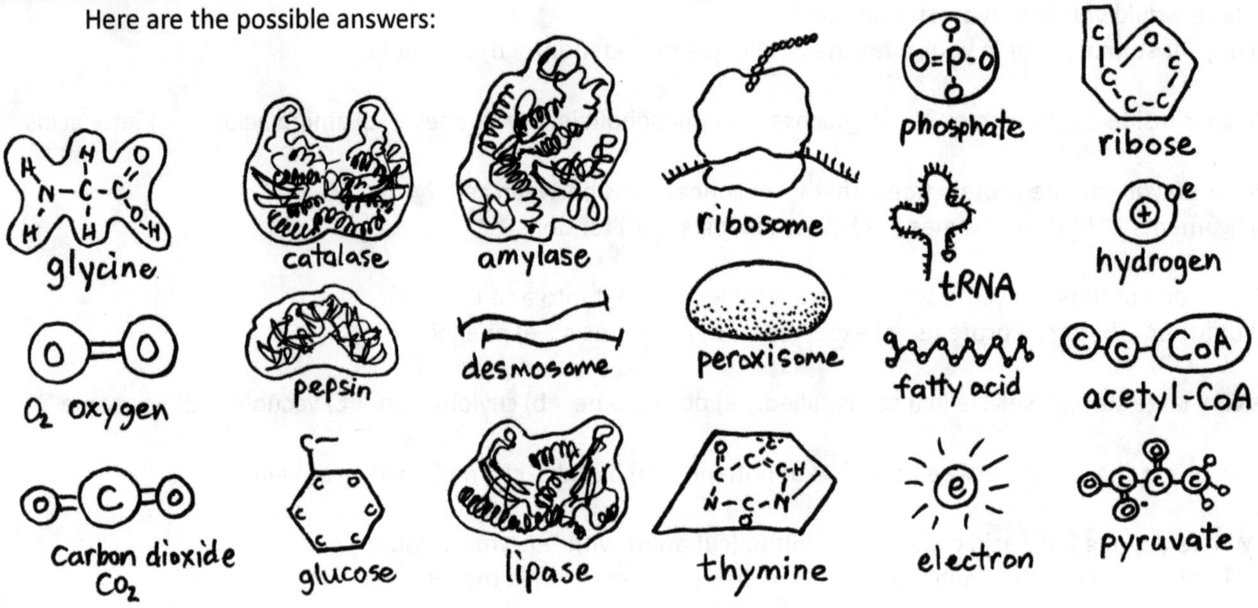

1) I am in DNA but not RNA. I am not a sugar. ________________

2) I am found in your mouth where I break apart chains of glucose molecules that make starch. ____________

3) I am in RNA but not DNA. I am not a base. ____________

4) I am found in liver cells.. I break apart hydrogen peroxide, H_2O_2, into water and oxygen. ____________

5) I break apart fat molecules. ______________

6) I am the fuel that gets burned in the Krebs "factory." ______________

7) I am 3-carbon molecule. ______________

8) I am the smallest amino acid. ______________

9) I break apart the proteins in meat, eggs, fish, nuts and beans. ____________

10) I am a primary site for anabolism because I build proteins. ______________

11) I am found in ATP, phospholipids, and DNA. ________________

12) My chemical formula is $C_6H_{12}O_6$. ______________

13) There are lots of me in liver cells. One of my jobs is to break apart toxic molecules. ____________

14) The NADH shuttle carries me to the first pump in the Electron Transport Chain. ______________

15) Animal cells produce me as waste, but plants use me as a raw material. ________________

16) WIthout me, the process of cellular respiration comes to a halt. ______________

17) I ride through the blood inside a chylomicron. ______________

18) I hold cells together. ________________

19) I bring amino acids to ribosomes. __________

20) I am the smallest atom. My nucleus is just one proton. ____________

CHAPTER 9: MITOSIS AND MEIOSIS

Like all living things, cells need to reproduce. There's a constant need for new cells, whether for growth or repair. The normal method of cell reproduction in your body is called ***mitosis***. The word mitosis comes from the Greek word "mitos," meaning "thread." This is the same "mitos" that begins the word "mitochondria." The "thread" being referred to is chromatin (DNA).

PAUSE

Okay—how about if we backtrack for a minute talk about mitochondria again?

They are so complicated we just couldn't fit all the information into chapter 3.

PLAY

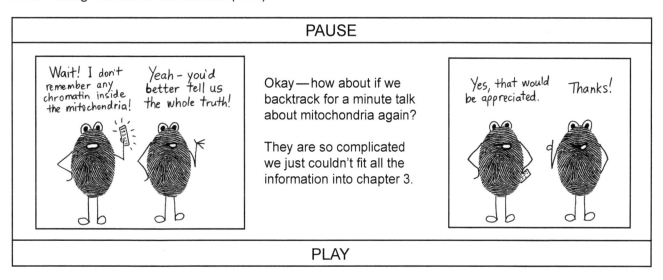

The ring-shaped chromatin (DNA) in the mitochondria is called ***mitochondrial DNA (mtDNA)***. Your mtDNA came almost exclusively from your mother. A microscopic egg cell has about 100,000 mitochondria; a sperm cell only has about 100. When they join, the sperm's mitochondria will be massively outnumbered by the egg's mitochondria. As the cells of the growing embryo begin to divide, the mitochondria divide as well. The mitochondria inherited from the mother will be, for all intents and purposes, the source of all the baby's mitochondria. Because of this direct inheritance of mitochondria from the mother, mitochondrial DNA is very useful for things like tracing the ancestry of people groups. It has also been used to positively identify the remains of famous people such as the famous American outlaw Jesse James, and the last czar of Russia, Nicolas II.

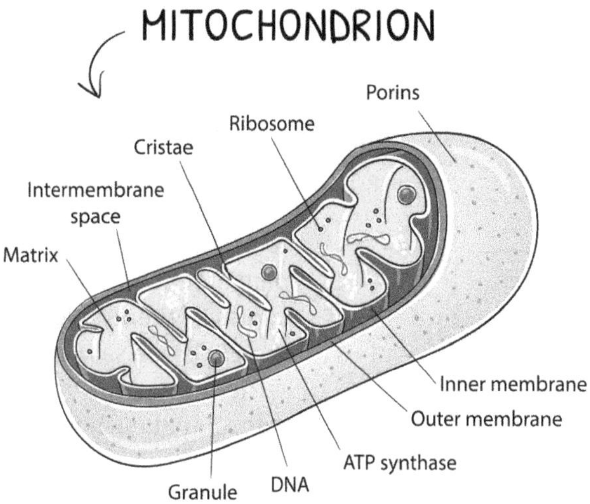

What instructions do you think are encoded in the mtDNA? A mitochondrion doesn't need information about things like lysosomal enzymes, or hemoglobin, or how to make mucus proteins. It only needs information that pertains to itself, such as directions for how to make the parts for the electron transport chain.

This circular diagram shows what each section of mtDNA encodes. Notice that there are also instructions for making ribosomes (rRNA), since they are needed to make protein parts. Strangely enough, not all of the information on how to make ion pumps is stored in mtDNA. The mitocondrion still relies on nuclear DNA for the construction of pump parts. No one knows why this is so. It would seem more efficient to have all the instructions in the mitonchondrial DNA.

A mitochondrion can have up to 10 copies of its mtDNA ring.

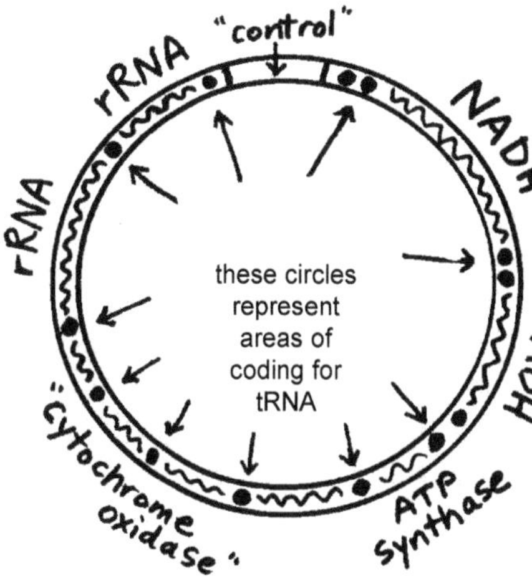

And now, back to mitosis...

Simply put, mitosis is when a cell splits in half. The original cell is called the "parent cell" and the two resulting cells are called the "daughter cells." This does not mean that the new cells are female. The term was chosen during the era when ships and hurricanes were always given feminine names.

To do mitosis, you must be a cell with a true nucleus—a nucleus with a membrane that envelopes all of the DNA and keeps it separate from the rest of the cell. Cells that have a true nucleus are called ***eukaryotic*** cells *(you-care-ee-ot-ic)*. "Eu" means "true" and "karyo" means "kernel" (referring to the nucleus). Animal and plant cells are all eukaryotic, and so are fungi and protozoa. Cells that are not eukaryotic are found in the bacterial kindgom. Bacteria cells do have DNA, but it is not contained within a nuclear envelope. Bacteria are called ***prokaryotic*** cells. "Pro" means first, or "before."

When prokaryotic cells split in half, it is called ***binary fission***. ("Bi" means "two," and "fission" means "split or divide.") Bacteria can accomplish binary fission in as little as 20 minutes, allowing their population to soar within a very short time. (Mitosis generally takes more time than binary fission, with the possible exception of cancer cells that are out of control.) Binary fission is also the term we use when an organelle such as a mitochondrion or a Golgi body splits in half to make a copy of itself. A cell carefully controls the number of mitochondria it has. It can multiply its mitochondria quickly to adjust to new conditions inside or outside the cell.

The words ***eukaryotic*** and ***prokaryotic*** are used frequently in cell biology, and you will see them again at some point in your education. We won't need to use these words in the rest of this book, but since these words are so important in biology, it is a good idea to learn them now.

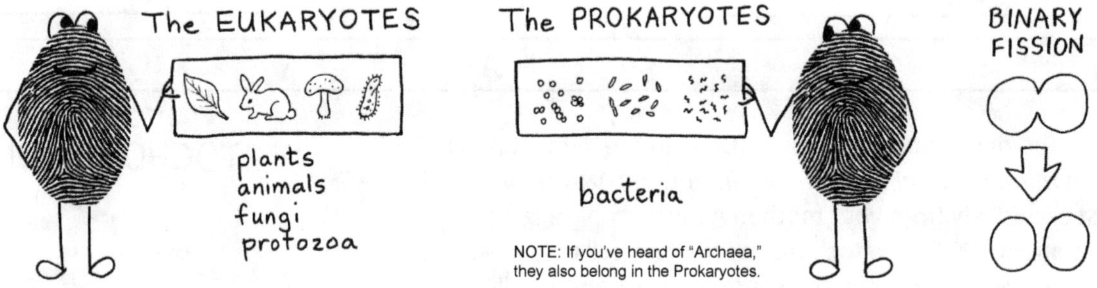

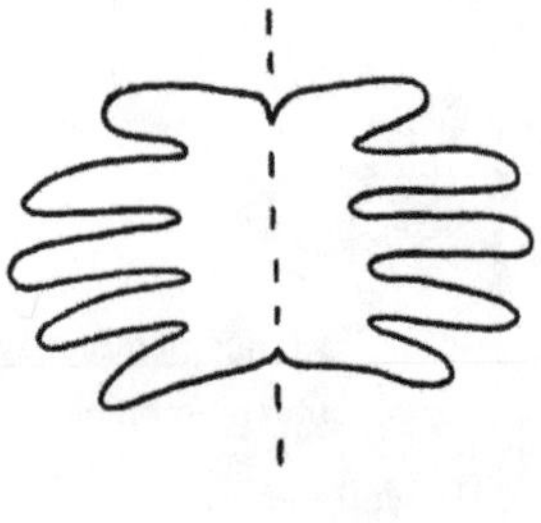

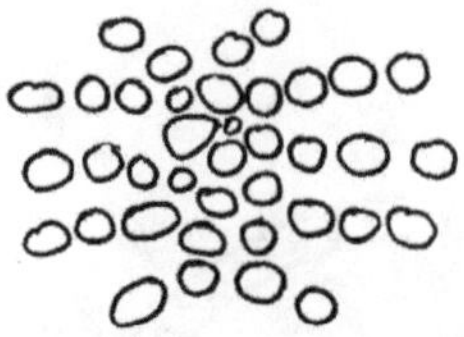

Before splitting in half, the parent cell has a lot of preparation to do. If it's going to become two cells, the original parent cell must have enough organelles for both daughter cells. The mitochondria and the peroxisomes can use binary fission to make copies of themselves until there are enough of them. Lysosomes can be generated by the ER, as we saw in chapter 6. Oodles of new ribosomes can be manufactured by the nucleosome area, although there are probably enough already floating around in the cytosol that each new cell would start out with quite a few. The ER can expand and make more of itself. Golgi bodies seem to have several ways they can multiply.

Sometimes a Golgi will go "super-size" and will double the diameter of its "discs." Then, this super Golgi can split in half the long way, creating two normal-sized bodies. Another method scientists have observed is the Golgi disintegrating itself into small blobs. Then, after the cell splits, each of these little blobs can grow into a whole new Golgi body. A cell's ER can probably manufacture new Golgi bodies, too, if necessary.

The most difficult replication task a cell faces is the duplication of the DNA in the nucleus. The entire DNA "library" has to be copied. We saw a copy of DNA being made back in chapter 5, as the RNA polymerase "sled" read and copied one small section of DNA. In this case, the entire length of the DNA (all 6 billion rungs) is copied by a "sled" called ***DNA polymerase***. The cell must also make a whole new set of spools (about 300 million histones), so that each copy of the DNA can be wound up tightly and neatly. And don't forget all those chaperone proteins that surround the DNA. What a massive undertaking!

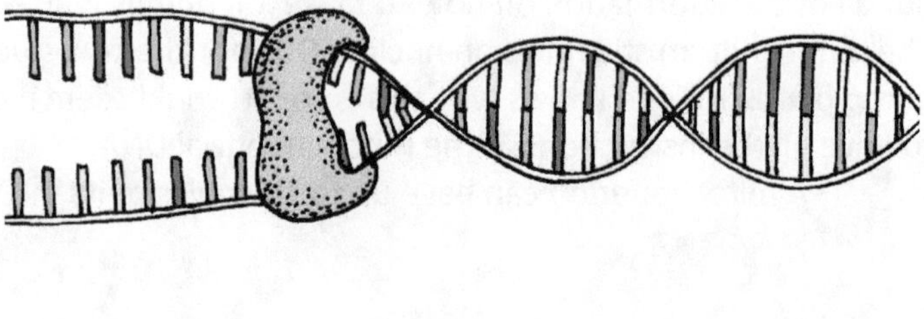

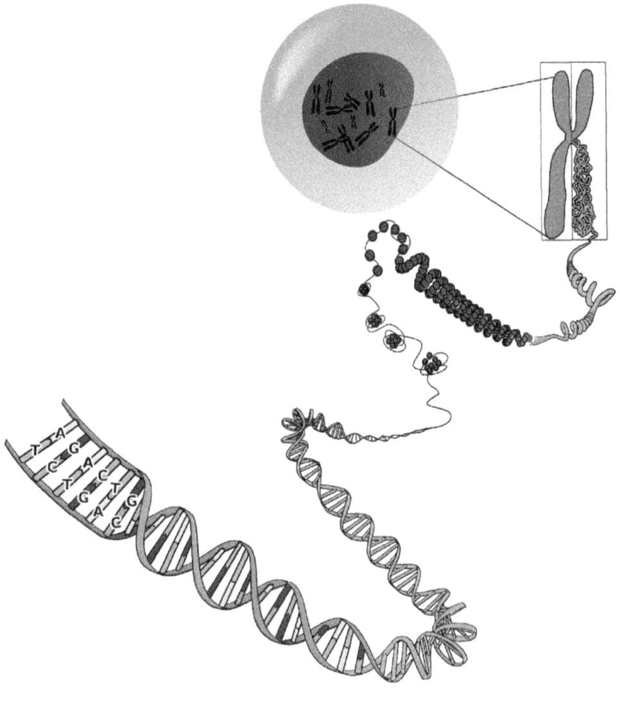

This illustration shows how DNA is organized during mitosis. The string of spools is curled up into a large "X" shape, which we call a ***chromosome***. There are 23 pairs of chromosomes in a (human) cell, giving a total of 46. Each pair consists of two almost identical chromosomes, one inherited from the mother and the other inherited from the father.

You only see chromosomes in their X shape if the cell is preparing to divide. Under normal circumstances, the DNA looks more like a plate of spaghetti. The chromosomes are still there, but they are loose, not coiled up. While the DNA is tightly coiled into these X shapes, it is not availble for making mRNA.

NOTE: Scientists will often use the word "***chromatid***" to refer to the chromosome during mitosis. After mitosis is over, they go back to using "chromosome" again.

After the DNA is duplicated, the nucleus has 46 pairs of chromatids, for a total of 92 chromatids. Keeping with the family theme, the two identical chromatids (the original and its copy) are called ***sister chromatids***. The sister chromatids are joined at the middle by a little dot called a ***centromere*** that is just sticky enough to keep the chromatid sisters together, but not so strong that the bond can't be broken, as we shall see shortly.

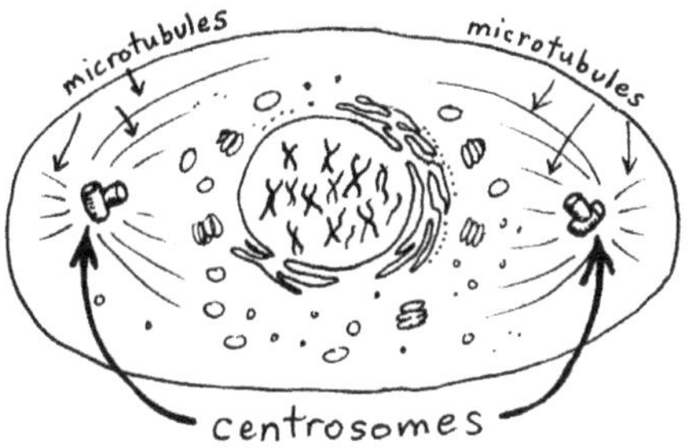

This is when the little centrosome begins to play its biggest role. The centrosome now duplicates itself. Then each centrosome starts making a long strand of microtubule, stretching it out toward one end of the cell. Then motor proteins come along and drag the centrosomes along those microtubule "roads" they just made, until the centrosomes are on opposite sides of the cell. Then the centrosomes must wait until the next step is accomplished.

The next step is to dissolve that double-thick nuclear membrane. The nuclear membrane completely disappears, leaving the chromatids sitting there out in the open, unprotected! But this step is necessary, and the cell will build new nuclear membranes after mitosis is complete. Fortunately, cells are very good at rebuilding their parts quickly.

Now that the nucleus is gone, the centrosomes go to work making more microtubules. These tubules stretch out forming what is called a ***spindle***. Some of the spindle fibers attach directly to the chromatids. The equal pulling force on each side of the pairs of chromatids cause them to line up along the middle of the cell. Once the spindle is in place, the centrosomes start pulling even harder. They pull so hard that the pairs of chromatids snap apart right at the centromere and begin going toward opposite sides of the cell. When the centrosomes have finished pulling, each side of the cell has one complete set of chromatids. (And now we go back to calling them chromosomes.)

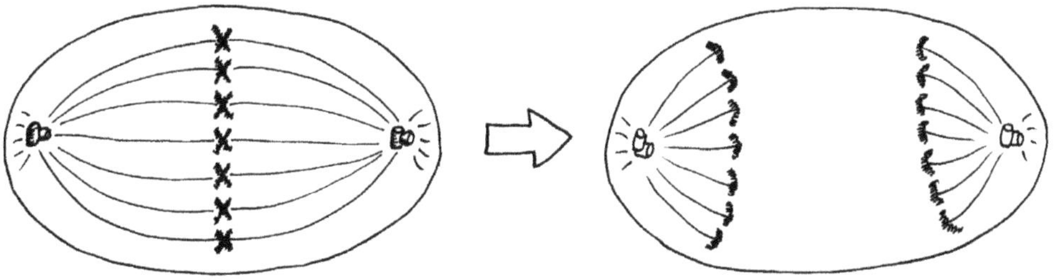

Now for the last part of mitosis. Each side of the cell has a full set of chromosomes, plus a full set of organelles. The cell now begins to get longer. The chromosomes at each end of the cell cluster together and go back to being a blob of chromatin. Then the ER gets busy re-organizing itself and re-building the nuclear membrane around the DNA. Soon, each end of this long cell has its own nucleus. This is technically where mitosis ends. The final splitting process is called ***cytokinesis*** ("cell movement"). However, most of the time when people talk about mitosis, they mean to include cytokinesis as well.

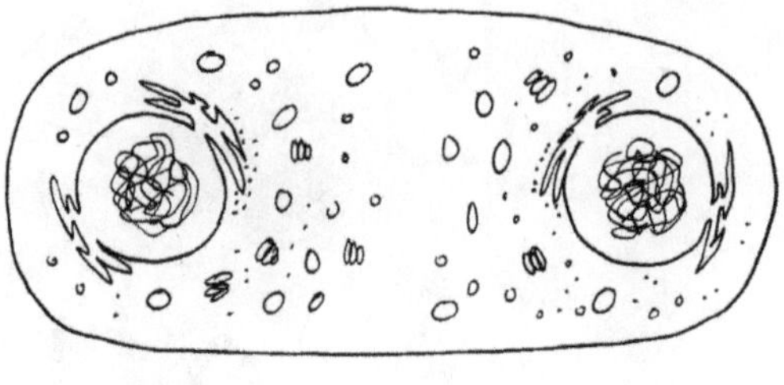

Now the cell is ready for "cytokinesis."

Cytokinesis is when the cell splits in half. It starts out as a pinched area on either side, then the pinch grows and grows until it's all the way down the middle. Interestingly enough (and just to show you how complicated cells are) vesicles created by the Golgi bodies seem to play a role at this point. Vesicles line up along the middle line and help to create the new membranes. In plant cells, these vesicles are even more important than in animal cells, as they will help to create a new cell wall, not just a new membrane. (Walls are much thicker and sturdier than membranes.) Once the pinch is complete down the middle, the final separation is made, and you've got two daughter cells.

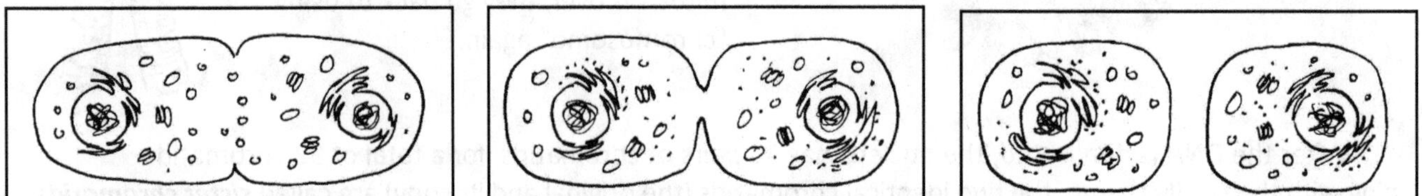

Scientists like to assign names to the steps of mitosis, to make it easier to for them to have discussions about it. They can say things like "at the beginning of prophase" or "at the end of telophase," and all the other scientists know exactly what stage of mitosis they are talking about. So, just in case you ever need to know, here are the names they assign to the stages. (You can also use key words "cell cycle" in a video platform such as YouTube and find lots of videos that have animations explaining each stage.)

INTERPHASE: The "normal" state of a cell. Most of the time, a cell is in interphase. Towards the end of this phase, the organelles start to replicate more than usual, to make sure that each future daughter cell will have enough. ("Inter" means "between.")
PROPHASE: The chromatin duplicates and then organizes into chromosomes. The centrosomes duplicate and go to opposite sides of the nucleus. ("Pro" can mean "first," so this is the first step.)
METAPHASE: The nuclear membrane dissolves, the spindle forms, and the chromosomes line up down the middle. (To remember this phase, think M for Middle, since they line up in the middle.)
ANAPHASE: The chromatids are pulled to opposite sides of the cell. Also, all the organelles have gone to one side or the other. (In this phase the spindles look like letter As on each end of the cell.)
TELOPHASE: The cell elongates and new membranes are formed around the DNA, which has gone back to being chromatin. ("Telo" means "far away," as in "teloscope." The chromosomes are far away now.)
CYTOKINESIS: The cell splits in half, making two daughter cells. ("Kine" means "motion.")

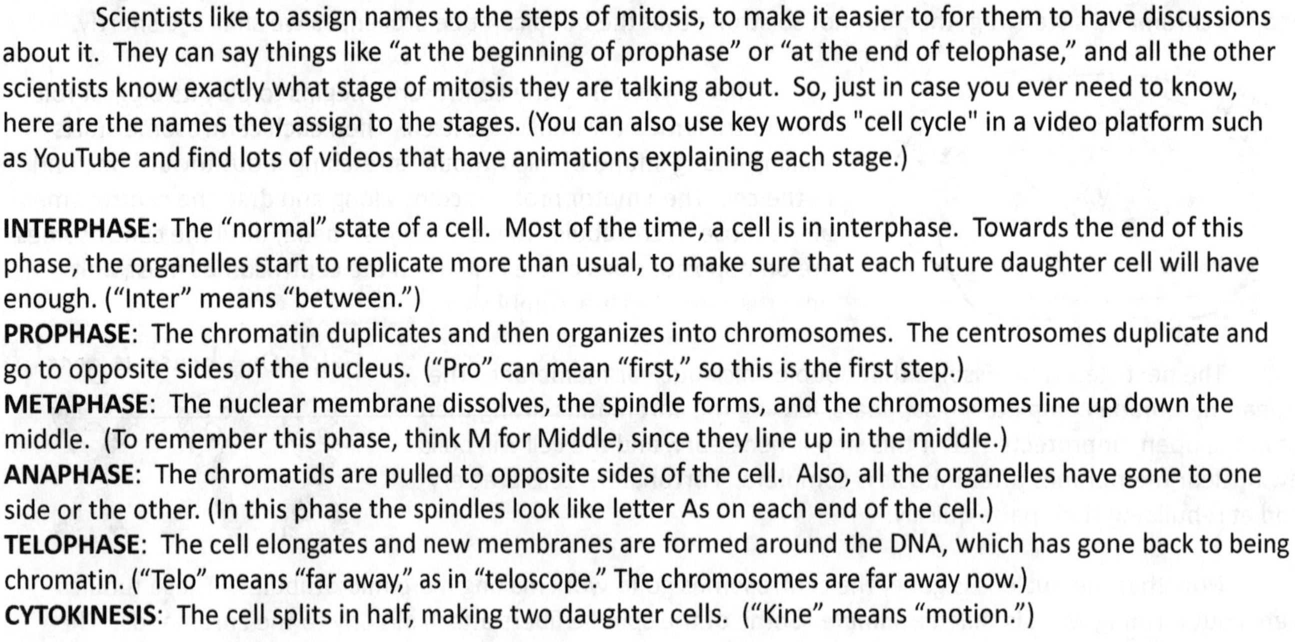

Eukaryotes have another other kind of cell division, in addition to mitosis. There is one situation in which you don't want a complete set of DNA in each daughter cell. In reproductive cells (eggs and sperm) you only want <u>half</u> of a set of DNA, not a full set. Why? Because an egg and a sperm join together to make a new set of DNA. One half plus one half equals a whole. If an egg and a sperm each had a full set of DNA, the embryo would end up with double DNA, creating a real mess!

The special process of making an egg or sperm cell is called ***meiosis*** (*my-O-sis, or mee-O-sis*). In meiosis, cell duplication takes place twice instead of just once. To make sure everyone knows which duplication is being discussed, scientists decided to call the first one "meiosis 1" and the second one "meiosis 2." Sometimes you will see Roman numerals used instead of numbers: "meiosis I" and "meiosis II."

The first step of meiosis 1 is the same as in mitosis: the cell duplicates its chromosomes into pairs of sister chromatids. It does this duplicating before the chromosomes get all bunched into little sticks, so for a short time the cell has double the normal "spaghetti mess" in the nucleus.

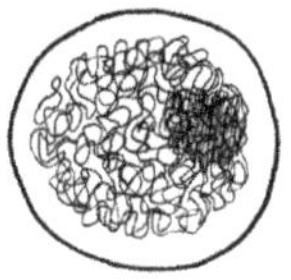

The next step is different from mitosis. Amidst this semi-organized chaos in the nucleus, the pairs of chromatids have the unbelievable task of having to go find their duplicate "counterparts." When they do, they form a temporary attachment, creating two pairs of sister chromatids.

Wait—this is getting so confusing! Let's try to make this more understandable by drawing the original pair of chromatids as a set of girl/boy twins. The girl twin will be the chromosome inherited from the mother and the boy twin will be the chromosome inherited from the father. So they each make a copy of themselves, giving us two sets of girl/boy twins. They are joined in the middle by a little sticky dot (centromere) that will keep them together until the microtubules pull them apart.

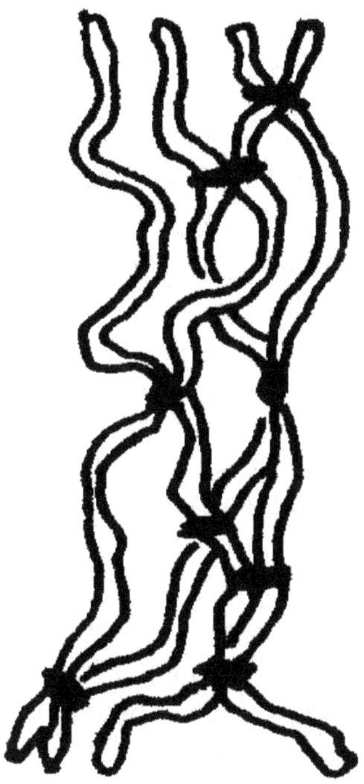

Real "crossing over" would look more like this. The dark spots are where the chromatids have traded their DNA.

Now something bizarre happens. Those four chromatids trade some of their DNA back and forth. In our mental picture of the two sets of twins standing next to each other, it would be as if they all traded some body parts. Imagine them as plastic dolls made of individual body parts that can be snapped on and off. They could pop off their body parts, mix them up, and them pop them back on randomly. Bobby might end up with Sally's feet. Sally might end up with Betty's hair or with her left hand. Betty might end up with Jimmy's left thumb. If chromatids were people it would be a truly bizarre scene! However, chromatids look pretty boring under the microscope, and this event isn't nearly as humorous in real life as it sounds when we imagine the chromatids as people. The chromatids simply trade some DNA back and forth. Why? The purpose of meiosis is to make reproductive cells that will create offspring with brand new combinations of DNA. Children are not clones of their parents; they are their own unique selves. It's partly this step in meiosis, called "***crossing over,***" that helps to mix up the genetic information and create offspring that are different from either parent. (In large populations, such as insects, these variations can help a species to survive.)

Now the cell is ready to divide. Alas—those couples are going to have to split up! The rules of the meiosis game say that the groups of four (the girl twins and boy twins), have to split up so that the girls go off together and the boys go off together. Just like in mitosis, the centrosomes form a spindle of microtubules. Then the chromosome pairs line up along the middle and are pulled apart.

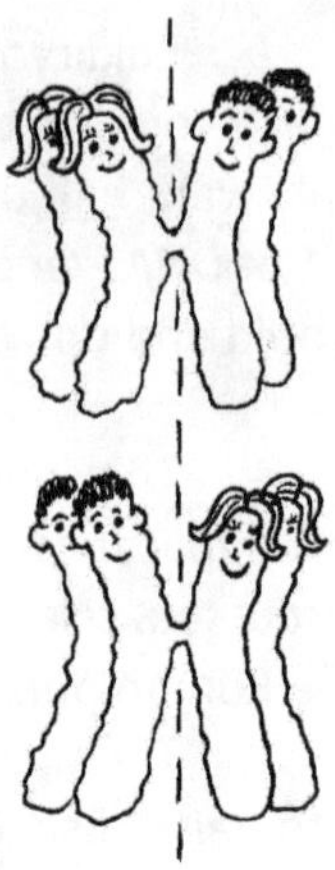

To be more accurate, the girls (or boys) wouldn't all be facing the same side. The "couples" would be arranged randomly, with some boys and some girls on each side. We only show one set of double twins here, but remember, real cells would have 23 of these sets of four chromatids. Theoretically, it would be possible for a daughter cell to end up with all 23 girl twins or all 23 boy twins, but the odds of this happening are very small. And even if it did, it wouldn't really matter because the twins have mixed-up body parts, so the daughter cells will get some DNA from both parents no matter what happens.

Now for the second part of meiosis. It's a good thing that chromatids don't have feelings, because now the "twins" must be separated. It's time for another cell division. This time, there is no duplication of DNA. The twins simply line up along the middle and are pulled apart by the microtubules in the spindle. It's a sad goodbye as the twins say "Adieu" and never see each other again. Once the individual chromatids are separated, the cell splits in half, forming two daughter cells that have only half the normal amount of DNA.

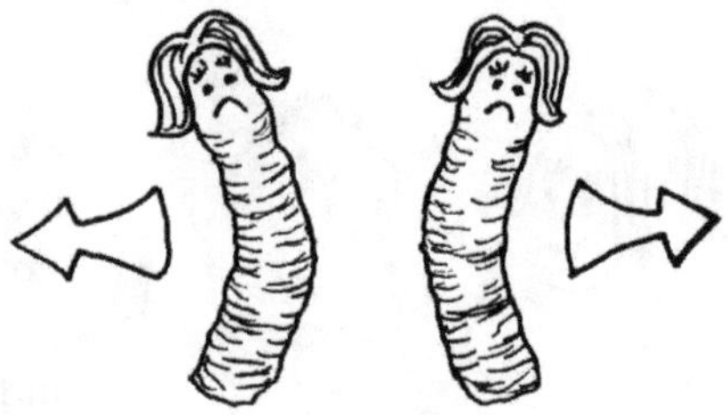

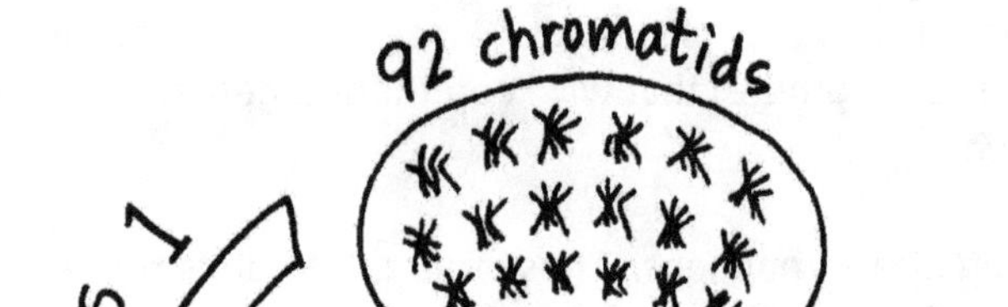

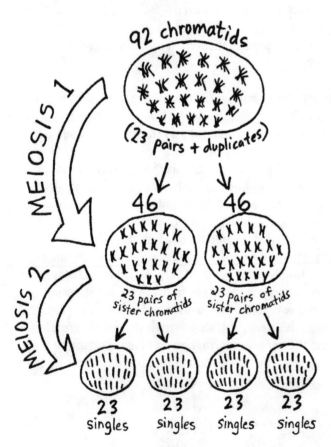

On the left is a diagram showing how many chromatids are in human cells during the stages of meiosis. You can see that the original cell starts out with 46 chromosomes (23 pairs). They duplicate themselves, making a grand total of 92 chromatids. After they do the "crossing over" routine to mix up the DNA, they then split apart to make daughter cells that have 23 pairs of sister chromatids (46 chromatids in all). In the second stage, the chromosomes don't duplicate, they just separate into daughter cells that have only 23 chromosomes, which is half of the "normal" amount of DNA.

What is the fate of those sad "ex-twins" that end up in those cells on the bottom row? These "haploid" cells (called ***gametes***) will form into either egg cells or sperm cells, depending on whether they are in a male or female body. There are a few more twists and turns in the plot; not all female daughter cells end up as eggs. But the rest of the story is... another story. We are going to say goodbye to these gamete cells and let you learn more about them in another course.

NOTE: These numbers are for human cells. Most other animals, even other mammals, have more or less than 23 pairs of chromosomes. The only mammals that have 23 pairs are the Reeve's muntjac (a small deer-like animal), and the nilgia (a type of cow). Oddly enough, another muntjac, the Indian muntjac, has only 3 pairs! Apes have 24 pairs, dogs have 39 pairs, and cats have 19 pairs.

If you want to know more about chromosome counts in various living things, including many plants and insects, look up this Wikipedia article: "List of organisms by chromosome count."

Can you remember what you read?

1) TRUE or FALSE? Mitochondrial DNA is inherited almost entirely from the mother.

2) TRUE OR FALSE? Mitochondrial DNA contains all the information a mitochondrion needs.

3) TRUE or FALSE? Mitochondrial DNA looks just like the DNA in the nucleus.

4) TRUE or FALSE? A mitochondrion can have more than one set of mtDNA. (bottom of page 79)

5) TRUE or FALSE? Prokaryotic cells do not have a nucleus.

6) TRUE OR FALSE? Only eukaryotic cells can do mitosis.

7) Prokaryotic cells mainly belong to this kingdom:
a) animals b) bacteria c) fungi d) protozoa e) plants

8) When a cell duplicates itself, the resulting cells are called the ______ cells.
a) sister b) daughter c) children d) clone

9) When a chromatid duplicates, the duplicate is called its:
a) sister b) daughter c) twin d) clone

10) Which organelle creates the spindle? ________________

11) What is the spindle made of? __________________

12) When the cell actually splits in half, this action is called:
a) mitosis b) binary fission c) cloning d) cytokinesis

13) Which one of these doesn't do binary fission?
a) bacteria b) mitochondria c) eukaryotic cells d) peroxisomes

14) In which process (mitosis or meiosis) will you find "crossing over"? ______________

15) What does "crossing over" do?
a) Helps to "mix up" the DNA b) Sorts out which chromatids will go to which daughter cells
c) Causes mutations d) Determine the gender of gamete

16) How many single chromatids end up in a (human) "gamete" cell (egg or sperm)? ______

17) TRUE or FALSE? The nucleus disappears during mitosis.

18) Which one of these is technically not part of mitosis?
a) prophase b) metaphase c) anaphase d) telophase e) cytokinesis

19) TRUE or FALSE? DNA is always organized into chromatids.

20) Which of these is responsible for making a duplicate copy of the cell's DNA?
a) centrosome b) RNA polymerase c) DNA polymerase d) ribosomes

ACTIVITY 9.1 Look at real cells in various stages of mitosis

If you do an Internet image search using key words "onion mitosis" you will find an endless supply of pictures showing microscopic views of cells from the tips of onion roots, a place where mitosis takes place at a very rapid rate. You'll see cells in just about every stage of mitosis.

ACTIVITY 9.2 Watch those supplemental videos on the playlist!

Don't miss the meiosis square dance!

ACTIVITY 9.3 Cell number puzzle

Try to match the correct number to each clue.

1 2 3 4 5 6 7 8 9 10 11 13 14 20 23 36 40 92 146 500

1) The number of ATPs that a cell can get out of one glucose molecule. ____
2) The number of histone proteins in a nucleosome spool. ____
3) The number of steps in glycolysis. ____
4) The number of electrons in a hydrogen atom. ____
5) The number of amino acids in a codon. ____
6) The number of tubulin dimers it takes to make one ring of a microtubule. ____
7) There are approximately this many different enzymes in a lysosome. ____
8) The number of centrioles in a centrosome. ____
9) The highest number on the pH scale. ____
10) If we could enlarge a strand of human DNA so that it was as thick as a thread, it would be _____ miles long.
11) There are this many types of amino acids. ____
12) There are this many pairs of chromosomes in a human cell. ____
13) There are this many types of nucleotide bases (including both DNA and RNA). ____
14) There are this many layers of phospholipids in the nuclear membrane. ____
15) Number of chromatids in a human cell right during mitosis or meiosis right before it splits in half. ____
16) This number is "neutral" on the pH scale. ____
17) This is the number of carbon atoms in a glucose molecule. _____
18) Number of non-essential amino acids. (check table on page 39) ____
19) There are about this many DNA rungs wound around each nucleosome spool. ____
20) The number of essential anino acids (check table on page 39) _____

ACTIVITY 9.4 Chromosome trivia: Is it possible to live with just one set of chromosomes?

Yes. Male honey bees have only one set of chromosomes. The queen bee can choose whether or not to fertilize (put sperm into) an egg. If she deposits sperm into the egg she has laid, the egg will then have two sets of chromosomes, one from herself and one from a drone (male) bee. The fertilized eggs will always develop into female worker bees. If she does not fertilize the egg, the egg will still devolop into a full grown bee, but it will be a male who is "haploid," meaning having half the normal number of chromosomes. Organisms with two sets of chromosomes are called "diploid."

CHAPTER 10: TYPES OF CELLS

Everything we have been learning about cells is true for most eukaryotic cells. All eukaryotic cells (with just a few exceptions) have the same types of organelles and use the same basic metabolic processes. However, cells can come in many different shapes and sizes and some have amazing adaptations that make them suitable for a particular function. In this chapter we are going to look at a selection of cells found in humans and mammals.

SKIN CELLS

Skin cells belong to a group of cells called ***epithelial*** cells. ("Epi" means "outside" and "thelia" means "tissue.") Epithelial cells can be found not only in the skin but also in the lungs, intestines, eye, blood vessels, and many glands.

Skin cells form what we call the ***epidermis*** (there's that word "epi" again, plus "dermis" for skin). The epidermis is made of layers of skin cells piled on top of each other. Their shapes are very simple, like blocks. They are packed together very tightly, connected by desmosomes. In fact, they are so tightly packed that there aren't any capillaries running through them to nourish them. They have to rely on diffusion from below. The skin cells up at the surface are too far away from the capillaries to get any nutrients, but they don't care because... they are dead. These dead cells no longer have any organelles. They are filled with a substance called ***keratin***. Keratin is made of long, thin strands of protein and is classified as an intermediate cytoskeletal filament.

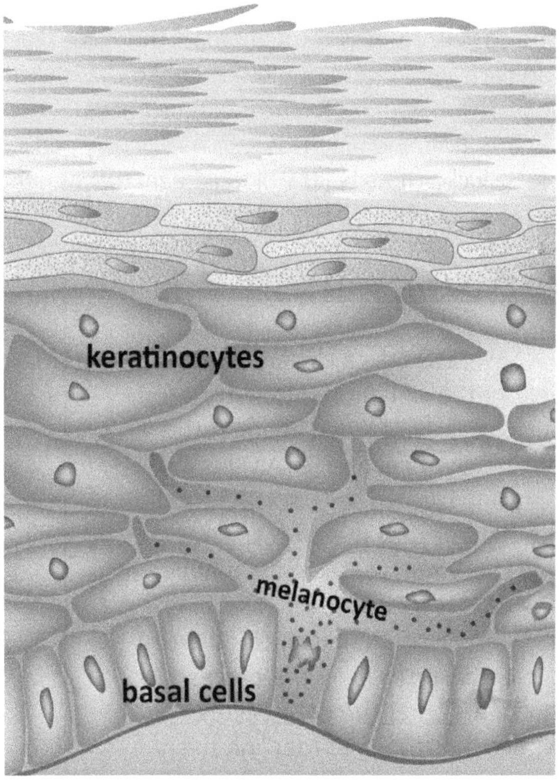

EPIDERMIS

The ***basal cells*** at the bottom of the epidermis are constantly undergoing mitosis. As new cells arise, the older ones above them get pushed upward. As they go up they gradually get filled with more and more keratin and are called ***keratinocytes***. By the time they reach the top, they are totally filled with keratin. It's sort of like the cytoskeleton takes over and kills off the organelles. But there's a bright side to the death of all these cells—keratin is tough and waterproof. All that keratin (plus some oil made by oil glands under the epidermis) gives your body a fairly waterproof surface.

Mixed in with the keratinocytes are a few ***melanocytes***—cells that make ***melanin***, the brown pigment that gives color to our skin. The difference in skin color is due to the number, and activity level, of the melanocytes. Melanocytes contain a special organelle called a ***melanosome***, which uses the animo acid tyrosine to manufacture the melanin pigment molecules. The melanin is put into vesicles that are both transported throughout the melanocyte cell and also exported outside the cell so that keratinocytes can take them using the process of endocytosis. Melanin is a molecule that can absorb ultraviolet light which might harm the cell. When we expose our skin to sunlight (which contains ultraviolet), the melanosomes will become more active and will change shape, putting out pseudopods that will shade and protect the nucleus. This is the mechanism behind "getting a tan."

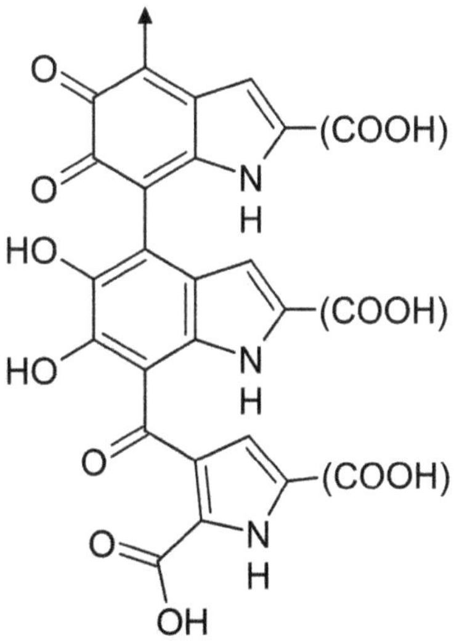

MELANIN

CAPILLARY CELLS

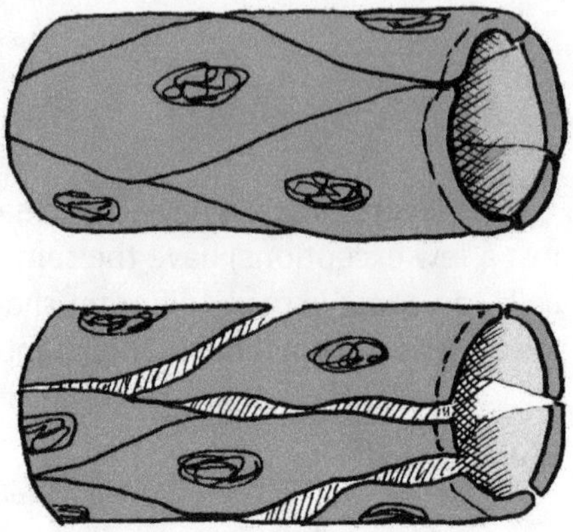

Another place where you will find epithelial cells is in your capillaries. Hopefully, you will remember this cell from chapter 8. It looked different when we saw it in that chapter because we were looking at a cross section, not the whole cell. These "**endothelial**" cells are pancake shaped; they curl around and together they form a tube shape. The cells are held together tightly by desmosomes. In chapter 8, we saw them transfering nutrients to cells.

In some places in the body, such as the liver and the spleen, red blood cells need to be able to enter and exit the capillaries. In these places, you will find larger cracks between the cells—cracks large enough for red cells to pass through. Also, there is a place in the brain where the capillaries are intentionally "leaky" so that the brain can sample what is floating in the blood and send out signals to other body parts to correct imbalances.

Even though these cells have a very odd shape, they are still cells, and have a nucleus and organelles. They are pretty lucky, though—the bloodstream rushes right past them all the time. No problem about getting enough nutrition!

CELLS OF THE VILLI

Let's take a closer look at the cells of the villi, those little finger-like structures we learned about at the beginning of chapter 7. A villus has only one layer of cells, so all the cells are right next to the tiny capillaries and the lymph vessel in the middle. The cells responsible for absorbing nutrients from the mushy food are the ***enterocytes*** *(en-TARE-o-sites)*. Nutrients <u>enter</u> the <u>entero</u>cytes. Notice the "fringe border" on the enterocytes. (It really is called the fringe border.) The ***microvilli*** are tiny finger-like projections of the cell's plasma membrane. They help to increase the surface area of the cell so that there are more places where the cell will come into contact with the mushy food passing by.

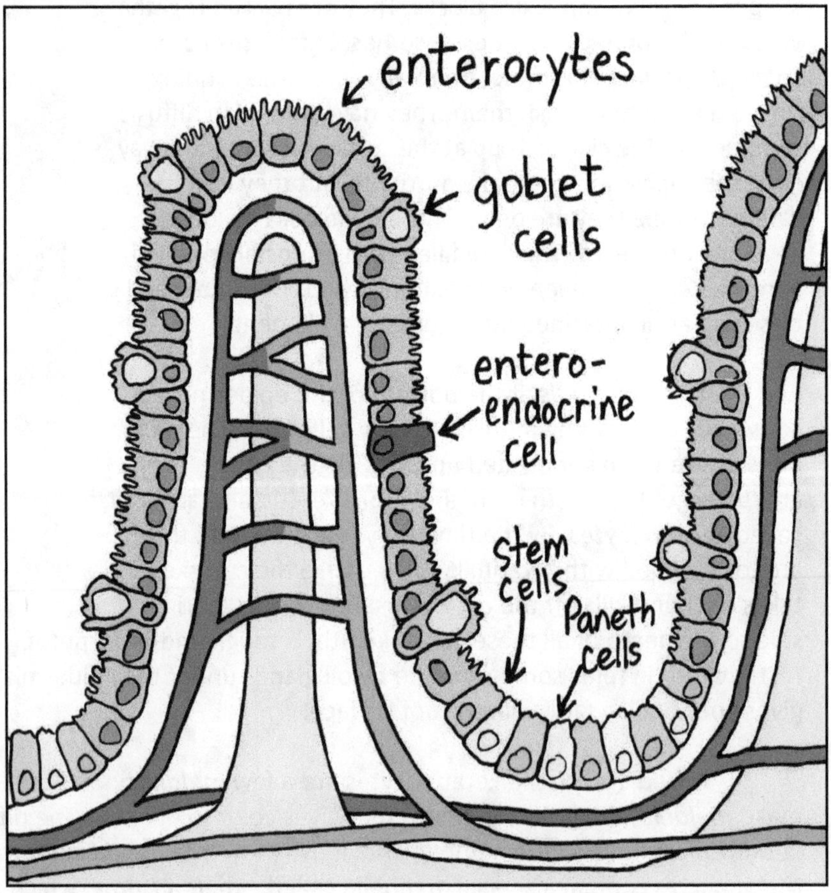

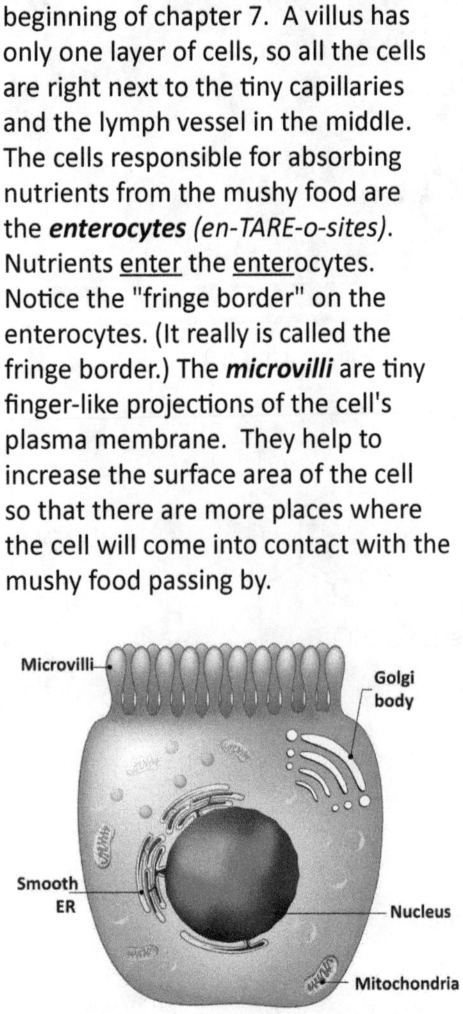

There are many types of ***enteroendocrine*** cells, though only one is shown here. They act as controllers and coordinators, monitoring what is passing through and sending out signals to other body parts such as the brain, the pancreas, the liver, and the muscles around the intestines. For example, if you suddenly change your diet and start eating a lot more or less of certain foods, your digestive organs will have to ramp up production of some chemicals and cut back on the production of others.

Stem cells are like baby cells that don't know what they want to be when they grow up. They can turn into any of the villi cells. Most of them will become new enterocytes because these cells live a rough and tumble life out of there on the edge. Enterocytes are a lot like skin cells and get replaced constantly.

Paneth cells were named after the scientist who discovered them. They sense when harmful bacteria are present and release protein weapons called ***definsins***. This illustration shows the structure of the two most common defensin molecules. These molecules are designed to damage the outer membrane of bacteria cells so that the cell collapses. Paneth cells are "smart" cells, that learn to tell the difference between the "good" bacteria that is supposed to live in our intestines and "bad" bacteria that has invaded.

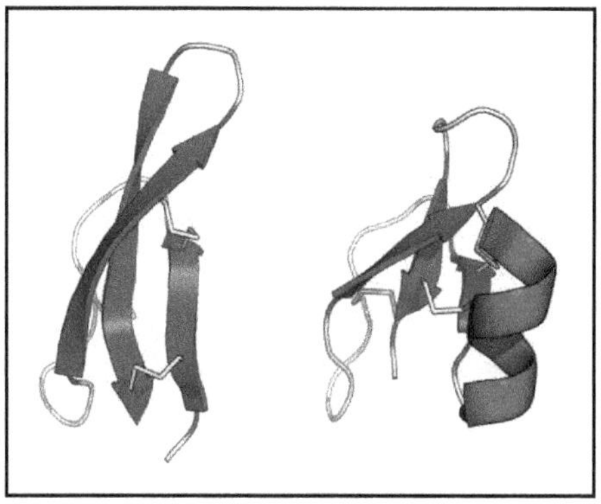

alpha defensin *beta defensin*

Goblet cells make mucus. These cells are found not only in your intestines but also in your respiratory tract. In the intestines, mucus provides lubrication so that things slide through easily, as well as acting as a protective coating for the enterocytes, keeping bacteria, viruses and fungal cells away from them. In the lungs, trachea and nasal passages, mucus catches dust particles and washes them out. It also provides a barrier that is hard for bacteria to cross. As much as we don't like too much mucus in our sinuses when we are sick, without mucus we would lose an essential form of protection from infections.

The goblet cells get their name from their shape: narrow on the bottom with a large "cup" on the top, like a goblet. The narrow bottom "pedestal" part of the cell contains the nucleus. The top "cup" part of the cell contains the storage vesicles full of mucin waiting to be released. Like the enterocytes, the top part of goblet cells has microvilli. When the microvilli are irritated or stimulated, they trigger the release of the vesicles through the process of exocytosis.

The main component of mucus is a protein called ***mucin***. Like all proteins, it is made by ribosomes and then processed by the endoplasmic reticulum and Golgi bodies, which add many sugars to it. The strings of sugars can attract and hold many water molecules, giving mucus its gooey, sticky texture. The mucin proteins are folded into very compact shapes so that they can fit into the storage vesicles. When the mucin is released, the protein unfolds and mixes with water, causing the mucus to expand to 600 times its previous size!

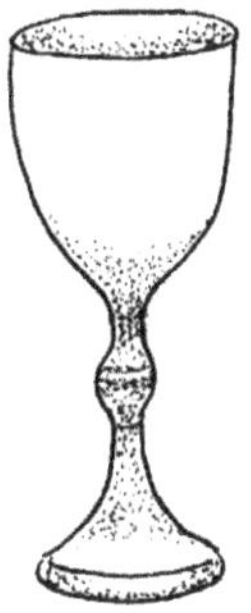

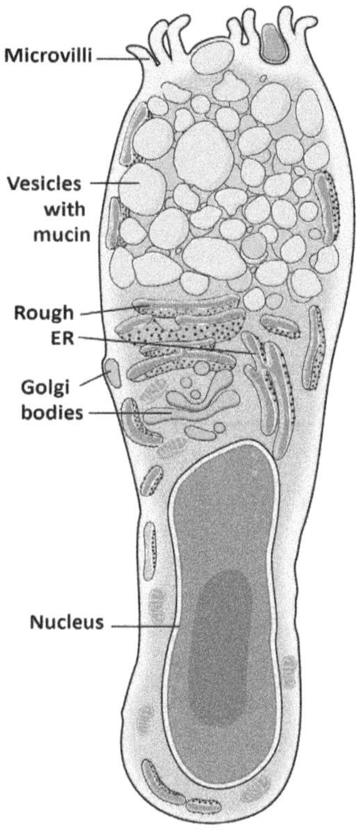

Goblet cells belong to a group of cells called ***secretory*** cells. *(SEH-creh-tore-ee)* Secretory cells "secrete" things like sweat, tears, saliva, milk or hormones. There are many types of secretory cells in your body that don't have the goblet shape, but they function in a similar manner: they make their products, store them in vesicles, then release them at just the right time.

You have two types of sweat glands, both with obnoxious technical names: apocrine and eccrine. Never mind the names, the interesting fact is that you have two types. The eccrine glands are found in most parts of your skin, even on your head. They respond to the temperature around you and make you sweat when it gets hot. The aporcine glands are only found in certain places such as the armpits, eyelids, and the ear canal. (The smell we associate with sweat comes not from the glands, but from bacteria.) The skin also has sebaceous glands that produce an oily substance that helps to make the surface more waterproof.

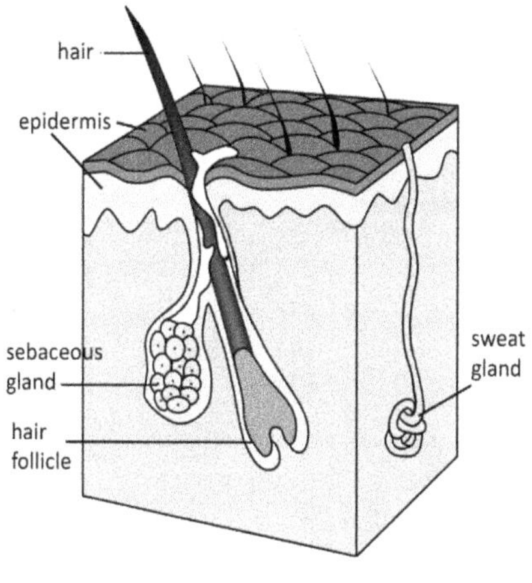

BONE CELLS

We all know that bones help us to stand up and keep our shape. But did you know that ounce for ounce your bones are stronger than steel? Your femur (thigh bone) is stronger than a steel bar weighing 4 to 5 times more than the bone. The average bone can withstand the weight of five trucks (don't try this at home!). How can your bones be so strong and so light? Let's look at the cells that make our bones.

Cells called ***osteoblasts*** ("osteo" means "bone" and "blast" means an immature cell) make a large interwoven network using a tough but flexible protein called ***collagen***. These cells have a large Golgi complex and endoplasmic reticulum so that they can process the large amounts of collagen protein they need to make. Once they have made this large spiderweb structure of collagen protein (called a ***matrix***), they begin to mineralize it using mostly calcium and phosphorus, plus magnesium and a few other minerals. These minerals bind to the flexible collagen matrix and give it extra strength. The hollow spaces of the matrix keep the bone light, the protein network gives it flexibility, and the minerals give it strength.

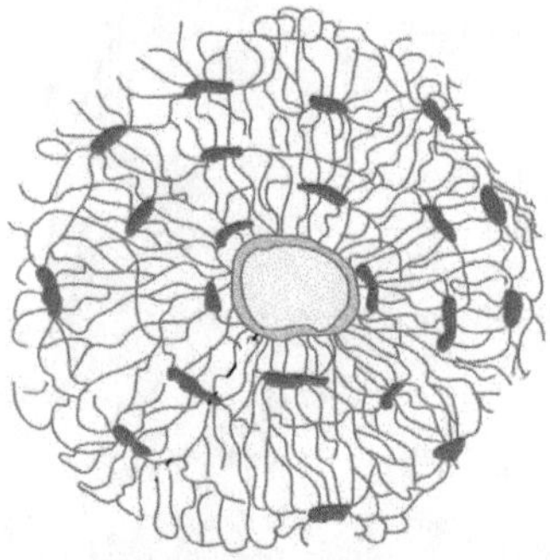

These osteocytes are forming a circular osteon. The red blobs are the cell bodies; the thin lines are the cells' "arms" (filopodia).

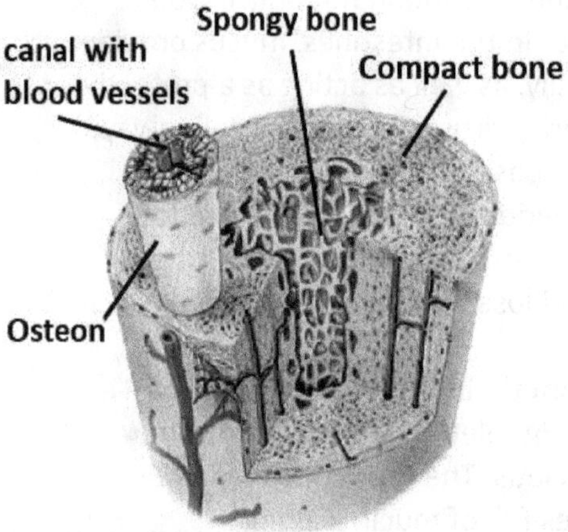

Eventually, the osteoblast will be stuck inside the bone "cage" it has made around itself. After this happens, we call the cell an ***osteocyte.*** The osteocytes are arranged into a circular shape called the ***osteon***. The osteocytes leave a hole in the center of the osteon for tiny blood vessels. Running along with the blood vessels are lymph vessels and nerves.

The osteocytes have very long and extremely thin "arms" called ***filopodia*** ("thread feet"). The filopodia reach out to touch the filopodia of other cells. (Imagine a group of your friends standing just far enough apart so that they can barely touch each other using their fingertips.) The cells that are close to the blood vessels get nutrients passed along to them from the endothelial cells of the capillaries. Those osteocytes must pass nutrients to other osteocytes that are further away from the blood vessels. This system only works at short range, so there is a limit to how large an osteon can be (as shown in the osteon image above). The osteocytes also use the filopodia to communicate with each other. Chemical messages can be passed from cell to cell.

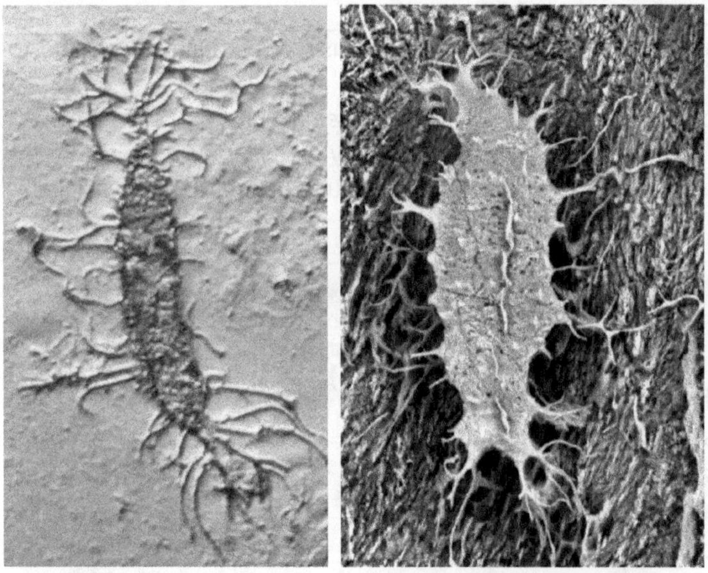

Triceratops osteocyte under regular microscope

Triceratops osteocye SEM image

Photos by Dr. Mark Armitage

These photographs show osteocytes taken from dinosaur bones. Your osteocytes look very much like those of other animals, including reptiles. In the image on the left, the bone's minerals were dissolved away, releasing the osteocytes to float freely in the solution.

Osteocytes can live for several decades. Eventually, the body's recycling system will dissolve all the older sections of bone and replace them with new osteons. The cells that dissolve the old bone are called ***osteoclasts.*** The bones can also be used as an emergency supply of calcium and other minerals. If the level of calcium in your blood gets too low, the osteoclasts will go to work releasing minerals. When more calcium becomes available, the osteoblasts will then repair the bone.

MUSCLE CELLS

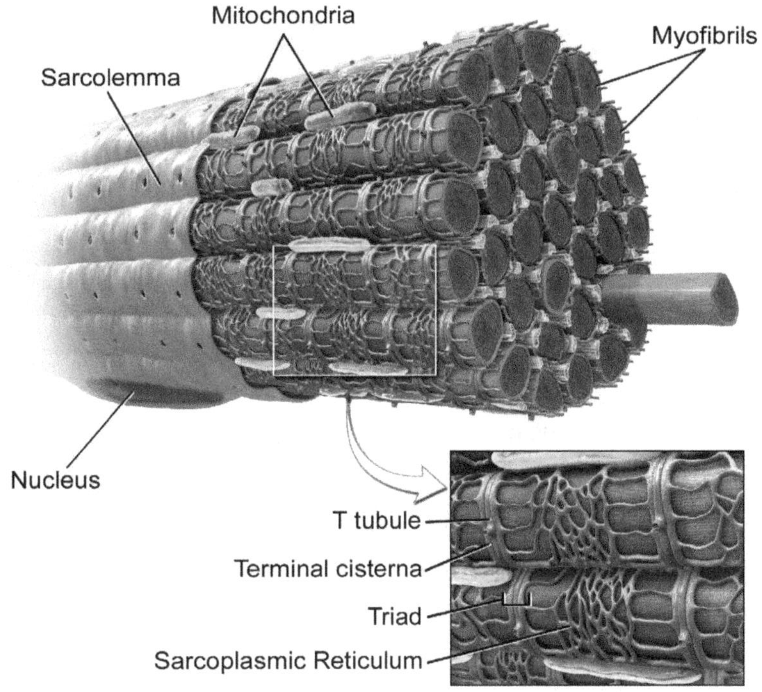

Muscles cells are weird. Individual cells fuse together to form a very long cell that has lots of nuclei. This cooperative cell is so strange that we don't even call it a cell—we call it a ***fiber***. The cells started out as single "baby" cells, then, as they matured into "adult" cells, they merged together to form one long fiber. The fiber has all of the organelles from the original cells, so it ends up with a whole bunch of everything, including nuclei. A cell with more than one nucleus? How can this work? Which one is in charge?

Having multiple nuclei is an advantage to the muscle fiber (cell) because it is so long. For example, if there was only one nucleus, the messenger RNAs would have to travel long distances to get to the ribosomes at the far ends of the fiber. This would slow down the process of making proteins. However, with multiple nuclei, the mRNA can simply go to the nearest nucleus. All the nuclei have the same set of instructions for repairing muscles, so it doesn't matter which nucleus the mRNA chooses.

Muscle cells have a special phospholipid membrane surrounding them called the ***sarcolemma***. This membrane is special because it actually forms tunnel-like tubes that pass all the way through the diameter of the cell and out to the other side of the sarcolemma. These tunnels house nerve endings that carry electrical signals from your brain. An electrical signal causes an organelle inside the cell to initiate the contraction process. This special organelle is called the ***sarcoplasmic reticulum***. It is similar in shape and structure to an endoplasmic reticulum, but its phospholipid membrane network extends through the whole muscle fiber. Its job is to store calcium ions. When the electrical signal comes down the nerve endings through the sarcolemma tunnels, the "electrical shock" causes the sarcoplasmic reticulum to release its calcium ions into the cytoplasm of the muscle cell. These calcium ions bind with myosin, causing it to slide across the actin, resulting in contraction.

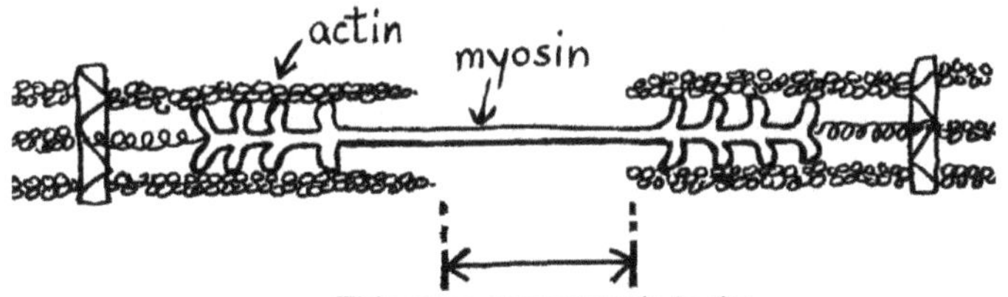

This area corresponds to the stripes on the myofibrils

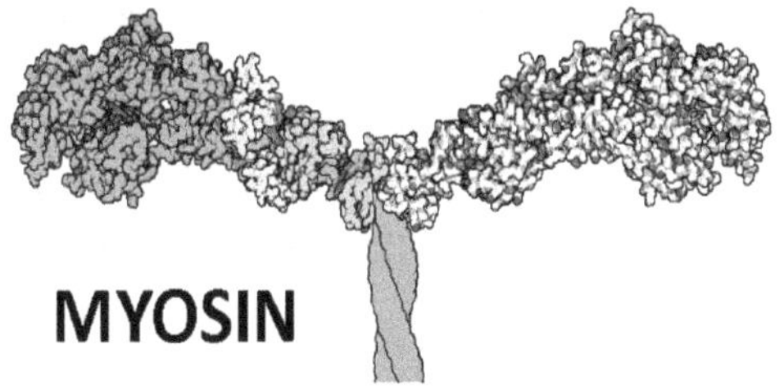

The bundles inside the muscle cell are called ***myofibrils***. They are made of two stringy proteins: ***actin*** and ***myosin***. Actin is a type of microfilament. We mentioned it briefly at the top of page 18. Myosin has tiny "oars" along its length that push against actin when calcium is present, causing the two proteins to slide past each other. The myosin "oars" should remind you of the heads of motor proteins that walk along microtubules. To help you visualize the motion caused by action and myosin, imagine a radio antenna, and think about how the tubes slide into each other, allowing the antenna to become longer or shorter. The sliding of actin and myosin causes the myofibrils to shrink in length. Then, the myosin lets go and the myofibril relaxes and gets longer again. ATP plays a critical role in this process, as it binds and unbinds to the myosin "oars." As you might guess, muscle cells have many mitochondria to supply a large and steady source of ATPs.

RODS AND CONES

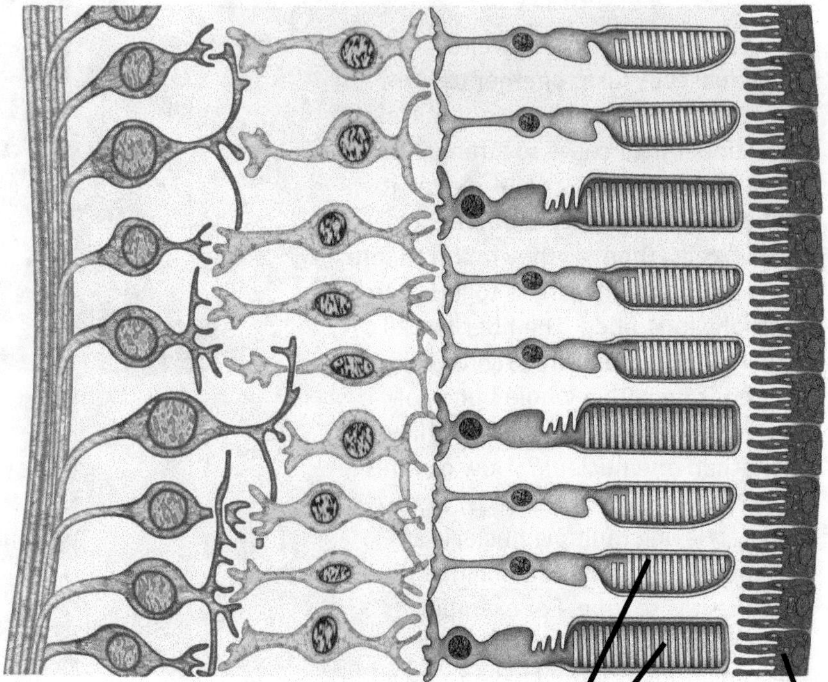

These cells act like electrical wires.

The cells with square ends are rods; the more pointed ones are the cones.

These cells nourish and protect the rod and cone cells.

The cells at the back of your eye are a marvel of engineering. Several types of highly specialized cells work together to capture the light that comes into your eyes through your pupil.

This picture shows only the very back of your eye, an area that we call the ***retina***. The cells that look a bit like feathers are called rods and cones. (The cones are the more pointed ones.) All those lines represent layers of membrane in which light sensitive pigments are embedded. Only the cones can sense color, but the rods play an important role when the level of light is very low. Thanks to your rods you can see just well enough in the dark to avoid stubbing your toe when you have to get up in the middle of the might.

When a photon of light is captured by the light-sensitive pigments, the cell generates an electrical pulse. The green and purples cells (which, of course, are not green and purple in real life) act as electrical wires and pass the electrical signal on to the yellow area, which become a large "wire" that eventually leads to the brain.

ADIPOCYTES

Adipocytes *(AD-i-poe-sites, or, ad-i-PO-sites)* are "fat cells." Fat plays an important role in the body, as it provides insulation and protection for internal organs, helps to give our skin a pleasant texture, and provides a source of stored energy in case the body is unable to take in food for a period of time. The fat is stored in a very large vesicle which takes up most of the space in the cell. Inside it are ***triglyceride*** molecules. "Tri" means "three." There are three fatty acid chains on a tryglyceride molecule. The "glyceride" is the glycerol "clip" that holds the fatty acid chains. We saw a glycerol molecule back in chapter 2, in the middle of the phospholipid molecule. If you clip off the phosphate, you can replace it with another fatty acid.

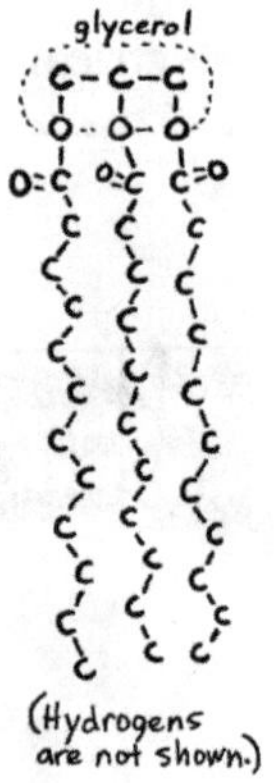

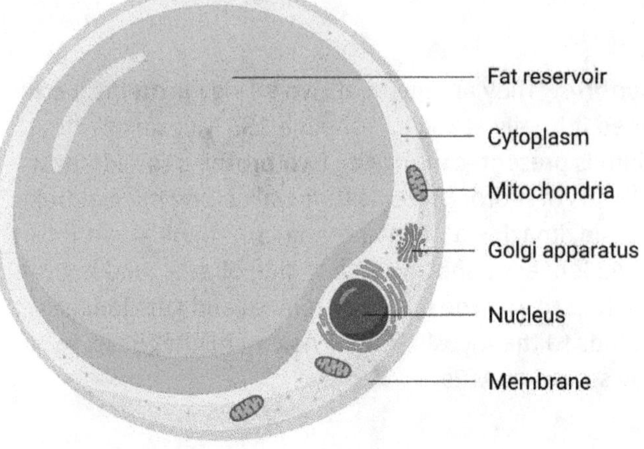

NERVE CELLS

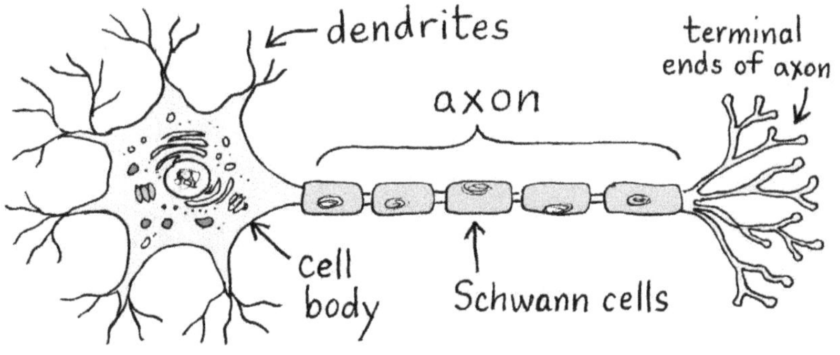

Real neurons don't have any color. Diagrams are colored to make them look nice.

Nerve cells are called ***neurons***. Their job is to send and receive electrical signals throughout the body. The three main parts of a neuron are the cell body (soma), the axon, and the dendrites. The cell body contains the nucleus and the organelles.

The dendrites stretch out from the cell body like branches from a tree. ("Dendron" is Greek for "tree.") The dendrites detect electrical signals coming from other neurons. When they pick up an incoming signal, they pass it along to the axon. The signal passes through the axon very quickly and then goes out through the terminal ends. The terminal ends of the axon reach out to almost touch the dendrites of other neurons. There is a tiny gap between neurons, called the ***synapse***. This gap is bridged by special chemicals called neurotransmitters. Once the chemicals carry the signal across the gap, it goes back to being an electrical signal.

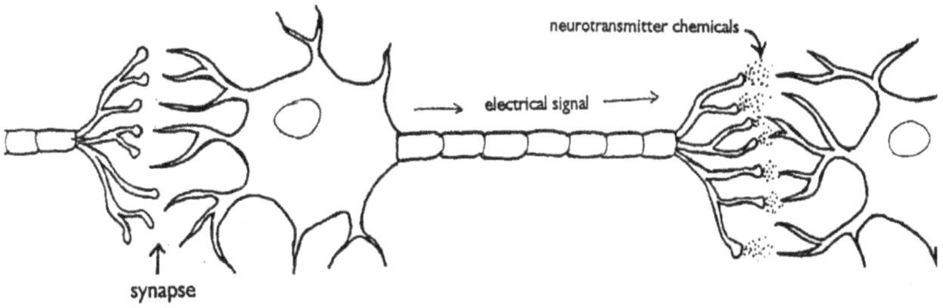

The ***Schwann cells*** (after discoverer Theodore Schwann, a scientist we met in chapter 1) provide insulation for the axon, like the rubber around the copper wire in an electrical cord. The Schwann cells are separate cells, but they are never found apart from neurons. They are by far the thinnest and flattest cells of the body. Their plasma membranes are made of special phospholipids that are good at attracting and holding other lipid molecules. Imagine Schwann cells as greasy pancakes wrapped around an axon "wire."

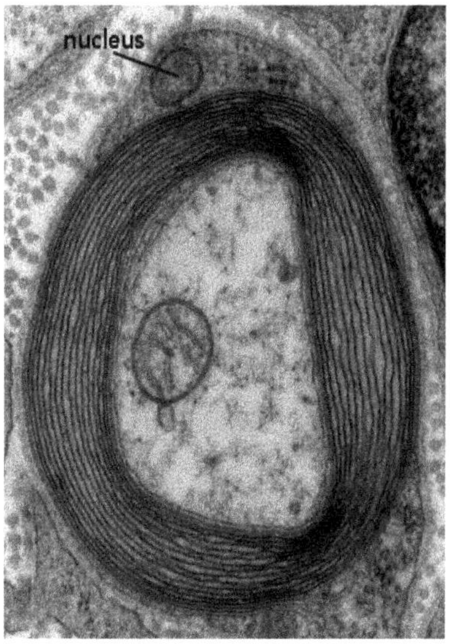

The Schwann cells become almost unrecognizable as cells and are thus often thought of as a continuous covering called the ***myelin sheath***. Myelin is made of cholesterol, phospholipids, and at least one other type of fat (with a very long name we won't bother you with). The photograph shown here is a cross section view of a Schwann cell. (Imagine cutting a cardboard tube in half and looking at the end that you just cut.) Imagine how wide the cell must be to form all those layers! The light colored center is the axon of a neuron. The large dot could be a mitochondria. Notice the location of the Schwann cell's nucleus, at the top of the image.

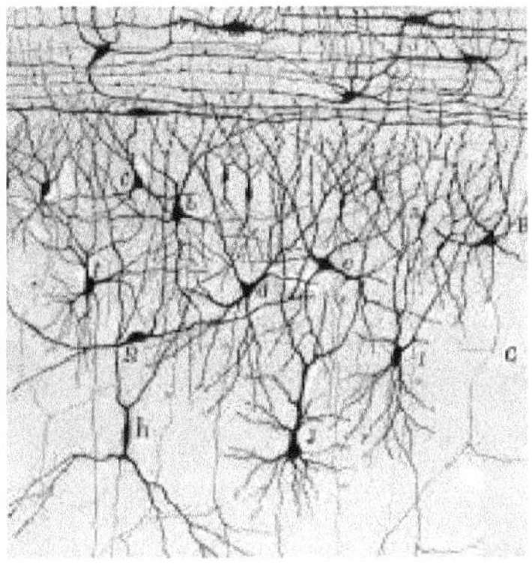

Neurons drawn by Ramon y Cajal, who discovered them over a century ago.

Inside the brain, we find other types of nerve cells, besides neurons. Neurons are so busy doing their job that they need help with basic tasks like getting enough food and water. They are surrounded by "helping cells" called ***glial cells***. ("Glia" is Greek for "glue.")

The most abundant type of glial cell is the ***astrocyte***. ("Astro" means "star.") The astrocyte's job is to feed and protect the neurons. They regulate the chemical environment around the neuron and get rid of excess amounts of minerals. They form a protective barrier that scientists called the "blood-brain barrier." The upside to this barrier is that the brain is protected from toxic chemicals and invaders such as bacteria. The downside to this barrier is that it makes it difficult to get medicines into the brain, should the need arise.

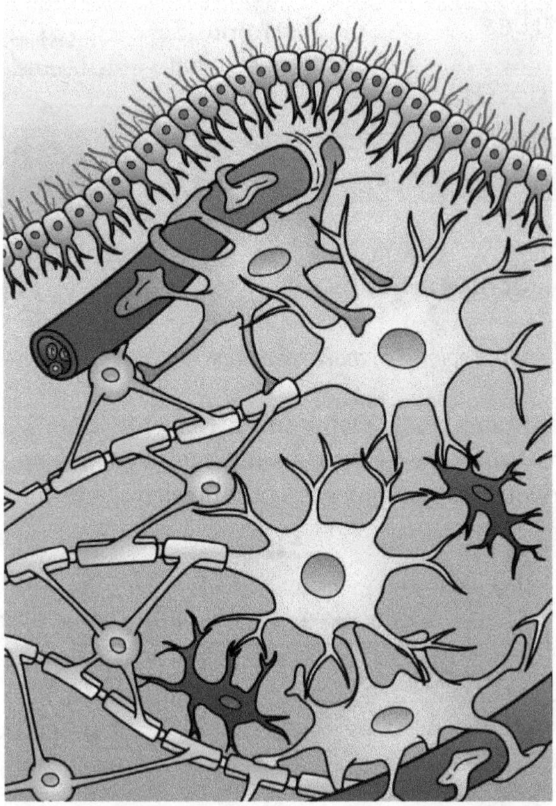

Artwork by Holly Fischer - http://open.umich.edu/education/med/resources/second-look-series/materials - CNS Slide 4, CC BY 3.0, https://commons.wikimedia.org/w/index.php?curid=24367125

The green cells in this diagram represent astrocytes. You can see them holding on to (red) capillaries, absorbing nutrition through the capillary cells. They are also touching the cell bodies of neurons (light yellowish tan), and passing the nutrients to them.

The blue cells in the diagram are ***oligodendrocytes***. "Oligo" means "few," and we know that "dendron" means "branch" and "cyte" means "cell." So oligodendrocytes are cells that have relatively few dendrites compared to other glial cells. The ends of their dendrites wrap around the axon of a neuron and act like Schwann cells, providing an insulating coat of myelin. We don't find Schwann cells in the brain.

The smaller red cells are called ***microglia***. They are the immune cells of the brain, attacking any viruses or bacteria they find. Of course, on rare occasions, invaders get past the microglia and brain infections occur. However, this happens quite infrequently because the astrocytes do such a good job of forming that tight blood-brain barrier.

The line of small, pink cells along the top are only found lining the edges of places in the brain and spine that are supposed to be filled with fluid. These cells make the fluid.

BLOOD CELLS

There are many types of blood cells. They all float in a clear, watery fluid called ***plasma***. The plasma carries the nutrients that go to our cells: proteins, fats, sugars, vitamins, and minerals. These nutrients are incredibly small—far too small to see with a regular microscope. The blood cells are large in comparison and can easily be seen even with an average microscope, the kind you might use at home or in school.

Studying blood cells can be a bit confusing because there are many technical names for them and some cells start out being one thing and end up as another. Also, they can be classified several different ways, which makes it difficult to make the subject simple for beginning students. First, we will divide blood cells into two categories—red and white—and then we'll take a more in-depth look at five kinds of white cells.

ERYTHROCYTES

This is an SEM (scanning electron (microscope) image of an erythrocyte.

Red blood cells are also known as ***erythrocytes*** *(ee-RITH-ro-sites)*. "Erythro" means "red." Red blood cells are very different from other cells because they have no nucleus and no organelles. (Of course, the top layer of your skin also has cells with no nucleus or organelles, but those are dead cells. Red blood cells are alive.) This means that erythrocytes can't make proteins, so they can't repair themselves or reproduce, and they can't metabolize, so they can't "feed" themselves. So what CAN they do? Just one thing—carry oxygen.

Erythrocytes cells are "born" in the ***marrow*** (center) of your bones. A "baby" erythrocyte has a nucleus, but as it "grows up" (in only one week), the nucleus disappears. As an "adult" the erythrocyte looks sort of like a donut, but without the hole in the center. The cytoplasm is filled with molecules called ***hemoglobin***. Hemoglobin is rich in iron, and it's the iron that gives the cells their red color. The hemoglobin molecule can grab and hold on to oxygen molecules. As the red blood cells pass through the capillaries in the lungs, the hemoglobin molecules get loaded up with oxygen molecules. Then, as the red blood cells circulate throughout the body, the oxygen molecules get distributed to cells that are in need of oxygen.

What would happen if a red blood cell had mitochondria? Those mitochondria would be constantly stealing oxygen molecules to use for their electron transport chains. The red cell would then be much less efficient at delivering oxygen. But the red cells don't need mitochondria because they live in a nutrient-rich environment all the time. They get enough glucose from the plasma around them so that they don't need mitochondria like other cells do. However, the lack of organelles does reduce their life span. They only live for a few months.

Erythrocytes are flexible, like rubber, and can bend to fit through very, very small capillaries. The ***concave*** (donut-like) shape of the cell helps it to flex. When red cells have to go through capillaries that are almost too small for them to get through, they release some ATPs, which act as a signal to the capillary (endothelial) cells, causing the capillary to expand a bit so they can get through.

Your body has over 20 trillion red blood cells, making 2 million every second! This high production rate is necessary because of the erythrocytes' short life span. When they die, they end up in the liver where a special type of white blood cell (a macrophage) comes and digests them. The hemoglobin molecules are disassembled and the cell parts are recycled into new cells.

LEUKOCYTES *(NOTE: If this next section is too much information for you, skip it, or come back to it later.)*

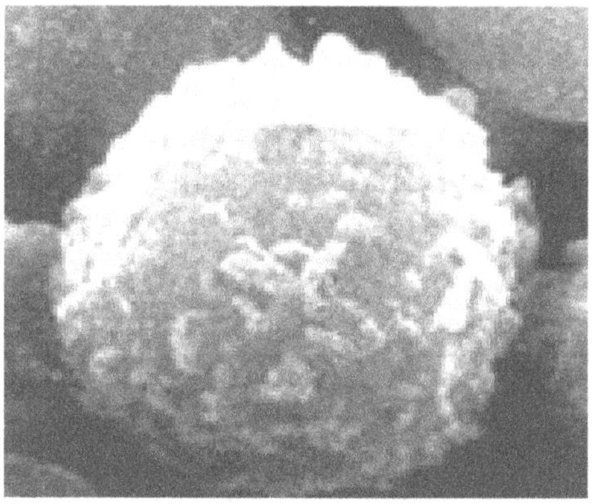

An SEM image of a white cell. This type of image shows you the external texture. It does not tell you anything about the inside.

White blood cells are called ***leukocytes*** *(LU-koh-sites)*. "Leuko" is Greek for "white." White blood cells contain no hemoglobin so they don't carry oxygen. The white cells form a microscopic army that fights harmful invaders such as bacteria and viruses. Just like a real army is a complex society made of soldiers, officers, pilots, mechanics, sailors, cooks, doctors, nurses, truck drivers, and communications specialists, so the immune system is also very complex and has different cells that do a variety of jobs. Strangely enough, this army of white cells looks surprisingly sparse when viewed under a microscope. For every one white cell, there are 600 to 700 red cells! This is a bit shocking considering how important white cells are. However, even though they are outnumbered by red cells, you still have billions of white cells.

There are five basic kinds of white blood cells. Three of them have names that end in "phil" and the other two end in "cyte."

The three "phils" are ***basophils, eosinophils and neutrophils***. We've seen the root word "phil" before, in the word "hydrophilic." Since "phil" means "to like or love," these cells must like something. It's not anything very interesting, though, just stains. Yeah, big deal. Basophils like blue stain, eosinophils like eosin (red) stain and neutrophils like being neutral and not taking either stain. Remember, everything looks pretty clear and colorless under the microscope unless you apply stains. Stains really are important, they just aren't very exciting to talk about.

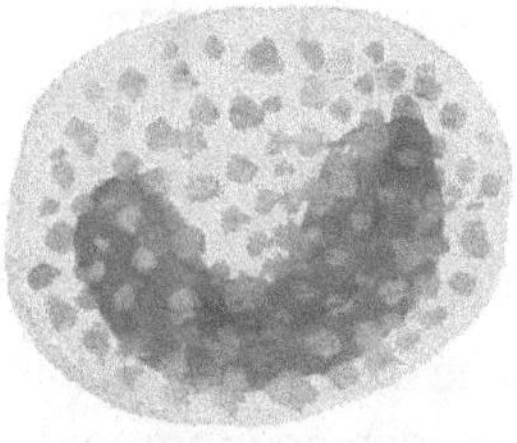

realistic basophil
The blue thing is the nucleus.

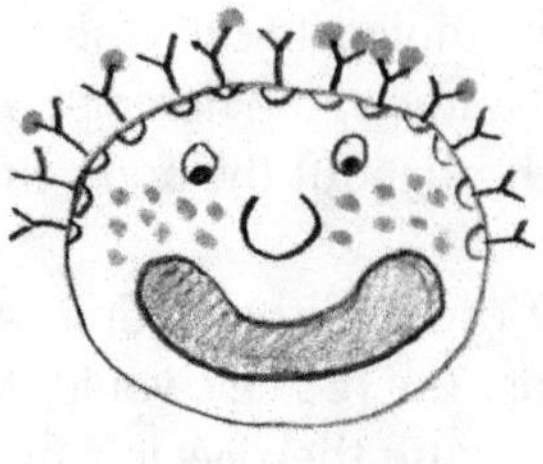

cartoon basophil
The Y's are allergy antibodies

Basophils *(BASE-o-fills)* are the least common type of white blood cell. They account for less than 1 percent of your total white cell population. The blue stain they love has a "basic" pH (as opposed to acidic), thus the name "basophil." Basophils freak out and burst open when they encounter things like venom, poisons, certain chemicals, allergens, tissue damage or even heat. They release a chemical called ***histamine,*** which causes swelling and itching. They also release chemicals that open capillaries and keep blood from clotting. Basophils might sound like nothing but trouble makers that we'd be better off without. Who wants itching and swelling? However, every cell in the immune system is there for a reason, even if it produces side effects we don't enjoy. A certain amount of swelling around an infection or a bee sting can help keep that area "quarantined." Also, the basophil's chemicals (even the unpleasant ones) are important messengers that signal other immune system cells to come and help. Those Y-shaped things that look like hair on our cartoon basophil are called antibodies. The "prong" end catches invading particles. When enough particles are collected, they trigger the release of histamine. The blue freckles represent vesicles filled with histamine.

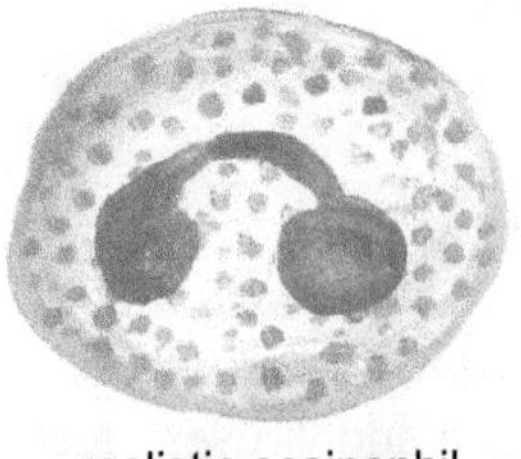

realistic eosinophil
Notice the 2-lobed nucleus..

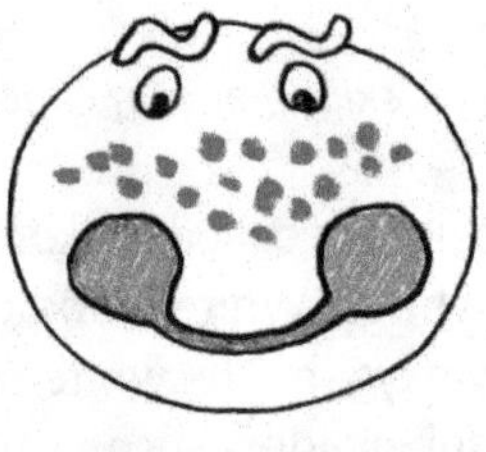

cartoon eosinophil
The eyebrows remind us of worms

Eosinophils *(EE-o-SIN-o-fills)*, which "like" the eosin stain that turns them shades of red or pink, are able to clean up the mess that basophils leave behind after they burst. They are slightly more abundant than basophils, and make up about 2 to 4 percent of your total white blood cell count. Eosinophils are multi-taskers, able to do a variety of jobs. Perhaps their most notorious role is to fight parasites, especially infestations of parasitic worms. Eosinophils are identified not only by their stained color but also by their oddly shaped 2-lobed nucleus. The freckles on our cartoon eosinophil are filled with the chemicals the cell needs to do its jobs. The wavy cartoon eyebrows are designed to remind us of its ability to fight worms.

The number of eosinophils you have in your bloodstream varies throughout the day, with the afternoon generally being the peak.

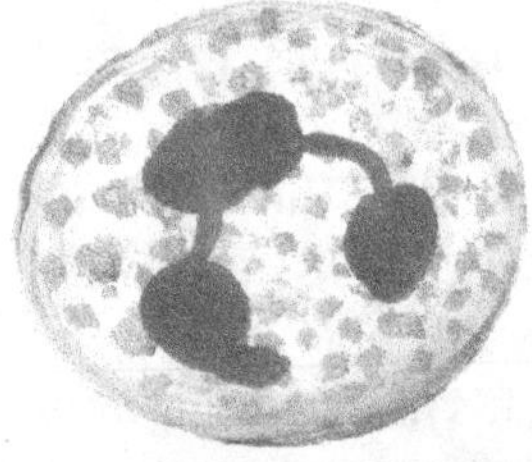

realistic eosinophil
This neutrophil is from a female.

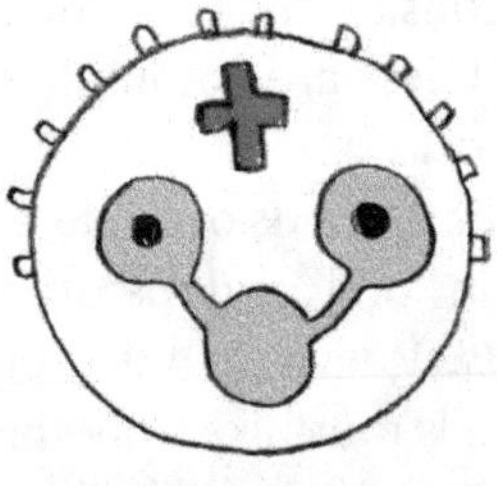

cartoon eosinophil
Red cross = "first responder"

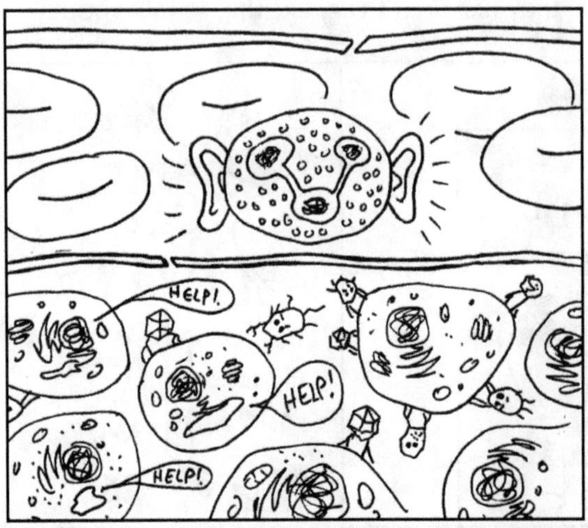

Neutrophils are your first line of defense against infection. These cells can sense foreign invaders and "eat" them using the process of phagocytosis. (The little bumps on the head of our cartoon cell represent the sensors.) Neutrophils only live for 1 to 2 days before they are replaced. The upside to this is that parasites can't live inside of them. The tiny protozoans that cause malaria like to crawl into red blood cells, making them into cozy little homes for several months. If a parasite tries to live inside a neutrophil, it will find its home disintegrating soon after it moves in (certainly before it has had time to buy furniture or stock the refrigerator). Neutrophils are your body's "first responders" and will be the first immune system cells to show up at the scene of an accident (cuts and scrapes).

Another important function of neutrophils is that they can "hear" distress signals coming from cells that are under attack. If bacteria get in, for example, certain chemicals are released, which react with proteins on the capillary cells, causing a chain of events that ends with the neutrophil being pulled out of the blood stream. There are already thin slits between the endothelial capillary cells, but they become wider as the connecting fibers are temporarily dissolved. The odd, 3-lobed shape of its nucleus helps the neutrophil to become very flat so it can slip through the newly opened cracks between capillaries and get out into the interstitial space between the cells. Once they are in the interstitial space, the neutrophils can start gobbling up the bacteria. So many neutrophils gather at the infection site that you can begin to see them. The white stuff we call "pus" is white because of the neutrophils.

Neutrophils are the most abundant type of white blood cell. They make up about 60 percent of your white cell population. Forensic scientists have a special interest in neutrophils because they look different in males and females. In this picture, you can see a long bulge (nicknamed the "drumstick") protruding from one lobe of the nucleus. This is one of the X chromosomes in the XX pair found only in females.

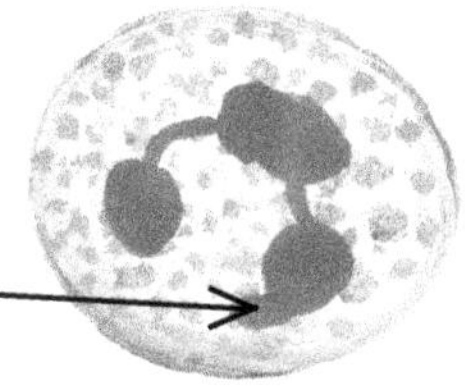

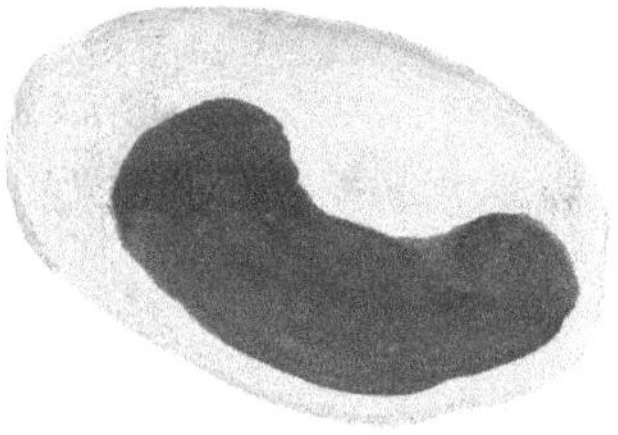

Monocytes float around in the blood until they are called into action. Like neutrophils, they can squeeze through the cracks between the capillary cells and get out into the interstitial spaces between the cells. Once they have left the capillaries, they are called ***macrophages*** ("big eaters"). They go around gobbling up all sorts of nasty stuff that you don't want in your body, including your own cells that have become infected with viruses. They are the ultimate garbage collectors and recyclers (thanks to their lysosomes!) Monocytes make up only about 5 percent of your white cell population. Half of them are stored in your spleen and are mobilized in about 8 hours if you come down with an infection.

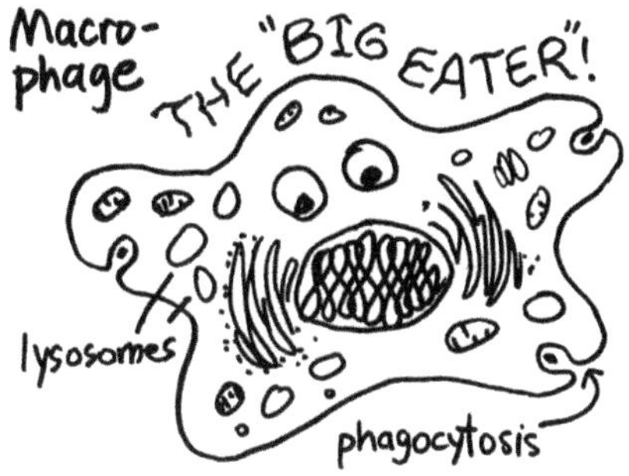

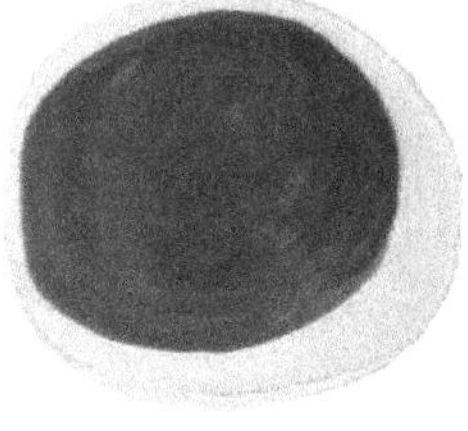

The nucleus of a lymphocyte is very large, filling the cell.

About 30 percent of your white cells are ***lymphocytes.*** There are three kinds of lymphocytes: ***T cells, B cells and NK cells***. All lymphocytes are "born" in the bone marrow (like all blood cells) but the T cells mature in the thymus (a small organ in front of your heart and lungs) and the B cells mature in the bones. "Natural killer" (NK) cells, mature in the tonsils and in lymph nodes. The lymphocytes all look the same when viewed under the microscope; you can't tell the difference between a T cell and a B cell. Lymphocytes work together to give you your "aquired" immune system that learns how to fight the germs you come into contact with during your lifetime.

This B cell has been activated and is now making antibodies.

The ***B cell's*** main job is to make ***antibodies***. Antibodies are Y-shaped "tags" that will stick to things that need to be destroyed. The antibody tags attract the attention of macrophage cells, which will eat anything that has been tagged. Some B cells function as memory cells after your illness is over. They will remember which tag was needed for that disease and if you are exposed to that germ again, they will be able to quickly gear up and immediately begin producing millions of those tags. (Back in chapter 3, antibody tags were used to discover the cytoskeleton. Scientists used antibodies that stuck not to germs, but to microtubules.)

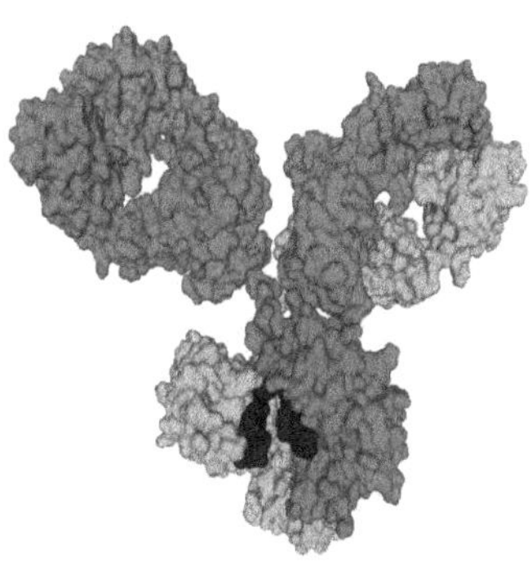

The protein structure of an antibody

There are ***three basic kinds of T-cells: helper cells, suppressor cells and cytotoxic cells***. The ***T helper cell's*** primary job is to direct the actions of the B cells. When B cells pick up particles, they "present" them to T cells to "ask" if the particle is anything to be concerned about. The T cell will determine if the particle represents a threat to the body. If the particle is a piece of a virus or bacteria, the answer will be "yes." The T cell then "tells" the B cell to begin making its antibodies (which match the shape of the particle).

These cartoons are taken from the author's video course called "Mapping the Body with Art." There are 10 (video) lessons about the cells of the immune system. You can find the course at www.ellenjmchenry.com.

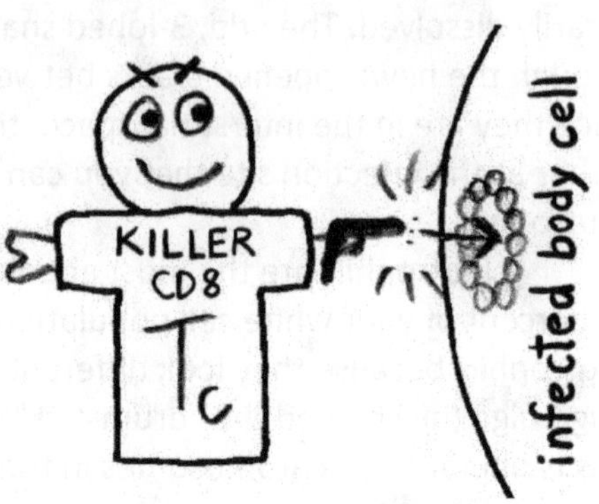

Another type of T cells is called the "killer" (or "cytotoxic") T cell. A the name implies, this type of T cell is able to kill cells that have already become infected or damaged. This type of T cell has a "weapon" that can make holes in the cell's plasma membrane, causing it to collapse. A macrophage will likely come and clean up the mess afterwards, recycling all parts in to raw materials.

The ***natural killer cells*** have the same type of weapon as the killer T cells. They go around killing anything that has an antibody tag on it. They will attack bacteria, virus-infected body cells and tumor cells. (But those things have to be tagged by the B-cells first.) In this cartoon you can see the NK cell feeling the surface of the cell to find the ID flags we first mentioned way back in chapter 2. The NK cell also has "sensors" that can "feel" particles that the cell has clipped to those ID flags. The particles will give the NK clues about what could be going on inside the cell. If the NK cell detects bits of virus, for example, it will kill the cell so that the virus will not have a chance to reproduce itself and thus spread the infection to other cells.

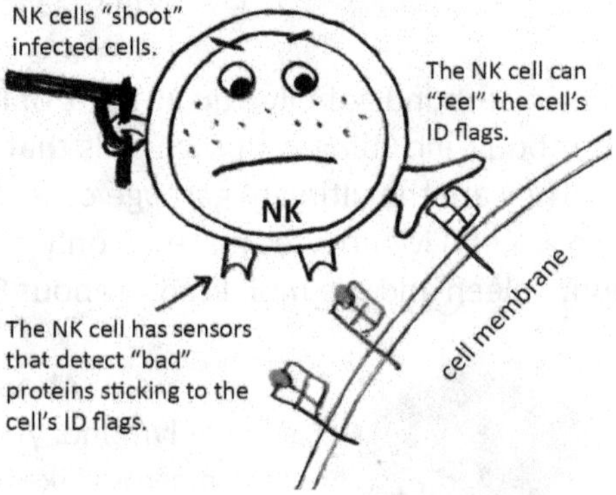

The ***T "suppressor" cells*** are the "brakes" for the immune system. Once the level of infection drops, the suppressor T cells make chemicals that send messages to all these killers to stop killing.

The types of cells we have looked at in this chapter are not the only types of cells in the body. There are many other types of cells that secrete various fluids or chemicals, such as salivary glands (saliva), mammary glands (milk), lacrimal glands (tears), and gastric glands (digestive enzymes). There are specialized cells in the hormonal glands such as the thyroid and pituitary. There are chemical sensors in the nose, and taste buds on the tongue. There's liver cells, kidney cells... but we can't fit them all into this chapter. Now that you know the basics of how cells work, you can do more reading on your own and be able to understand how these cells are suited to their role in the body.

Can you remember what you read?

1) What does "epi" mean? ______________

2) Which of these cells gives your skin its color? a) keratinocyte b) melanocyte c) basal cells d) goblet cells

3) TRUE or FALSE? The top layer of your skin is made of dead cells.

4) Which of these is NOT a type of epithelial cell?
a) endothelial b) keratinocyte c) enterocyte d) melanocyte e) osteocyte

5) What do goblet cells make? a) saliva b) blood c) mucus d) keratin e) enzymes

6) Which of these will you NOT find in the villi of the intestines?
a) goblet cells b) melanocytes c) enterocytes d) stem cells e) Paneth cells

7) Which of these can change its identity and become one of several different types of cells?
a) Paneth cell b) enteroendocrine cell c) stem cell d) enterocyte e) goblet cells

8) Which type of cell has filopodia ("thread feet")?
a) osteocyte b) neuron c) Schwann cells d) goblet cells e) enterocytes

9) Which builds bone? a) osteoblasts b) osteoclasts c) osteocytes

10) TRUE or FALSE? Human bone cells (osteocytes) look very similar to those found in animals.

11) TRUE or FALSE? Bone cells die once they are locked into the mineral matrix of compact bone.

12) Which kind of cell loses its own identity and merges with other cells to become one large cell?
a) muscle cell b) goblet cell c) skin cell d) villi e) neuron f) bone cell

13) What is the primary function of the sarcoplasmic reticulum?
a) help ribosomes make proteins b) conduct electrical signals c) make vesicles d) store calcium

14) Which type of cell gives you your color vision? ______________

15) Which organelle is the largest in an adipocyte?
a) nucleus b) mitochondria c) Golgi bodies d) ER e) vesicle f) cytoplasm

16) Which of these does NOT generate any electrical signals?
a) rods b) cones c) Schwann cells d) neurons

17) What do you call the "arms" that branch off the cell body of a neuron? ________________

18) What do you call the tiny gap between neurons? ________________

19) Which of these is NOT something that glial cells do?
a) provide nourishment to neurons b) protect neurons from damage
c) make a barrier to keep out harmful substances d) pass electrical signals through axons

20) Which of these does a red blood cell not have?
a) nucleus b) Golgi bodies c) ribosomes d) ER e) lysosomes f) all of these

Answer these questions if you read the section on leukocytes (white blood cells).

21) How long do red blood cells live?
a) a few days b) a few weeks c) a few months d) a few years

22) TRUE or FALSE? White blood cells carry oxygen.

23) Which cell releases histamine, which creates an itchy feeling?
a) basophils b) eosinophils c) neutrophils

24) Which type of cells clean up the histamine mess after an allergic reaction?
a) basophils b) eosinophils c) neutrophils

25) Where are blood cells "born"?
a) in the blood b) in bones c) in lymph nodes d) in the spleen

26) Which type of cell makes Y-shaped antibodies? a) T cell b) B cell c) NK cell d) all of these

27) Which type of white cell needs lots of lysosomes in order to do its job?
a) macrophage b) basophil c) eosinophil d) T cell e) B cell

28) Which of these controls B cells? a) T cells b) NK cells c) macrophages d) none of these

29) Which type of cell is probably the first one to arrive at the site when you get a cut or scrape?
a) eosinophil b) neutrophil c) macrophage d) B cell

30) Which type of cell can help detectives figure out the gender of a victim (if they have a blood sample)?
a) basophil b) eosinophil c) neutrophil d) macrophage e) T cell f) B cell g) NK cell

****

Activity 10.1 **"It's Greek to me!"**

Can you match the Greek word root with its meaning? (These words are not just from this chapter.)

1) soma ________
2) dendron ____________
3) gluc/glyc ___________
4) chroma _____________
5) erythro __________
6) leuko ____________
7) lysis _______________
8) phil ____________
9) osteo ____________
10) trans ____________
11) script ____________
12) ana ____________
13) cata __________
14) hydro ____________
15) di _______
16) reticulatus ______________
17) phob ____________
18) phago ____________

Possible answers:
across
dissolve
like
eat
tree
red
body
water
sugar
net/network
up
white
two
down
fear
bone
color
write

<u>Activity 10.2</u>   My "cellular" day

Agh! I've been studying cells to much! Everything around me is starting to look like cells or parts of cells! Can you finish my sentences as I describe how my day went? (Use each word once.)

adipocyte	cytoskeleton	ER	lipid rafts	microvilli	nucleus	RNA polymerase
amino acids	dendrites	erythrocyte	lysosome	mitochondria	pinocytosis	Schwann cell
antibodies	desmosomes	goblet	macrophage	motor proteins	villi	vesicle
axons	DNA	Golgi body	melanocyte	mRNA	polypeptide	plasma membrane
chromosomes	enterocytes	keratinocytes	melanosomes	muscle fibers	portal protein	water molecule
cytoplasm	enzymes	left-handed DNA	microtubules	neurons	ribosomes	

This morning at breakfast, my fried egg reminded me of an __________________. I decided to roll my pancake around my sausage since the pancake looked to me like a ______________________. As I took little sips of juice from my glass, I thought of how cells "sip" through the process of ________________. I was extra hungry this morning, so I finished breakfast by eating a donut and imagined that I was eating an __________________. Wow, I felt like a "big eater." If I was a ________________, my stomach would be a ________________.

After breakfast I went to the library. I felt like I was walking through a ____________________ and into a nucleus. I was not allowed to borrow the book I wanted to read, but I was allowed to copy some of the pages. The copier reminded me of ______________________ and my copies seemed like ______________.
I walked to the post office to mail a package. I felt like I was carrying a ___________ to a _________________.
The workers inside the post office reminded me of _______________ doing all their various processing jobs.

At lunch, my soup made me think of ______________, so the bowl was the ________________ around it. The oval-shaped beans in the soup (which would be a source of energy for me today) reminded me of ____________. For dessert I had a chocolate chip cookie; the dark chocolate chips reminded me of the brown ______________ inside a _______________ cell.

After lunch the weather became chilly so I put on my coat and gloves. The fingers of my gloves reminded me of the ___________ in the intestines. I went for a walk to the park. I passed a pond with lily pads that made me think of ________________. In the park, the trees reminded me of _____________. I saw the branches as ______________, the trunks as __________ and the roots as terminal ends. Someone had been gathering firewood into bundles that looked to me like _____________________. I passed a playground and saw a tall slide with a ladder that reminded me of ___________. A child was jumping rope to the alphabet song and was singing the last phrase, "X, Y, Z." These letters made me think of _________________, ________________, and ______________________. People walking along the sidewalk looked to me like ___________________ walking along ___________________. A short brick wall signaled the end of the park. The stacked bricks reminded me of _________________ I had seen in skin tissue. I checked my map to make sure I would not get lost. The network of roads on the map look so much like a cell's ___________________.

I stopped at a restaurant for dinner. The waitress brought my ice water in a glass that looked too much like a ________ cell! She was wearing a bead necklace that I saw as a _______________ made of ________________.
I imagined that my plate was a cell. My giant meatball was the _____________ and the spaghetti noodles all around it was the _____________. The green peas dotted around the edges of the spaghetti were the _____________. I watched as someone mopped the floor on the other side of the room. The head of the mop, with all its strings soaking up the water reminded me of the stringy fringe borders called ________________, found on ___________________ that soak up nutrients in the intestines.

As my "cellular" day ended, I tucked myself into bed. The stitches between the squares of my quilt looked so much like the ________________ between the square fabric "cells." The teddy bear next to me made me laugh. His head and ears looked so much like a _____________________!

Activity 10.3 "Odd one out"

1) Which one is NOT made of protein?
a) kinesin b) microtubule c) enzyme d) plasma membrane

2) Which of these cells does NOT have a nucleus?
a) erythrocyte b) melanocyte c) osteocyte d) lymphocyte

3) Which of these is NOT a membrane-bound organelle?
a) ribosome b) Golgi body c) mitochondria d) lysosome

4) Which of these enzymes does NOT break something apart?
a) catalase b) lipase c) ATP synthase d) amylase

5) Which of these is NOT found in mitochondria?
a) myosin b) DNA c) ATP synthase d) ribosomes

6) Which of these is NOT found in your intestines?
a) stem cell b) goblet cell c) enterocyte d) osteocyte

7) Which of these does NOT have only one nucleus?
a) goblet cell b) osteocyte c) muscle cell d) neuron

8) Which of these is NOT a way that a cell can takes things in?
a) pinocytosis b) phagocytosis c) exocytosis d) diffusion

9) Which of these organelles does NOT contain enzymes?
a) lysosome b) centrosome c) peroxisome d) Golgi body

10) Which one of these is NOT critical to cell division (mitosis)?
a) lysosome b) centrosome c) microtubules d) DNA polymerase

Can you figure out which is the odd one out, even without a clue?

11) a) macrophage b) neutrophil c) erythrocyte d) lymphocyte

12) a) goblet cells b) basal cells c) melanocytes d) keratinocytes

13) a) Adenine b) Thymine c) Cytosine d) Guanine

14) a) astrocyte b) Schwann cell c) neuron d) oligodendrocyte

15) a) kinesin b) myosin c) collagen d) dynein

16) a) rods b) melanocytes c) cones d) neurons

17) a) alanine b) adenosine c) lysine d) glycine

18) a) oxygen b) nitrogen c) water d) phosphorus

19) a) cholesterol b) glycerol c) fatty acid d) phosphate

20) a) James Watson b) Rosalind Franklin c) Francis Crick d) Linus Pauling

INDEX

This section can help you find something if you can't remember where you read it.

ANSWER KEY

CHAPTER 1:
1) b, 2) d, 3) c, 4) F, 5) a, 6) a, 7) F, 8) T, 9) T, 10) b, 11) d, 12) bacteria, 13) c, 14) F, 15) F
16) a, 17) F, 18) Scanning Electron Microscope, 19) gold, 20) F

Activity 1.1:
1) M, 2) P, 3) J, 4) L, 5) O, 6) H, 7) E, 8) I, 9) D, 10) G, 11) F, 12) A, 13) B, 14) K, 15) N, 16) C
A, 6J and P are TEMs.

CHAPTER 2
1) b 2) b 3) 2 4) 5 5) a 6) b 7) d 8) b 9) c 10) T 11) F It is a single layer with tails pointing inward.
12) F You also find them surrounding organells and forming storage vessicles. 13) T 14) T 15) c 16) c 17) c
18) b 19) a 20) e

Crossword, page 16:
ACROSS: 2) sugar 3) mosaid 4) hooke 5) lipid 8) micelle 9) hexagon 11) blood 16) transmembrane 17) raft
19) cholesterol 20) phosphate 21) hydrophobic
DOWN: 1) Schwann 2) switch 6) peripheral 7) mailbox 9) hormones 10) polar 12) Leeuwenhoek
13) communicate 14) water 15) scanning 16) transmission 18) Brown

CHAPTER 3
1) cytoplasm or cytosol 2) b 3) b 4) F 5) b 6) a 7) c 8) a 9) c 10) ATP
11) a 12) tubulin 13) dimer 14) e 15) b 16) centrioles 17) e 18) a 19) a 20) d

CHAPTER 4
1) ATP 2) c 3) phosphate 4) T 5) b 6) make 7) a 8) F (matrix membrane!) 9) c 10) b
11) d 12) F 13) 3 14) a 15) b 16) diffusion 17) T 18) 2 19) a 20)c

4.2 Crossword puzzle with picture clues
ACROSS: 1) cork cells 7 mitochondria 11) glycan 13) adenosine triphosphate 14) glycerol
17) ATP synthase 18) micelle 19) peripheral 20) raft 21) glucose 22) tubulin 23) dynein
DOWN: 1) cholesterol 2) centriole 3) phospholipids 4) transmembrane 5) water 6) fatty acid
8) microtubule 9) phosphate 10) centrosome 12) Leeuwenhoek 16) nucleus 18) matrix 19) kinesin

CHAPTER 5
1) a 2) F 3) b 4) d 5) F 6) T 7) c 8) c 9) c 10) c 11) b 12) sense 13) d
14) F 15) b 16) b 17) a 18) F 19) c 20) b

5.5
The new vocabulary word is POLYPEPTIDE. The root "pep" means "protein." A "peptide" is a chain of amino acids of any length, though it is most often used for shorter chains, not longer. The enzyme "pepsin" is used in our stomachs to break down proteins.

CHAPTER 6
1) b 2) a 3) d 4) c 5) F 6) F 7) c 8) a 9) F 10) T 11) b 12) d 13) c
14) T 15) T 16) T 17) a 18) d 19) T 20) T

Review crossword puzzle on page 60:
Across: 2) uracil, 4) centrosome, 6) ribosome, 11) DNA, 12) chaperone, 14) microtubules, 17) acid
18) phospholipid, 19) transcription, 21) diffusion, 22) cytoplasm, 23) protein, 24) nucleus, 25) synthase
Down: 1) translation, 3) adenine, 5) ptotons, 7) membrane, 8) microfilaments, 9) helix,
10) mitochondria, 12) cytoskeleton, 13) hydrophilic, 15) ATP, 16) hydrophobic, 20) nitrogen

CHAPTER 7

1) endoplasmic reticulum 2) b 3) c 4) c 5) a 6) d 7) F 8) d 9) T 10) F 11) a
12) 8 (not including binding histone) 13) 2 14) T 15) c 16) d 17) F 18) F 19) F 20) F

CHAPTER 8

1) b 2) a 3) b 4) c 5) a 6) d 7) b 8) c 9) c 10) c 11) b 12) d 13) b 14) d
15) d 16) d 17) b 18) c 19) a 20) c

8.3

1) thymine 2) amylase enzyme 3) ribose sugar 4) catalase enzyem 5) lipase enzyme 6) Acetyl-CoA
7) pyruvate 8) glycine 9) pepsin enzyme 10) ribosome 11) phosphate 12) glucose 13) peroxisome
14) electron 15) carbon dioxide 16) oxygen 17) fatty acid 18) desmosomes 19) tRNA 20) hydrogen

CHAPTER 9

1) T 2) F 3) F 4) T 5) T 6) T 7) b 8) b 9) a 10) centrosome 11) microtubules
12) d 13) c 14) meiosis 15) a 16) 23 17) T 18) e 19) F 20) c

9.3

1) 36 2) 8 3) 10 4) 1 5) 3 6) 13 7) 40 8) 2 9) 14 10) 500
11) 20 12) 23 13) 5 14) 4 15) 92 16) 7 17) 6 18) 11 19) 146 20) 9

CHAPTER 10

1) "outside" or "on top of" 2) b 3) T 4) e 5) c 6) b 7) c 8) a 9) a 10) T 11) F 12) a
13) d 14) cones 15) e 16) c 17) dendrites 18) synapse 19) c 20) f
21) c 22) F 23) a 24) b 25) b 26) b 27) a 28) a 29) b 30) c

10.1

1) body 2) tree 3) sugar 4) color 5) red 6) white 7) dissolve 8) like 9) bone
10) across 11) write 12) up 13) down 14) water 15) two 16) net or netlike 17) fear 18) eat

10.2

adipocyte, Schwann cell, pinocytosis, erythrocyte, macrophage, lysosome
portal protein, RNA polymerase, mRNA, vesicle, Golgi body, enzymes
cytoplasm, plasma membrane, mitochondria, melanosomes, melanocyte
villi, lipid rafts, dendrites, axons, muscle fibers, DNA, chromosomes, antibodies, left-handed DNA
goblet, polypeptide, amino acids, nucleus, ER, ribosomes, microvilli, enterocytes
desmosomes, water molecule

10.3

1) d 2) a 3) a 4) c 5) a 6) d 7) c 8) c 9) b 10) A
11) c, erythrocyte is not a white blood cell 12) a, goblet cells are not found in skin
13) b, thymine is not found in RNA, the others are found in both DNA and RNA
14) b, Schwann cells are not found in the brain 15) c, collagen is a protein that does not move
16) d, neurons are not cells that make pigments 17) b, adenosine is not an amino acid
18) c, water is not an element found on the Periodic Table 19) a, cholesterol is not a component of a phospholipid molecule
20) d, Linus Pauling did not help to discover the helix shape of DNA

TEACHER'S SECTION

You can choose which of these activities are suitable for your particular situation. Skip any activities that are not appropriate for your student(s).

If you have a paperback copy of this book and need a digital copy of the craft patterns so you can print them out, go to www.ellenjmchenry.com, click on FREE DOWNLOADS, then on HUMAN BODY, then scroll down until you see "Printable patterns for Cells curriculum."

SUPPLEMENTAL ACTIVITIES

PLEASE NOTE: You can choose which of these activities are suitable for your situation. Skip any activities that are not appropriate for your student(s).

CHAPTER 1

ACTIVITY IDEA 1A: Observe cork cells and plant cells (like Robert Hooke and Robert Brown did)

This activity requires a microscope. If you don't have one, just skip this activity. If you have a microscope but don't know how to use it, search for a tutorial video on YouTube or another video streaming service.

You will need:

- a compound microscope and a glass slide
- a piece of cork (NOTE: Cutting cork can be tricky. If it is too difficult, just use the onion skin.)
- a piece of onion skin
- a razor blade (or sharp craft knife like X-Acto)
- pencil and paper if you want the students to draw what they see

TIP: A good place to order science supplies is www.homesciencetools.com

What to do:

An adult should do the cutting of the cork. Use the razor blade or craft knife to make very thin wedge-shaped slices. The tapered edge (as it tapers off to nothing) will be the best place to look for cells. Put the wedge onto a glass slide. View under the lowest power (shortest objective lens) first. After you get it in focus, you can then move up to the next highest power. You will have to refocus, but only slightly. You will only need to use the 4x, 10x or 40x objectives (giving you 40x, 100x and 400x, respectively). If you want your student(s) to draw the cells, provide paper and pencils.

To view the onion skin cells, slowly peel off the papery layers one by one. Try to find the very thin, almost transparent "membrane" layer that surrounds the bulb. Lay a piece of this moist, transparent layer onto the slide. If you want to do a traditional "wet mount" you can place one or two drops of water onto the onion, and the place a cover slip on top. (If you need instructions on how to place cover slips, use key words "how to make a wet mount" in YouTube or your favorite video streaming service.) However, I have done this lab without cover slips and it has worked just fine. The onion will dry up eventually, but will remain moist long enough to complete the lab.

Again, use your lowest power first and get it in focus, then move up to the next power. You should be able to see a dark dot in many of the long, box-like cells. This dot is the nucleus. You might also see some other lighter dots that could be other organelles, but usually you can't see other organelles without special stains.

ACTIVITY IDEA 1B: HOW BIG IS A CELL?

This interactive activity lets students zoom in and out on a series of microscopic things, starting with a coffee bean and a grain of rice and ending with a carbon atom. Cells and cell parts are shown along they way, along with bacteria and viruses. (This link was active as of November 2021. If you find that this link doesn't work for you, try using an Internet search engine with key words: "cells scale how small microscopic"

https://learn.genetics.utah.edu/content/cells/scale/

CHAPTER 2

ACTIVITY IDEA 2A: DEMONSTRATION of FLUID MOSAIC MODEL

This activity helps students to understand the fluid nature of the membrane. The text mentioned that the phospholipid molecules can move around, sort of like ping pong balls floating in a bathtub. This demonstration is a fun way to help them understand this concept. (If time is limited, you might want to choose between this activity and idea 2B.)

You will need:

- a large, shallow metal or plastic tray of some kind (a 9x13 cake pan would be adequate)
- a pitcher of water
- a bag of "miniature" marshmallows (these will represent the heads of the phospholipid molecules)
- Chunks of apple that are slightly larger than the mini-marshmallows (You could also use chunks of Styrofoam® or anything else that is waterproof and will float.) Cut some of the chunks into thick discs you can use to represent lipid rafts. Cut some oddly shaped chunks that you can use to represent surface proteins.
- a few toothpicks

What to tell the students:

This will be a demonstration to show what it means when scientists say that a phospholipid membrane is a "fluid mosaic." The word "mosaic" means a pattern made from small pieces. "Fluid" means "flowing." In this demonstration, we will imagine that we are looking down on the outer membrane of a cell. We will only be able to see the heads of the phospholipid molecules. The marshmallows will represent these phospholipid heads. We will see how lipid rafts and membrane-bound proteins can move around, going in and among the phospholipids.

What to do:

Pour at least an inch of water into the tray. Dump in marshmallows until they cover most of the surface of the water. Add your representations of lipid rafts and membrane-bound proteins. You might want to use a few toothpicks to create proteins that are sticking up above the surface. Fill in any remaining gaps with marshmallows.

Allow the students to gently push the rafts and the proteins around. (You could even provide very small chunks of "protein" that they could set on top of the lipid rafts.) Notice how the marshmallows won't allow an empty patch of water. They immediately fill any gaps you try to create. The surface of the water remains covered at all times even though the positions of the rafts and proteins are constantly changing.

Extra tips:

You might want to put a bath towel under your pan to absorb any slop-over accidents, and have plenty of paper towels on hand. (Also, just in case you have any thought of up-sizing this and using a small plastic pool, I tried this and found the results disappointing in several ways. Definitely not worth the time and money.)

ACTIVITY IDEA 2B: MAKE AN EDIBLE FLUID MOSAIC MODEL

This is a delicious way to review the amazingly complicated surface of the membrane. This activity is somewhat similar to idea 2A, so if time is limited you may want to chose one or the other.

You will need:

- a selection of healthy edibles such as grapes (excellent for representing phospholipid heads), strawberries, blueberries, carrots, cucumbers, celery, broccoli, pieces of melon, etc.
- a large paper plate for each student, or a large tray for a small group of students

- safe knives for students to use
- toothpicks
- paper towels

What to tell the students:

In this activity you will make an edible model of a membrane as seen from above. Imagine looking down on a membrane, so all you see are the heads of the phospholipid molecules, plus the various rafts or proteins floating among them. You don't need to worry about accurately portraying any particular structures, just the general idea that the surface of a cell is a very crowded place, with many strange-looking structures sticking out of it.

What to do:

Provide the students with the raw materials, safe tools to work with, and a plate or large tray on which to arrange their edible "sculpture." Photograph the final results, especially if the students are keeping a portfolio that documents their educational activities during the year.

ACTIVITY IDEA 2C: A HUMAN MODEL OF A MEMBRANE (20 or more students)

You will need:

- a fairly large space (but an average-sized classroom will do, if you move the tables and chairs)
- a few small balls (ping pong or tennis balls) and at least one large ball (basketball, beach ball)
- a long stick to be a "flag" (cell identification marker) You might even want to tape a piece of paper to the end, that says something like, "I belong" or "I am part of the body" or even the name of one of the students.

What to tell the students:

In this activity, each of you will represent a phospholipid molecule. You will imagine that your head is the water-loving head of the phospholipid, and you will stretch our your arms to be the water-hating tails. You'll have to ignore your feet for a while! Just like in a real membrane, your water-hating tails will face each other and your water-loving heads will be on the outside. After you have lined up and created your membrane, we will do some demonstrations that show how the membrane works.

What to do:

Line up the students in two rows, facing each other, but not too close. Tell them that their heads will represent phosphate heads. Then have them put their arms out straight in front of them to represent the two lipid tails. Their hands should come close but not touch. If one of the students asks what their feet represent, tell them to ignore their feet and pretend they just have heads and arms. Weird, yes, but we have to deal with the constraints of gravity and human anatomy. Human are not phospholipids.

With the students lined up and modeling a piece of membrane, show the students a few small balls and tell them the balls represent very small molecules such as water or oxygen or carbon dioxide. Then gently toss or roll the small balls between the students. Emphasize that this is to show that very small molecules can pass right through the membrane. Then show them the large ball. Say that this represents a very large molecule, such as a food molecule or a piece of protein. Demonstrate that the molecule will not be able to slip through. (The students should be standing close enough together that the large ball can't get through the space between their legs or bodies.) Then ask if anyone remembers how cells regulate the entry and exit of large molecules. (Answer: portal proteins) Volunteer one pair of students in the middle to be a portal protein. Have them turn to the side and put their arms out to their sides. Designate one side to be "outside" the cell and give the ball to the portal protein who is on that side. Have the outer protein pass the ball to the inner protein. The inner protein can then just let go of the ball or give it a gentle toss into the "inside of the cell." If time permits, you can let other students take turns being portals.

One student is playing the role of a portal protein. The ball is a molecule.

Volunteer a student on the outer side of the membrane to pull their hands in and pretend to be a protein instead of a phospholipid. Then give him/her a long stick to hold up. This will represent an identifying "flag" that will tell all other cells that come into contact with it that it's part of the body and not a foreign invader.

Something else you can do that was not mentioned in this chapter, is the flipping of phospholipids from the inside to the outside, or vice versa. (Special proteins called "flippase" and "floppase" help with this process. Assign a number to each pair of students and let them switch places when you call out their number.

<u>ACTIVITY IDEA 2D</u>: MAKE A PHOSPHOLIPID ORNAMENT

OPTION A: Use a clear plastic ornament
OPTION B: Use just paper and chenille stems

You will need:
FOR OPTION A:

- a clear plastic ornament for each student (The kind that has two halves that snap together is best, but you can make other hollow spheres work if you are able to cut a slit in the bottom. If clear plastic ornaments are not "in season" when you do this unit, they can still be ordered online through various websites. Search for "clear fillable ornaments.")

The blue tails have tiny beads for C atoms.

FOR BOTH OPTIONS:

- a copy of one of the following pattern pages

(If you need to resize the circles to fit into your ornaments, you can shrink or enlarge the patterns slightly using a copier or scanner, or tell your printer to print at a percent less than 100.)

- colorful chenille stems
- colorful string or yarn to hang the ornament
- FOR THE PATTERN WITH DOTS: colored pencils or markers for oxygen, nitrogen and phosphorus atoms **(Suggested colors: red for oxygen, green for nitrogen, blue for phosphorus)**
- scissors and glue (NOTE about glue: Don't use glue sticks or white school glue. Buy either PVA glue (outside of USA) or try Elmer's Craft Bond glue, or Aleen's Tacky Glue.)
- OPTIONAL: small beads that will slide onto the chenille stem (to represent carbon atoms)

SMALL
2” dia.

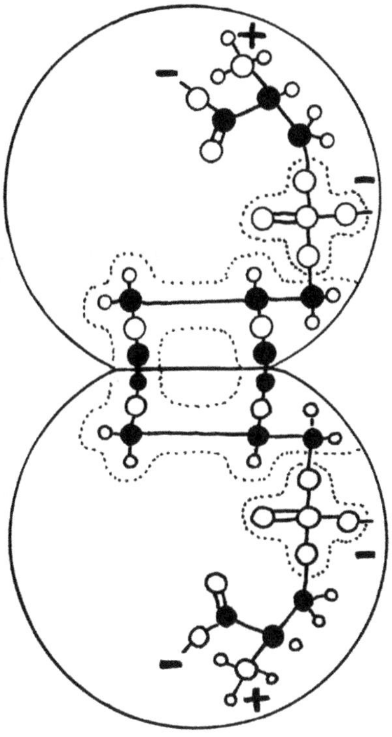
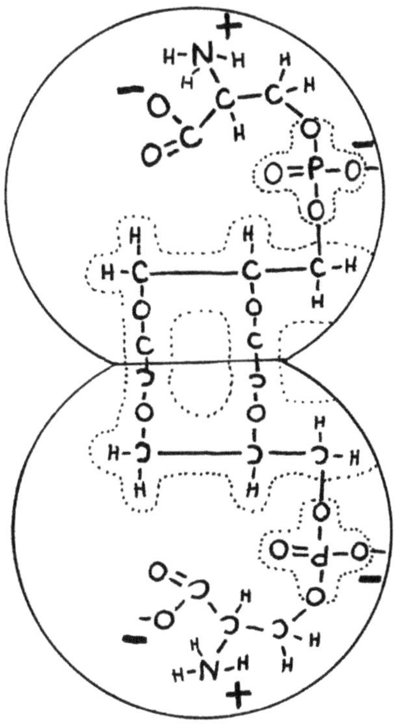
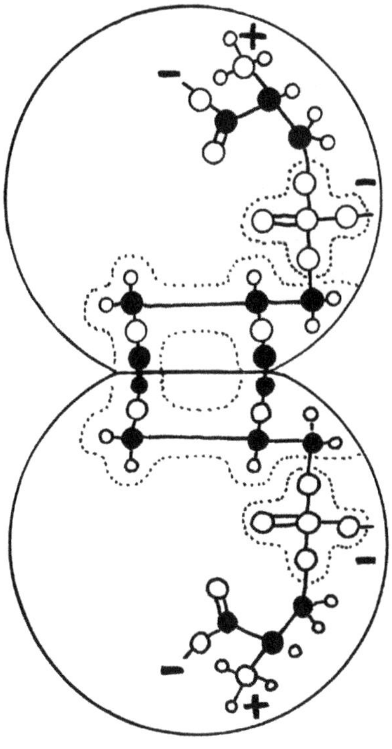
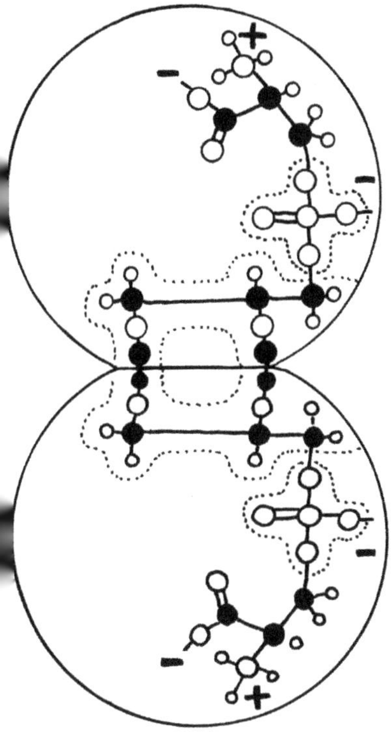
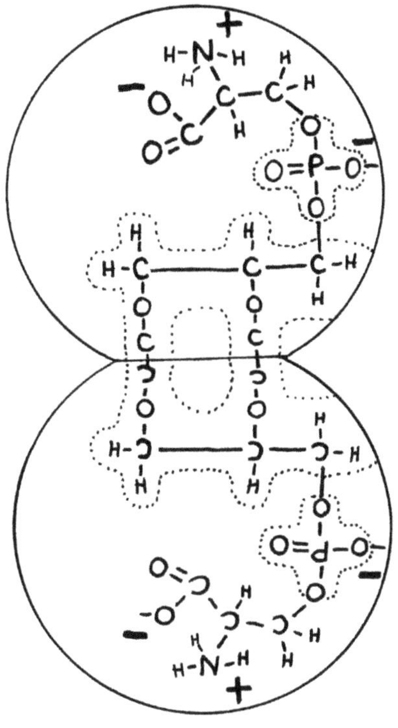
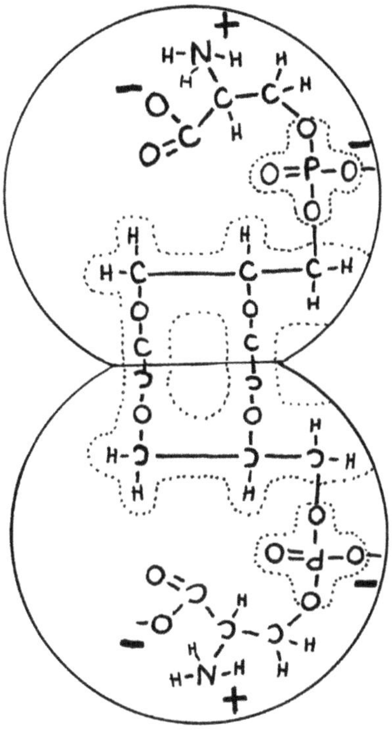
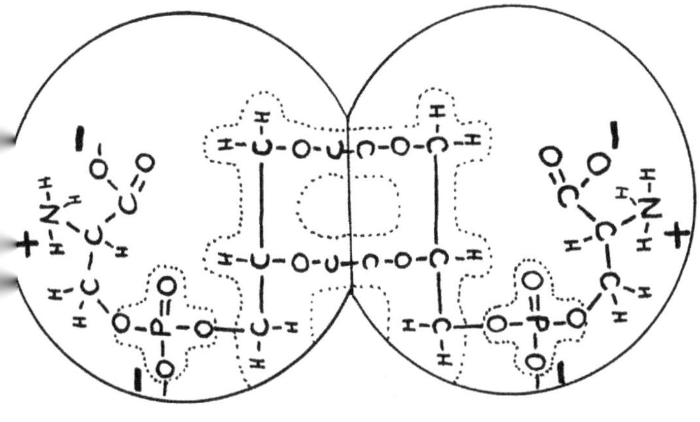
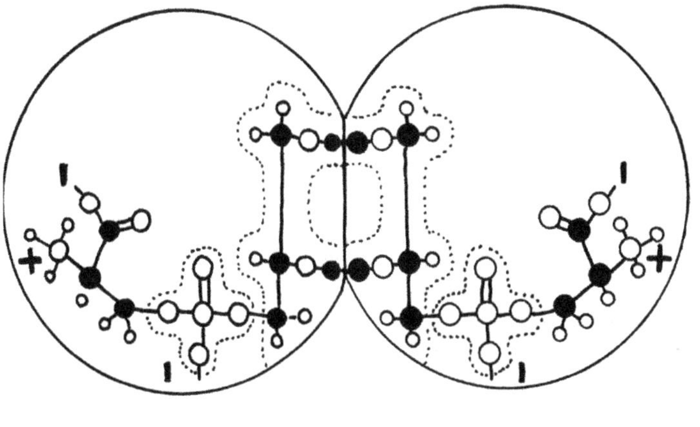

If you need a digital copy of this pattern page, go to www.ellenjmchenry.com, click on FREE DOWNLOADS, then on HUMAN BODY, then scroll down to find "Printable patterns for Cells curriculum."

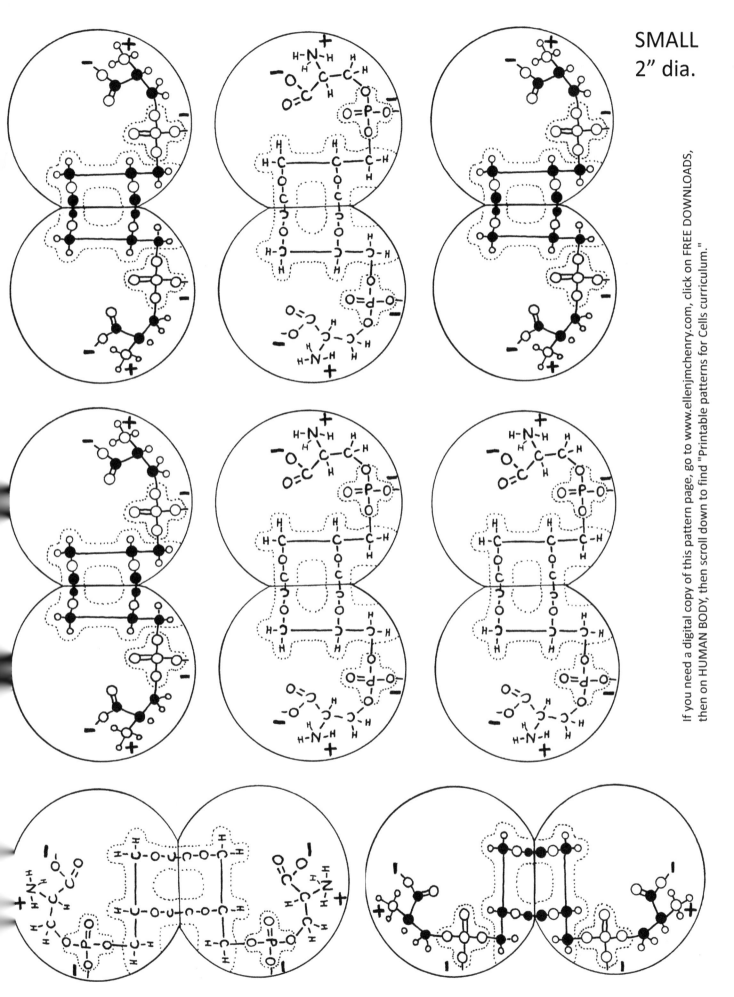

MEDIUM 2.5" dia.

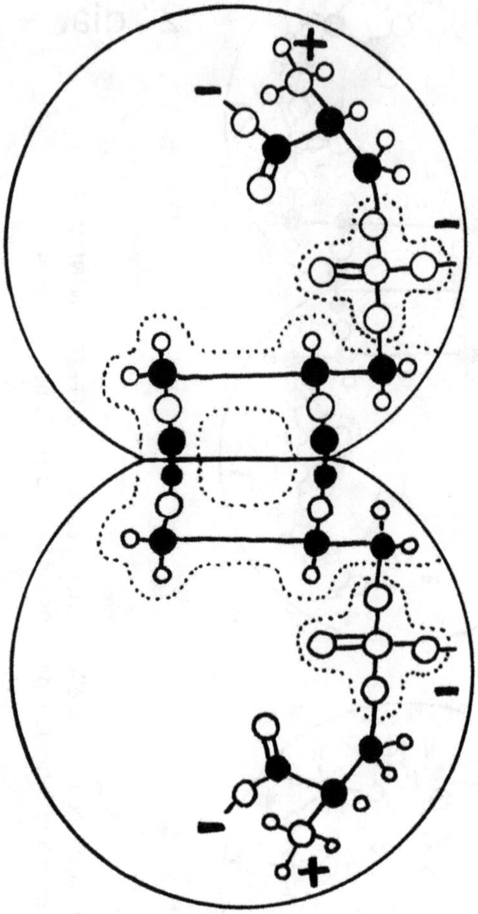

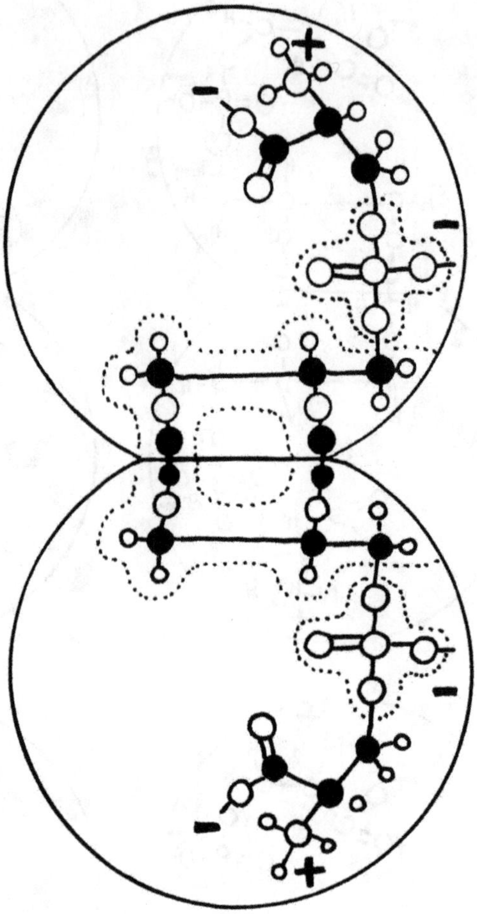

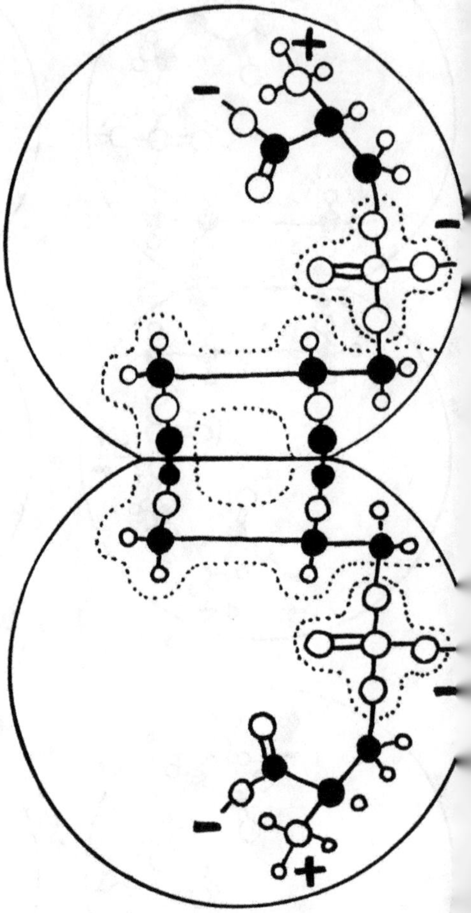

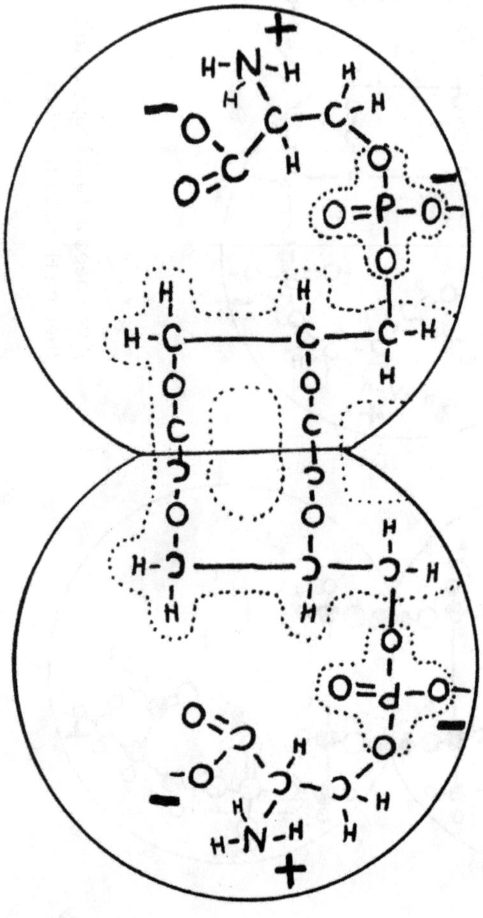

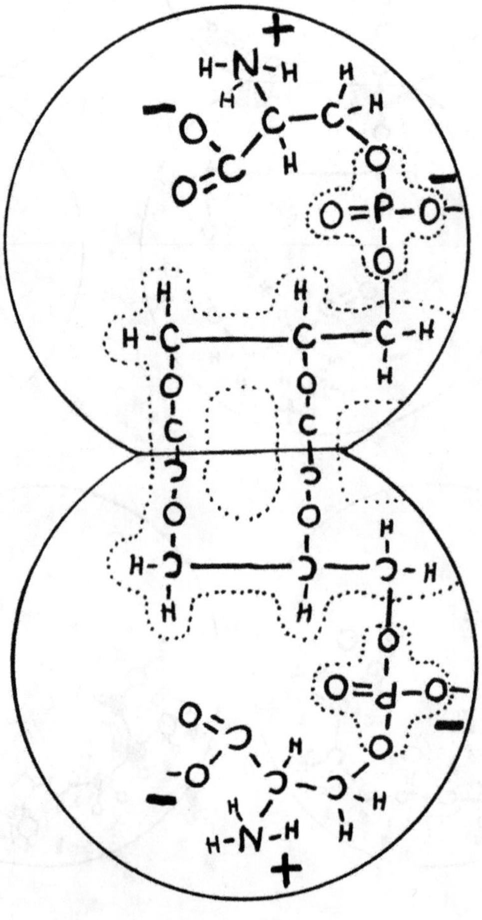

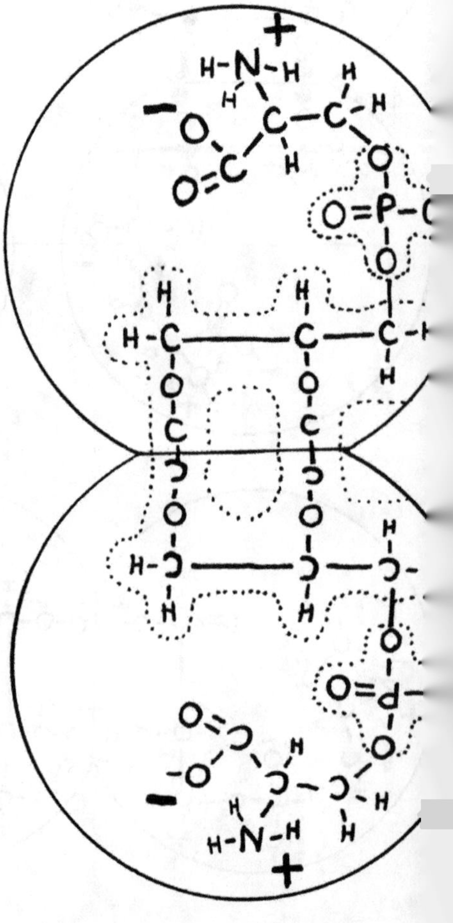

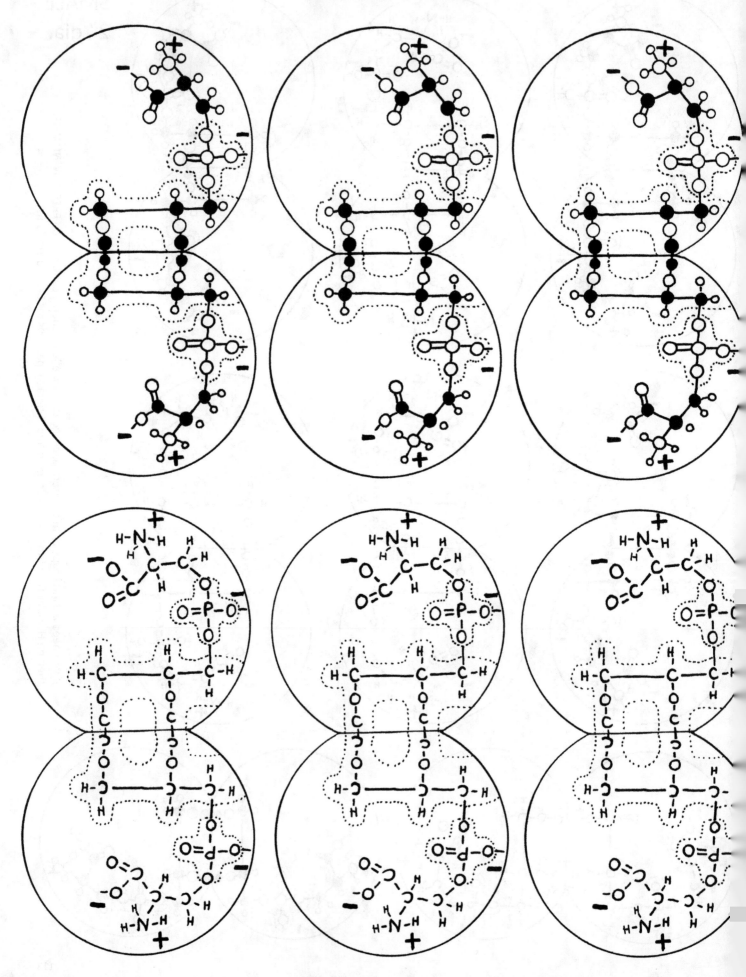

What to do:

The slit has to be just long enough so that you can maneuver the C-shaped molecule up through it.

FOR OPTION A:

1) If your ornaments have two halves that snap together you are ready to go. If your ornaments are one piece, you will need to make a slit in the bottom. I have tried various methods and found the easiest and most effective is to use a hack saw. Hack saw blades have very small teeth and will cut the slit in less than 30 seconds. If this option is not possible, you can try using a sharp craft knife (e.g. X-Acto) or a serrated bread knife, and be very careful to keep your fingers out of the way. (I set my balls in the top of a vice that was opened just enough to let the ball nestle down halfway.) Keep your fingers well out of the way and/or wear a rubber glove that will provide some protection while not being too slipper.

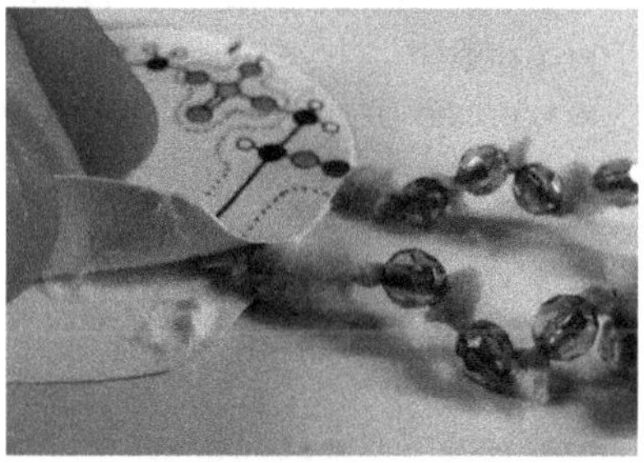

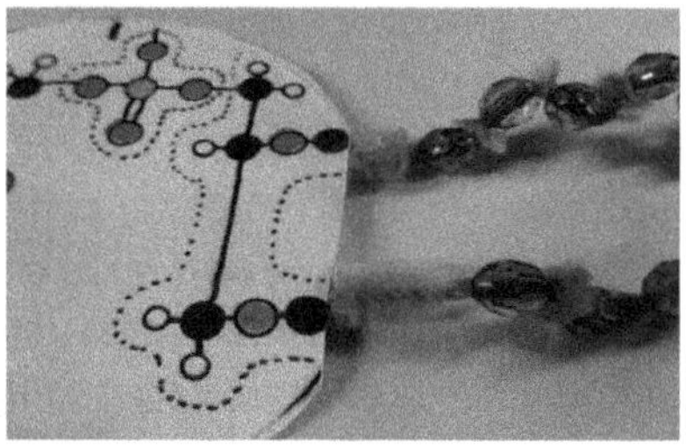

2) Cut out the two halves of your paper molecule. If you want do any coloring, do it now, before you glue the halves together. NOTE: You can mix and match halves, if you want to have the letters on one side and the dots on the other. Just be careful to get the appropriate images matched correctly so that you will be able to cut around the molecule on both sides.

3) Choose one chenille stem and cut it in half. If you happen to have small beads that fit onto the stem you can string on 8-12 beads to represent carbon atoms. You can leave the stems straight or bend them in a zigzag shape to simulate the shape of a real carbon chain.

4) Apply glue to one side of the paper molecule. Use enough, but not too much. Apply a few drops and then spread them out. You should not have glue oozing out when you press the papers together.

5) Make sure the ends of the chenille stems are tucked between the papers, right where the bottom carbon atoms are. You can cut along this bottom line, or you can make two tiny ships for the stems.

6) Press sides together, making sure that the stems will be glued into place when it dries. Let this dry for a few minutes.

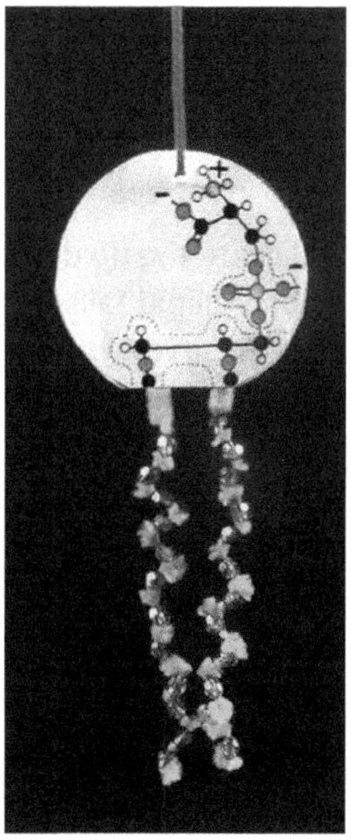

7) If you are using an ornament with two halves, simply put the paper into the ornament and snap shut. If you are using one that has a slit cut in the bottom, first cut carefully around the molecule, cutting away as much extra paper as you can without cutting off any atoms. The result will be kind of C-shaped (as shown in the examples on the previous page). You will be able to maneuver this shape up through the silt bit by bit until it is inside. Just be patient.

8) Add a decorative string, yarn, or ribbon so you can hang it up.

FOR OPTION B:

1) Choose your paper molecules and do any coloring you'd like to do.

2) Follow steps 3-6 above. You will not need to do any additional cutting. Leave the molecule as a circle.

3) Punch a small hole in the top of the paper so you can add a string or ribbon hanger.

<u>ACTIVITY IDEA 2E</u>: A PAPER MEMBRANE MODEL
Note: This activity can be used after chapter 3 if you are out of activity time for chapter 2.

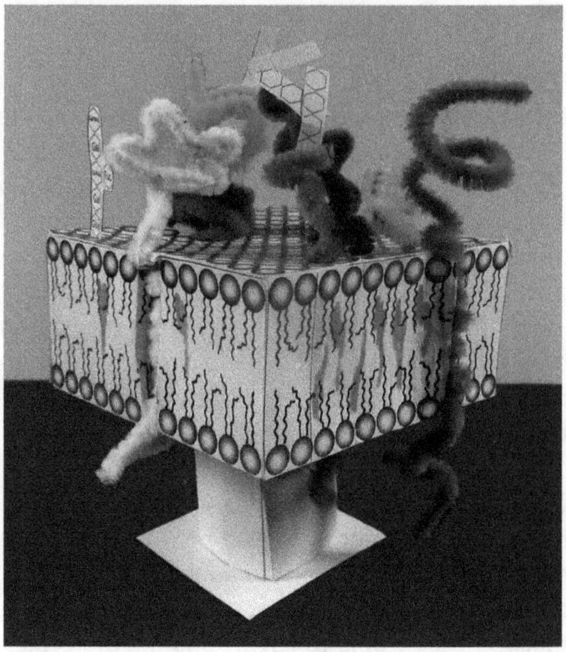

You will need:

- the following pattern page printed onto card stock

(See note on previous project if you need a digital pattern)

- scissors
- really good white glue (I order PVC glue from Amazon.co.uk.

You might also try a white craft glue, such as Elmer's Craft Bond (not regular Elmer's white glue), or Alene's Tacky Glue.)

- a tool for cutting the chenille stems (Scissors will cut them, but be aware that they will dull scissor blades.)
- a sharp craft knife (such as an X-Acto knife)
- a pencil
- colored pencils or markers, especially in yellow or orange (for cholesterol molecules)
- a ruler or straight edge
- a pointed object for scoring fold lines (compass point, nail, very large needle, etc.)

What to tell the students:

In this activity, you will make a paper module showing a tiny piece of a phospholipid membrane. You will add some protein structures to the membrane, both peripheral proteins and transmembrane proteins. One of these will be a real protein called MHC 1. It is the little flag that identifies the cell as part of the body. The other proteins will be examples of general types of proteins but won't be identifiable as any particular protein. For example, you will be putting in a channel that controls the flow of some type of molecule in or out of the cell but it won't be modeled after any particular channel. Also, you will add a number of receptors and "switches" but you can create your own squiggly patterns with the chenille stems. There so many protein shapes in a membrane, that whatever you create will probably look similar to one of them. There are still membrane proteins that scientists have not yet discovered!

What to do:

1) Before cutting out the pattern, score along fold lines. This step always seems like an extra bother, but it will make all the difference when you do the folds. The folds will be crisp and straight and the model will almost snap into place automatically. Use a ruler and something that will lightly scratch the paper without cutting it, such as the point of a compass, a nail, a large needle, or even a "dead" ballpoint pen. Scissors might be used successfully by an adult (I have done this) but can be tricky for many students. **If you are working with a younger group, you might want to consider doing this step ahead of time and give the students a pre-scored pattern page to cut out.**

2) Locate the lipid raft section (indicated by thin dotted lines). Add cholesterol molecules between the phospholipid tails in the lipid raft region. You can also add one or two cholesterols on other sections, if you find those few places that have more straight tails. Student might also want to add some light yellow to the phospholipid heads that belong to the lipid raft region, to make it more visible.

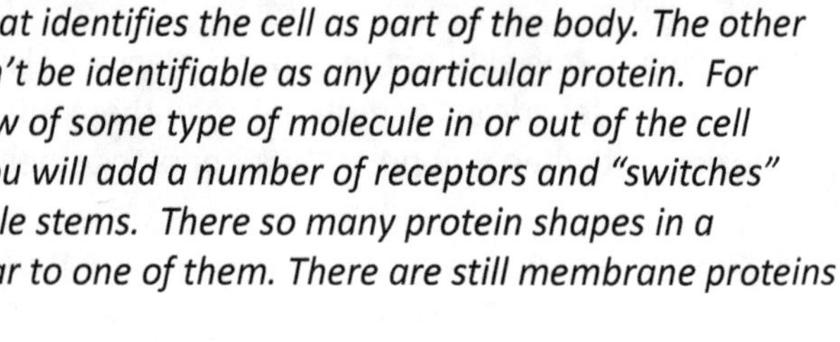

3) Cut around the edges of the paper pattern. Work carefully. Slopping cutting will make the assembly process much harder.

4) Fold on all the fold lines. After this step your project should similar to the one shown here.

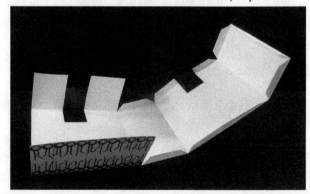

is a paper stand you can (as shown in sample photo). don't have to glue it to the el. The model can simply on top of it. Use this corner ne base but use it upside n so the words don't show,

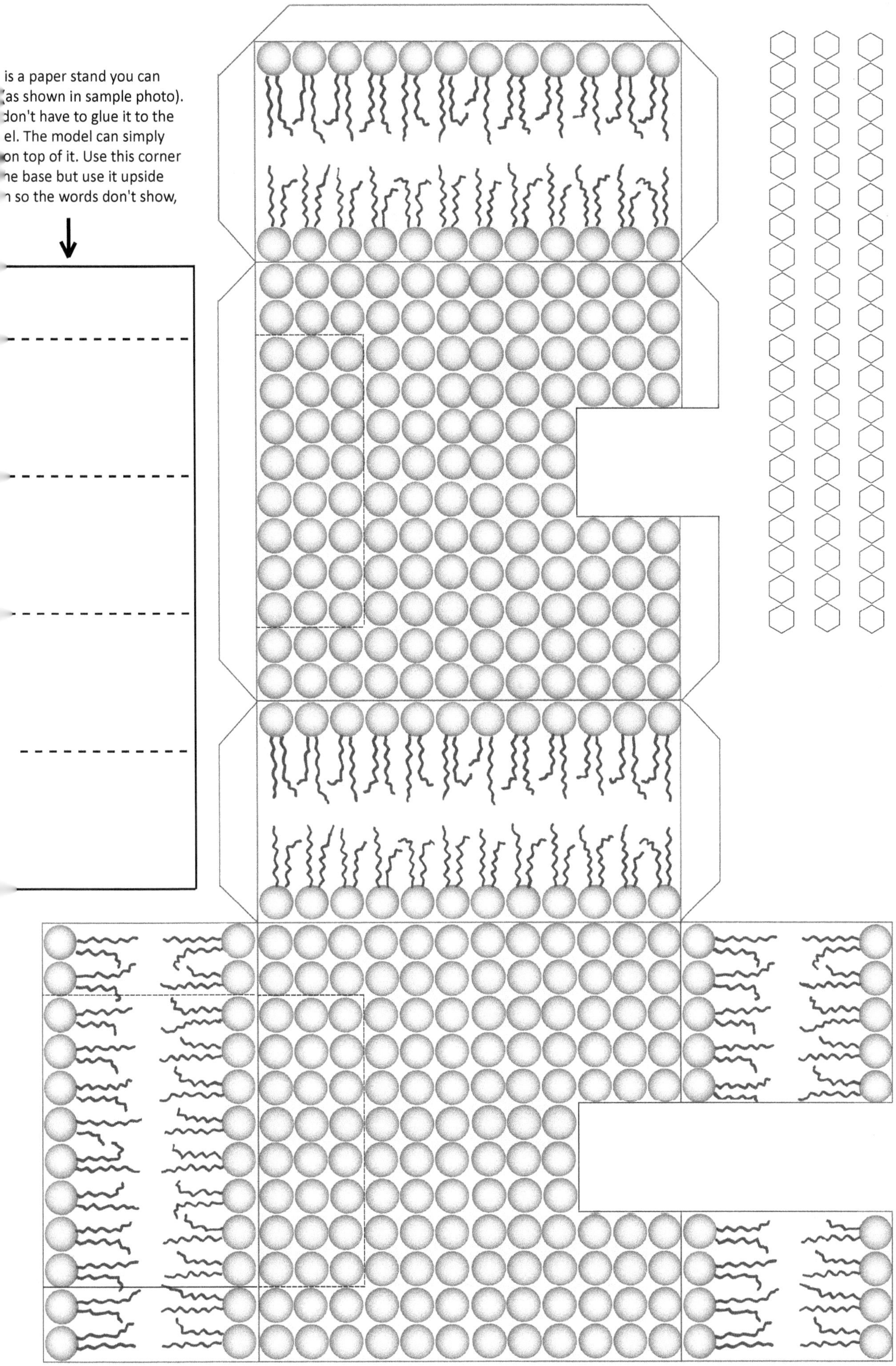

Optional:

ABO Blood antigens (made of sugars)

Type A

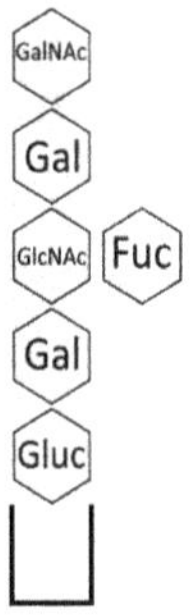

Type B

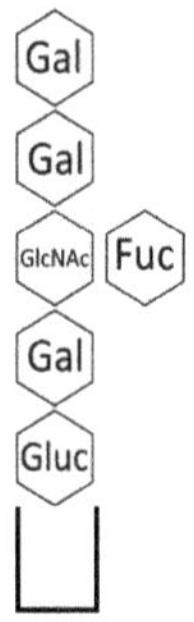

Type O

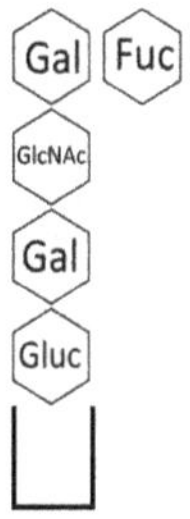

For type AB, use both A and B.

5) Begin gluing the glue tabs, perhaps in this order:

1

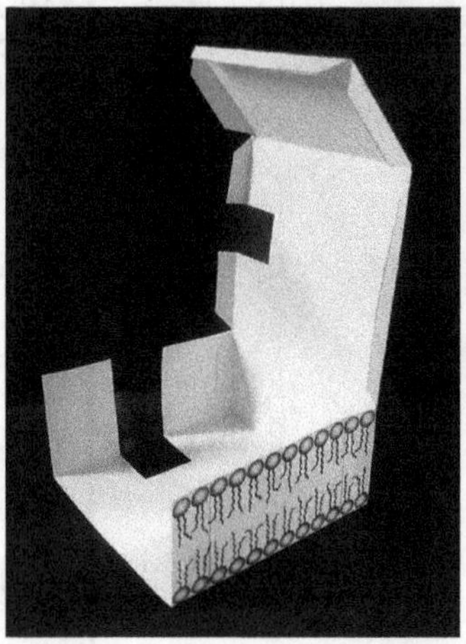

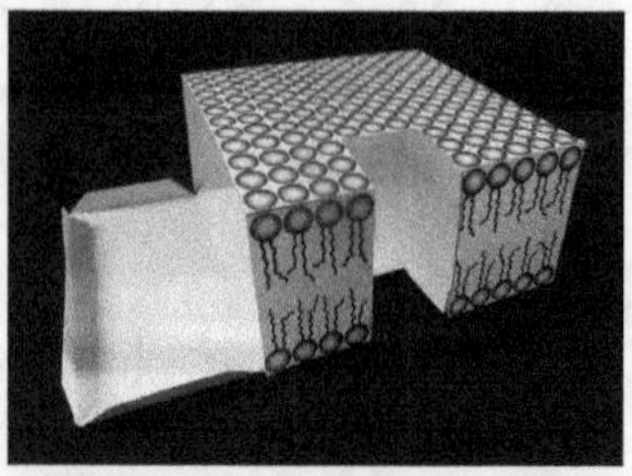

2

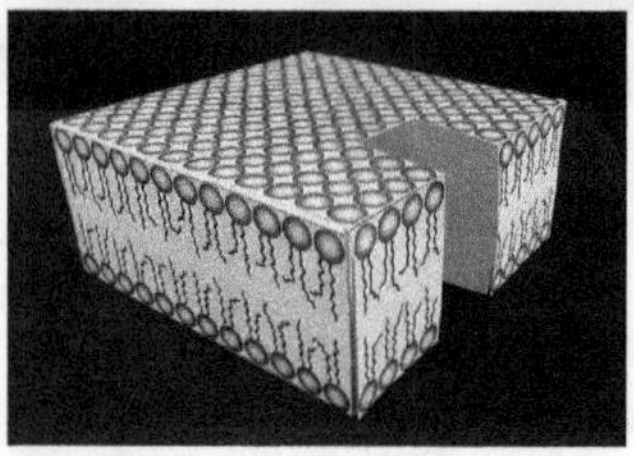

3

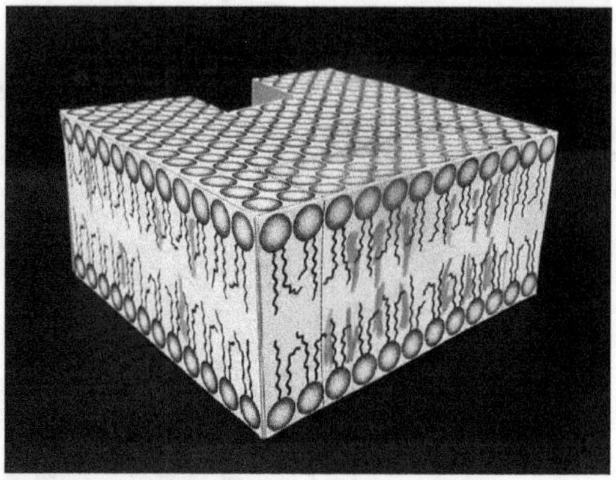

If you have colored the lipid raft area, your model will look something like this.

6) Use a chenille stem to make an "MHC 1" shape. (the ID flag)
You can use a pencil to create the round shapes.

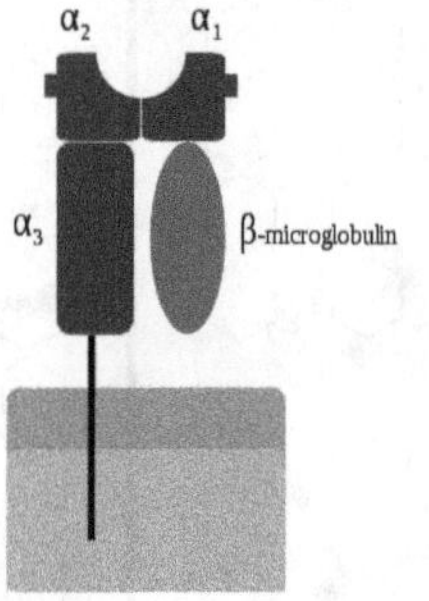

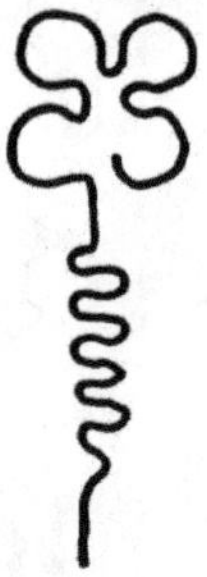

MHC 1 is shown many ways. Here are two typical diagrams. We'll imitate the one on the right.

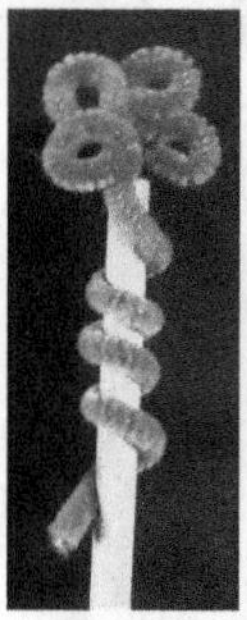

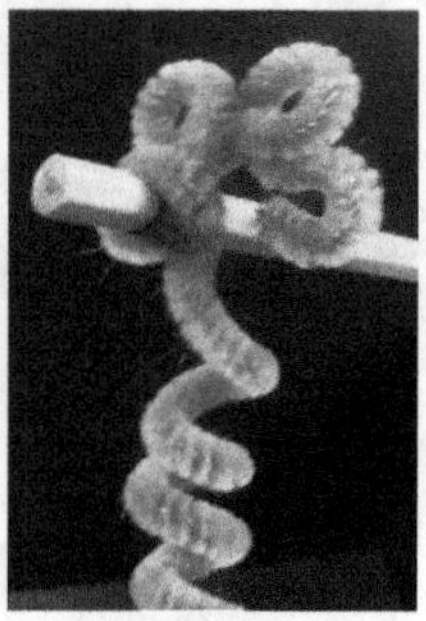

Your MHC 1 can be any color. My students found it helpful to use a pencil to make the curls.

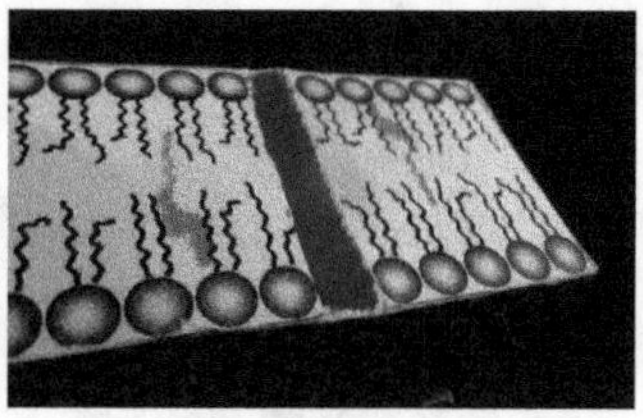

To insert MHC 1 (or any transmembrane protein) use a craft knife to cut out one phospholipid. Tuck in curled chenille. Add a tiny bit of glue to keep it in place.

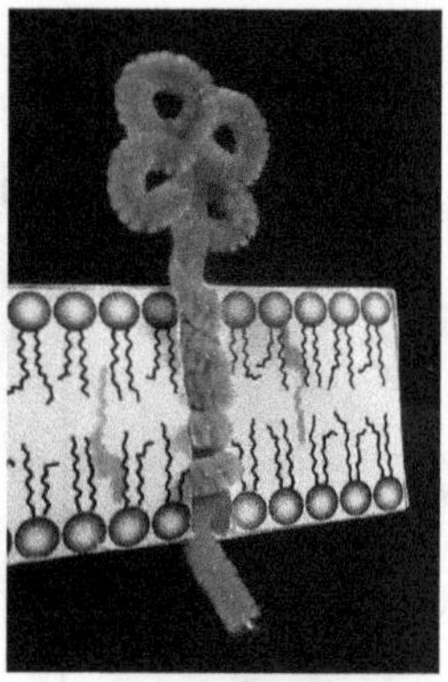

7) The large opening is for a beta barrel. Beta barrels are made of folded proteins called beta sheets. Beta barrel diagrams often look like a braided structure. We are going to just use two or more chenille stems and curl them in opposite directions around a finger. Try to make your barrel large enough to fit into the space in your membrane. Some barrels have extra curls on the top or bottom.

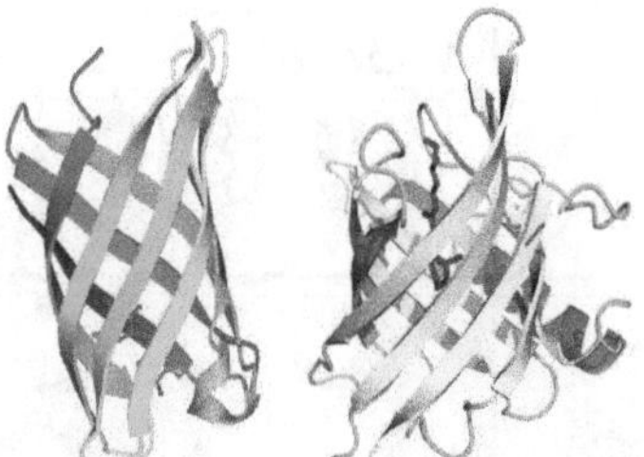

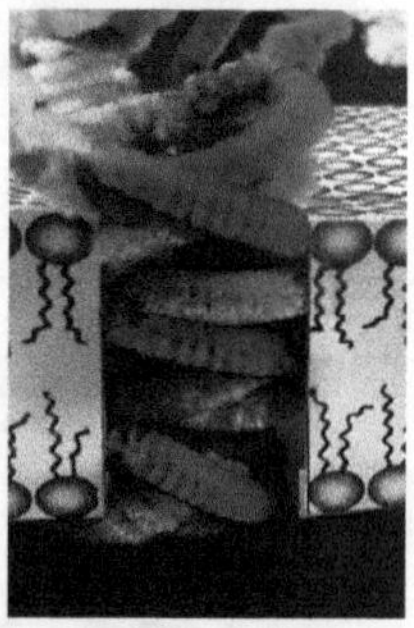

8) Add some peripheral proteins and some transmembrane proteins. They can be any shape. For the peripheral proteins, you can make a tiny hole, put a small blob of glue on the end of the chenille and stick it in (and let it dry). For the transmembrane protein you will need to cut a slot in the side of the model. Make sure you put several proteins in the lipid raft area.

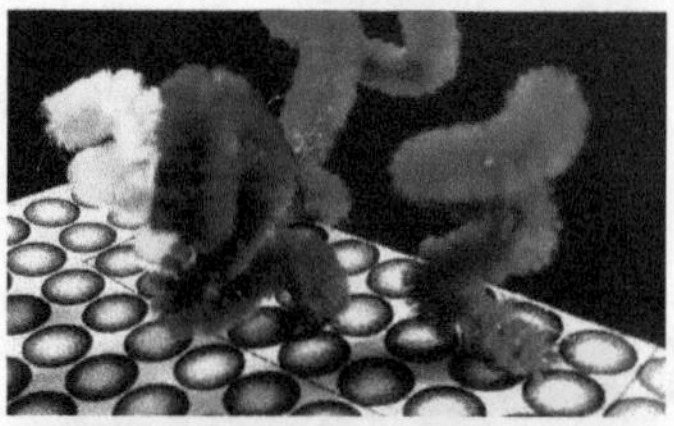

9) Add some glycan sugar strings to your model.

You have a number of options. You can add one of the famous "ABO" blood type markers, provided on the side of your pattern page. Only red blood cells will have these proteins on their surface, but we are not concerned with making this sample membrane accurate to any particular cell, so you can add these gylcans or leave them off, either way is fine. If you happen to know your blood type you can make your membrane match your own cells.

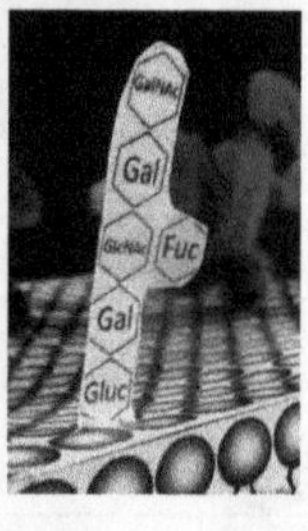

CHAPTER 3

ACTIVITY IDEA 3A: CRAFT: MOTOR PROTEIN PENS

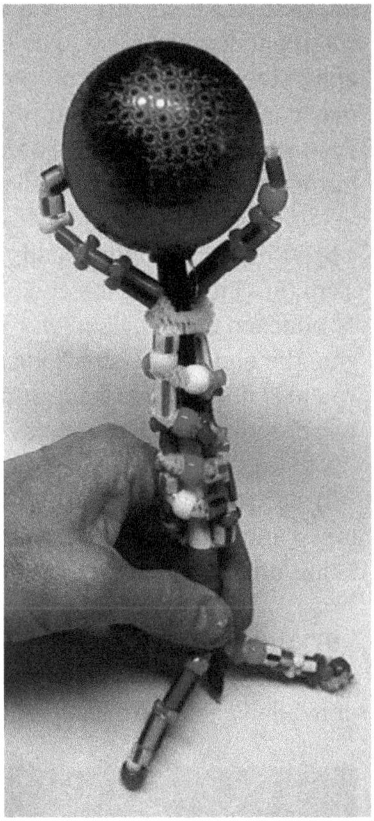

Even "craft-shy" students will probably like this craft because it is so bizarre. The students will be making a model of a motor protein that is also a functional pen they can write with.
NOTE: This craft has a lot of flexibility in how it is assembled. You can adapt the materials and/or construction to suit your situation. If you can find a better way to make it, please do so!

You will need the following for each student:

- a ballpoint pen (the kind with the cap that comes off)
- a handful of assorted colored beads for each student (miscellaneous sizes, shapes and colors)
- chenille stems (4 per student)
- tape: ideally, colored masking tape (available in craft stores) Regular masking tape or floral tape will also work. Avoid clear or duct tape.
- some kind of hollow ball (I used inexpensive plastic Christmas tree ornaments. You could also use a Styrofoam™ ball, or any lightweight plastic ball. If you can't find any suitable plastic balls, you can make one out of heavy card stock paper using the following dodecahedron pattern.
- hot glue gun (or fast-drying glue that sticks to plastic)
- drill to make small holes in side of ball

What to tell the students:

You will be making a model of a kinesin motor protein. It won't be a highly accurate model of a particular kinesin, as we have limitations due to the craft supplies we will be working with, but it will definitely be recognizable as a motor protein. The ball on the top will represent a vesicle that is being carried. At the bottom of the motor protein we will make the heads that walk along the microtubules. You will use plastic beads to represent amino acids that make the structural proteins. The nice thing about this model is that it is also useful as a pen you can write with. You can have a lot of fun making people guess what it is, then telling them what your learned about motor proteins.

How to assemble: (This is what I did, you can adapt where needed.)

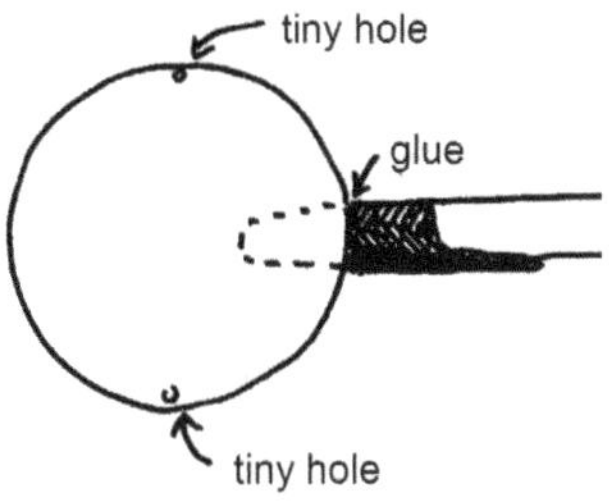

1) Take the cap off the pen. Use some hot blue to stick the cap firmly to the top. If using other glue, do this step ahead of time and let it dry. If no glue is available, just press the cap on as firmly as you can.

2) Drill or punch a hole in the bottom of the ball, if you are using a plastic ball. The hole must be just the right size so that it fits onto the cap at the halfway point. (see drawing at right) Use hot glue to make the ball stick firmly to the pen cap. (If using another type of glue, do this step ahead and let it dry.)

3) Drill or punch two tiny holes in the ball; the holes should be just large enough to accommodate the end of a chenille stem. Refer to photograph above; you can see where the chenille stem will go in. If you can't drill holes but you have a hot glue gun you can also glue them place.

4) Take two of the chenille stems and secure a bead to one end of each stem. Just loop the chenille stem around, give it a twist and tuck the end back into the bead. Make sure there isn't a sharp metal end sticking out. Then thread some beads onto them until you've covered about 5 cm (2 inches) of the chenille stem. The measurement does not have to be exact.

5) Lay the chenille stems alongside the pen so that the yet-to-beaded part is flush with the pen tip. (see drawing for clarification) Wrap some floral tape around at the tip, to secure them, then wind the tape up the pen about 4 cm (1.5 inches). The measurement does not have to be exact. Tear off the tape and continue wrapping until the end of the tape is sealed on. Press firmly. Floral tape won't seem very sticky so you may think it won't hold, but surprisingly, it will adhere very well and will stay in place even while the pen is being used to write with. The floral tape won't stick to fingers, just to itself. (For further clarification, look at the color pictures in the appendix.)

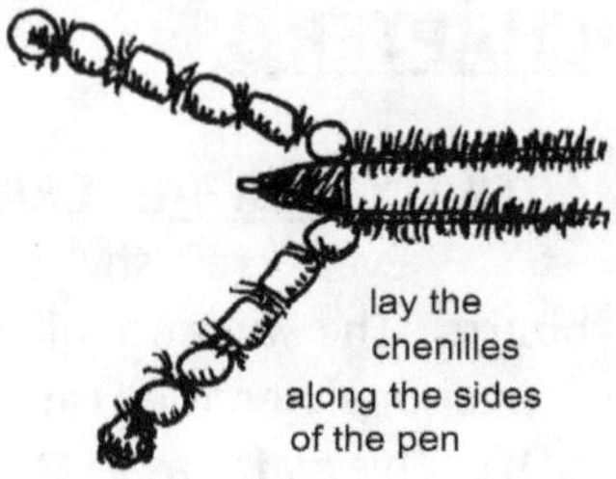
lay the
chenilles
along the sides
of the pen

6) Thread some more beads along those two chenille stems until you reach about 2 cm (3/4 inch) from the end. Those two ends will fit into the small holes you drilled or punched in the ball. (Refer to photograph on previous page.) You may want to glue the ends of the stems into these holes. (Ours stayed secure even without glue.) You may need to make adjustments for your particular situation, depending on what you are using as a ball and how large your beads are.

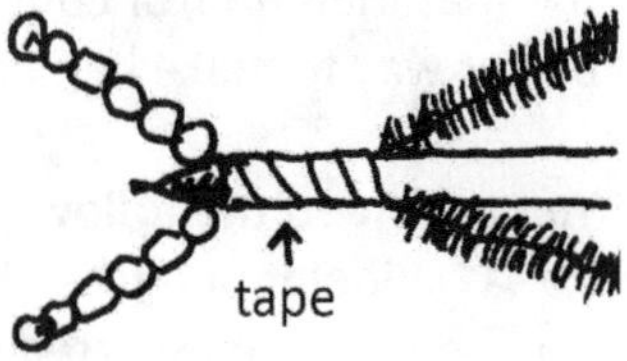
tape

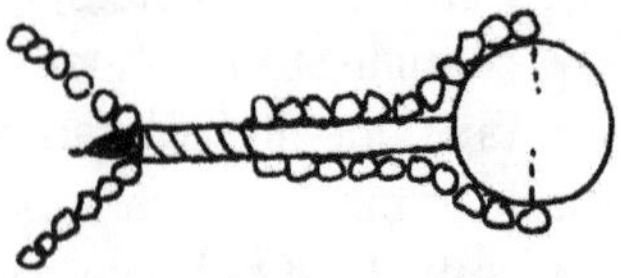

8) Secure a third chenille stem to one of the side stems right at the top of the taped area. Begin a pattern in which you alternate winding with threading a few beads on. Wind tightly.

9) After you have wound to the top, right under the ball, secure the chenille stem to one of the beads. You may also add a second beaded chenille if you'd like to have more beads on the model, and/or you can use an unbeaded chenille to wind around the middle, securing the others more tightly.

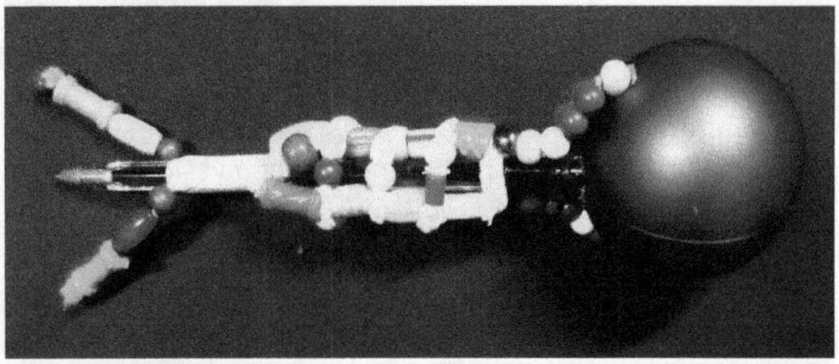

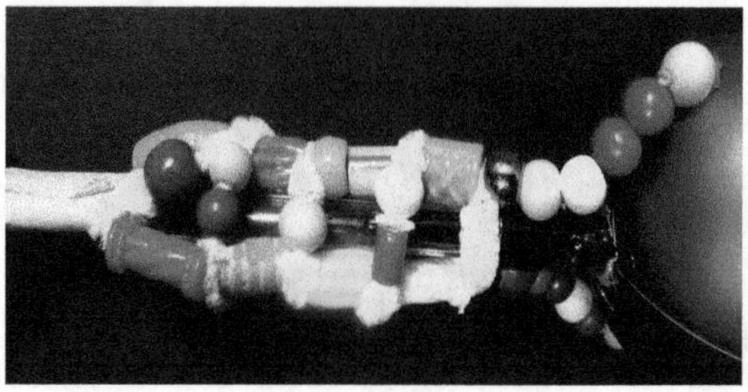

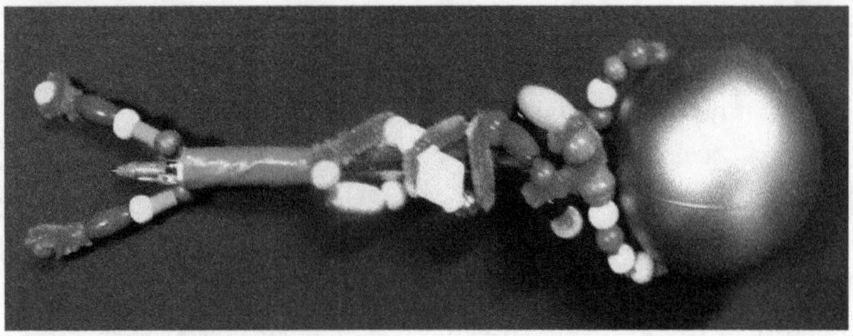

To use the pen, just bend the heads (which look like feet) up and out of the way. (Be careful not to bend them over and over again, as it is possible to "wear out" the metal in the chenille stems so that they break.)

<u>ACTIVITY IDEA 3B</u>: MOTOR PROTEIN CARTOON CHALLENGE

You will need:
- paper and pencils

Ask your students to look again at the kinesins on page 19 (and possibly supplement with some Internet images of both kinesins and dyneins?) and then make cartoon drawings of motor proteins. (Also, make sure they have seen the video "A Day in the Life of a Motor Protein" which feature animated cartoon motor proteins.) Post the drawings so they can be shared with the group. This activity will encourage artistic students whose abilities are not usually appreciated in science classes.

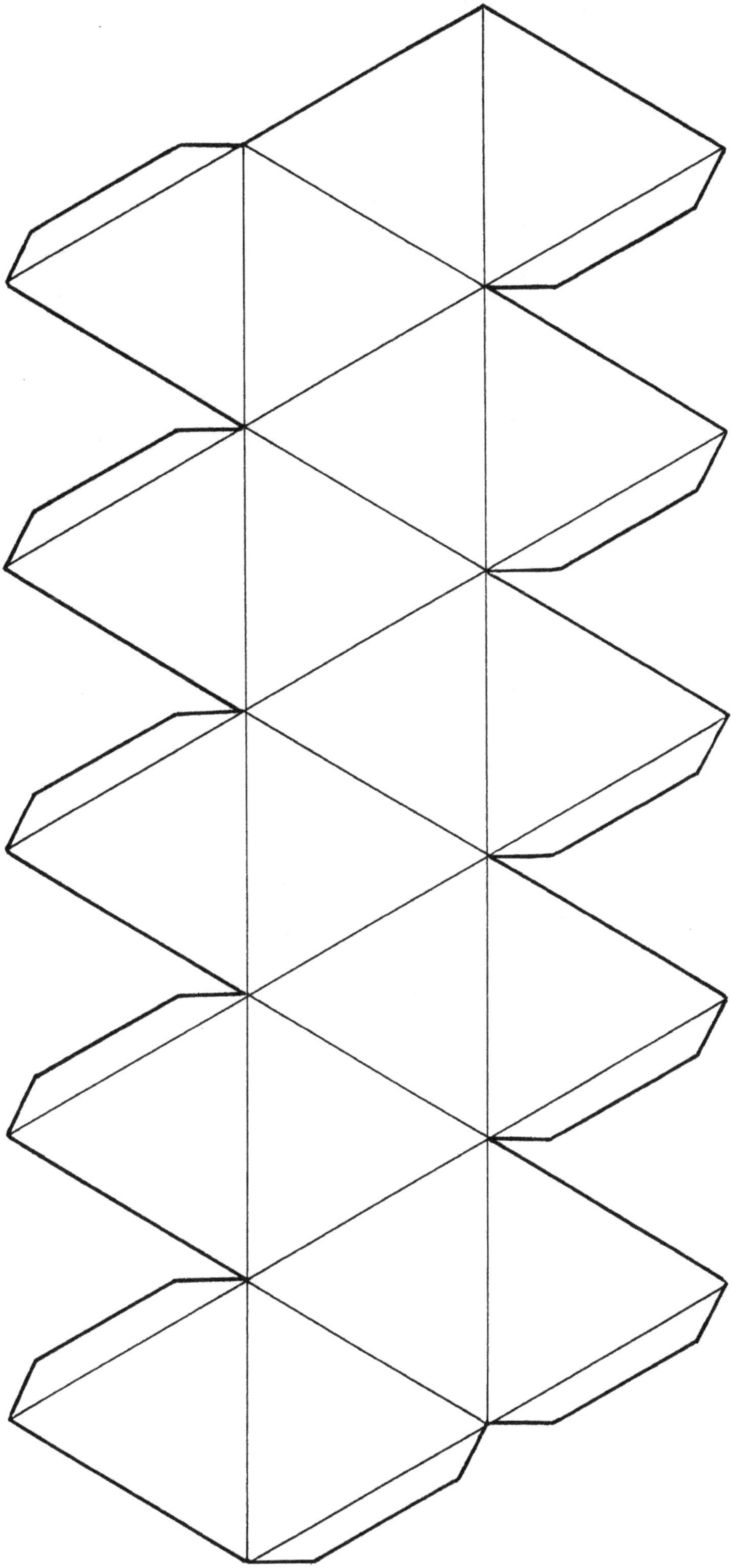

PATTERN for making a paper vesicle for motor protein craft if a hollow plastic ball is not available.

PRINT ONTO CARD STOCK

ACTIVITY IDEA 3C: MOTOR PROTEIN RELAY RACE

You will need:

- a fairly large space (long and narrow is fine, like a hallway)
- two very large, but lightweight, objects such as beach balls (or other large inflatables) or the "balance balls" used for exercise programs
- something to be your microtubules; I have used toilet paper and paper strips, but found blue painter's tape or masking tape to give best results. If you use paper of some kind, you'll need some tape to secure it to the floor at various points.
- two pieces of wide elastic, tied at the ends to make loops that will fit tightly but comfortably around the knees (notice elastic in bottom photo)

What to tell the students:

In this relay race, you will play the role of a kinesin or a dynein walking down a microtubule, carrying a vesicle. Walk carefully so you don't tear the microtubule! If you tear it, someone on your team will have to come and repair it before you continue. Cells often have to repair parts of their cytoskeleton and they can do it very quickly and efficiently.

How to set up for the race:

Prepare your microtubule "roads." If you are using paper, secure it in several places and make sure you have some extra left over (and some tape) for repairs. If you are using tape for the full length of the road, you probably won't need to worry so much about repairs.

Have a loop of elastic for each team at the starting line. Also, have the balls ready at the starting line.

How to run the relay race:

Divide the group into two teams. If you have an odd number of players, assign one player to go twice. Then divide those teams in half and put one half at each end of their road. Those on one side will be kinesins, and those at the other end will be dyneins. This will help to reinforce the idea that motor proteins only go one way. It won't matter which side goes first.

The first players stick their feet into the elastic loop and pull it up to just above their knees. They are only allowed to move the lower part of their legs, below the elastic. This will make them move more like a real motor protein. It will also add a humorous twist to the relay race and make it more fun. Then the players pick up their ball and hold it over their head. Now they are ready to start down the microtubule road. They must walk along the strip without tearing it. If it tears, one of the team members must take a piece of tape and fix the tear. The mending job doesn't have to be great. The minimum is that the two ends of the paper must be touching. (If using a tape road this step may might not apply, depending on how sturdy your tape is.)

When the runners reach the end, they put down the ball, take the elastic off their knees and hand it to the first player on that other side. This next player puts on the elastic, picks up the ball, and hobbles down the road to deliver the ball back to the side where it started. They hand both elastic and ball to the next player. Play continues like this until all the team members have been down the road. The first team to have all their members finish their run is the winner of that round. Since the race only takes a few minutes, several rounds can be played.

NOTE: I found that it takes a bit of adult supervision to make sure players are adhering to the rules and not "cutting corners" like putting the elastic too high or stepping off the road. It is up to you to determine how to handle things like this if and when they occur. (The main point is that they are having fun and learning!)

CHAPTER 4

ACTIVITY IDEA 4A: ACTIVE GAME: A TABLE-TOP RELAY RACE ABOUT THE ELECTRON TRANSPORT CHAIN

This relay race doesn't require a lot of space like a normal relay race does. It is designed to be a table-top game, but you can adapt the race to the space you are working with. We used several tables, but if you only have one table, you could put a relay each side, or make just one relay set-up and have the teams take turns using it (using a stop watch to time each team's race).

The set-up for this activity is bit involved. If you have enough class time, you may want to have the students work on making the parts ahead of time. If you have very limited class time, you'll have to make the parts yourself and spend the class time doing the race instead of setting up. It takes one person about 45 minutes to cut and assemble all the paper parts.

Helpful suggestion:

Before you begin preparing, watch the video demonstration where I explain how to play. I posted it on the "Cells" playlist at YouTube.com/TheBasementWorkshop.

You will need:

- card stock, colored if possible (8 sheets, 4 of one color and 4 of another--I used red and blue)
- the patterns on the following pages
- clear tape and/or glue
- scissors
- some small tokens to represent electrons, protons and oxygens (I used red and white dried beans for the electrons and protons, and large dried lima beans for the oxygens, but you can use whatever you have on hand. Just make sure they are small, approximately the size of dried beans.) For each relay set-up you will need a minimum of 4 tokens for electrons, 12 for protons and 2 for oxygens, but having lots of extras is recommended.
- plastic or paper cups (two per relay set-up)
- a cardboard tube that has a diameter smaller than the bottom of the cups (one tube per set-up)
- something to represent ATPs (large and small marshmallows on a toothpick, paper circles taped to a toothpick or whatever works well in your situation; I used Legos®) You will need at least 4 ATPs per set-up.
- black permanent marker

How to prepare:

1) Print or copy the following pattern pages onto card stock. (If you have only a paperback version of this curriculum and would like a digital copy of the patterns so you can print right from your computer, there is a free download of these patterns available online. Go to www.ellenjmchenry.com and click on FREE DOWNLOADS then on HUMAN BODY then scroll down to find "Printable pages for Cells curriculum."

For each relay set-up, print three copies of the first two pages (the one with the holes and the one with the two trays) and one copy of the other two pages.

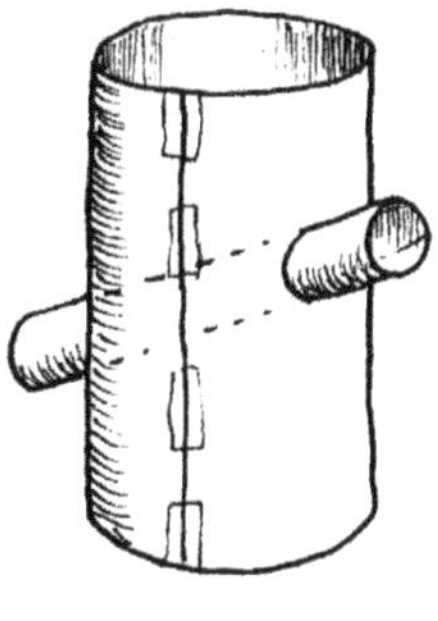

2) Cut out the holes on the pages with holes, and then roll the paper into a tube. Secure with tape.

3) Cut apart the rectangles (each one being exactly one fourth of a sheet of 8.5x11 paper) and roll each lengthwise so form a small tube. You don't need to tape these smaller tubes—you can just insert them into the large cylinders and then let the small tubes expand to fill the hole, thus holding them in place

4) Cut and assemble the large, shallow trays, cutting on the solid lines and folding on the dotted lines. Tape the corners so you have a square tray. Then tape a tray to the top of each large cylinder.

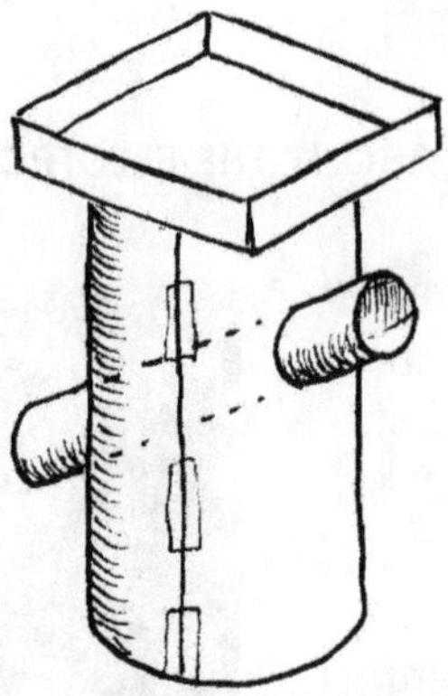

5) Cut and assemble the deeper rectangular trays, cutting on the solid lines and folding on the dotted lines. Bend the ends in to make a square corner and secure with tape. Mark one "NADH."

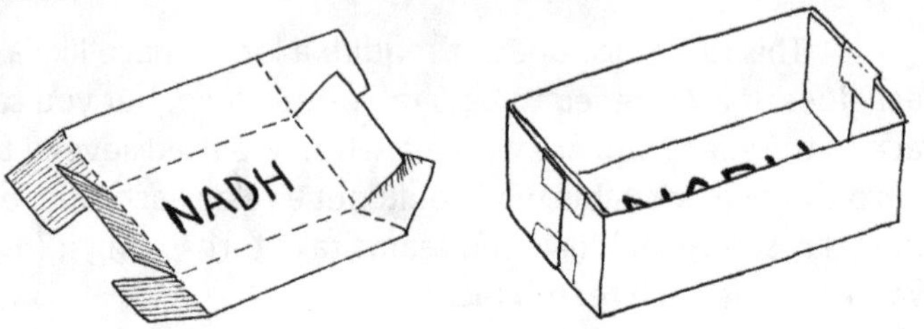

6) The other deep trays will be the shuttles between the ion pumps. The first shuttle can hold two electrons, the second only one. Write these numbers on the bottom of the trays so the players will see them and be reminded of how many each can hold. If you want to put the names of the shuttles on the sides of the trays, the first shuttle (between pumps 1 and 2) is called "Ubiquinone" *(you-BICK-wih-noan)* and the second (between pumps 2 and 3) is "Cytochrome C."

7) Cut out the squares with the H_2O molecules on them. For each H_2O, cut out two short strips and one long strip and roll them so the ends overlap until the strips are the same circumference as the printed circles. Secure with tape. Then secure the circles to the "platform" by taping the tabs down.

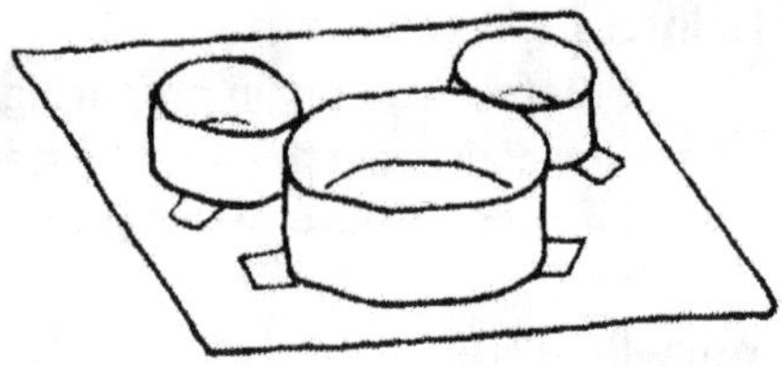

Assemble the O_2 molecule the same way, using two long strips.

8) Now you need to make the ATP synthase "machine." Cut holes in the bottoms of the cups, the same size as the diameter of your tube. If you are making a tube out of card stock you can just cut the holes whatever size you want, roll the card stock, stick it into the holes, then let it expand (just like you did for the small tubes in the sides of the cylinders). If your synthase machine is wobbly, add some tape to make it more sturdy. You may want to label the machine, either on the cup or by adding a paper label sticking up out of the top cup. You might also want to add a sign that says: **3 protons → 1 ATP** so the players don't forget how many proton tokens to put into the machine to get an ATP.

9) Prepare your ADPs and phosphates. Use whatever materials you think will work well in your situation. You may have some craft parts or food items on hand that work well. They don't even have to be round—you could use Lego® bricks. Just make sure you have three small items and one large one for each ATP. Assemble ADPs (one large and two smalls) and leave the third phosphates unattached. The phosphates will be hidden in the bottom of the synthase machine.

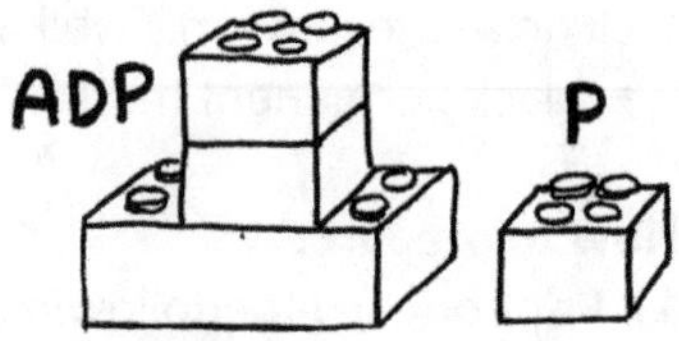

How to set up the relay:

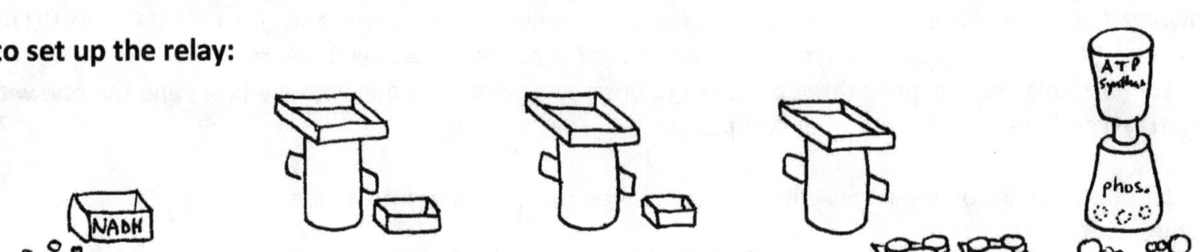

You can have any amount of space between the pumps. If you have a large playing area and want to make the players run, you could set these quite a distance apart. Put out plenty of proton and electron tokens. Put tokens in the oxygen circles. Individual phosphates are under the bottom of the ATP synthase. ADPs go in front of the synthase.

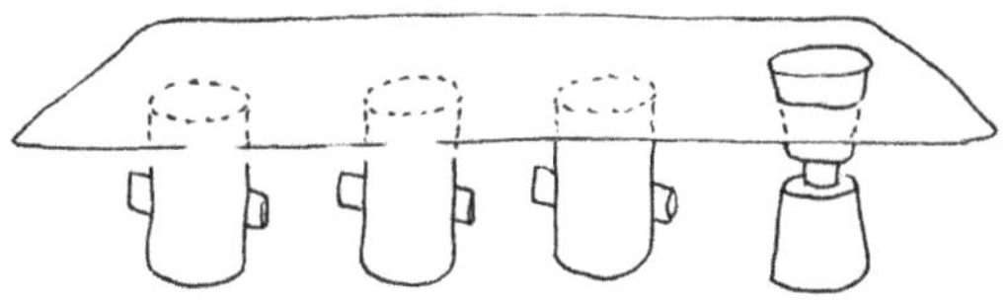

NOTE: An arrangement like this would be a more realistic representation because it would show that the pumps and the synthase machine are all connected to the same membrane, but this configuration would be more difficult to make as well as more difficult to play. Make sure the students understand that the pumps don't really have individual trays on top.

How to play:

(Reminder: Don't forget about that demo video on the YouTube channel!)

1) The first player comes to the NADH "shuttle bus" and puts two electrons into it. He then slides the NADH box across the table until it reaches the first ion pump.

2) The player then puts the electrons into the top of the small tube in the first pump. The electrons will slide through and come out the other side. (The player can position the next shuttle so that it catches the electrons as they come out of the tube.) After the electrons drop into the waiting shuttle, the player moves two protons from the table up to the tray on top of that pump.

3) The player then moves the shuttle to the second pump. This time he must put the electrons through one at a time because the shuttle waiting on the other side can only carry one electron. One electron goes down through the tube and then one proton is moved up to the top tray. Then he must empty this second shuttle before doing anything with the second electron. The player pushes the shuttle over to the third pump, puts the electron through, then the electron goes into one of the "H" circles on a water molecule. Then he can go back and repeat this for the second electron.

4) Now the cycle starts again. The player goes back to the beginning and puts another two electrons into the NADH shuttle. The whole process in repeated until you end up with four "tired" electrons, one in each of the "H" circles.

5) Now the player must complete the water molecules by picking up four protons from the table and putting a proton with each electron. Then the oxygen molecule is "split" and one oxygen token is put into each "O" circle on the waters. Now the waters are complete and the trays can be pushed out of the way.

6) The last step is to put protons down through the ATP synthase machine. There should be a total of 12 protons available, four on top of each pump. Pick up three protons and drop them into the top cup of the synthase machine, then tilt the machine and remove a phosphate from underneath it. Put this phosphate onto an ADP to make an ATP. Repeat this process until you have made 4 ATPs.

7) Now the turn is over. If you have limited tokens, the player will have to put all the protons back onto the table, put the electrons back at the start, and the shuttles back at their starting points. If you have limited ATPs, perhaps the adult helping with the race can dissemble them while the players is replacing all the electrons and protons. Having an adult help to reset the relay between players is recommended. The adult can also be in charge of recycling the parts in the water molecules, leaving the water circles empty and putting tokens back into the oxygen circles.

SHORTER VERSION:

To play a shorter version of the relay, you can make just one water molecule instead of two. Making two water molecules is more accurate because when you split an O_2 molecule, the O's have to go somewhere. But for the sake of time and attention span, it may work better in your situation to run through the cycle just once and make just one water molecule and 2 ATPs.

NOTE: Save the parts from this relay! You will get a chance to use them again in chapter 8. The relay suggested at the end of chapter 8 is slightly more complex than this one, but will show the students where those the electrons come from (the ones that go into the NADH).

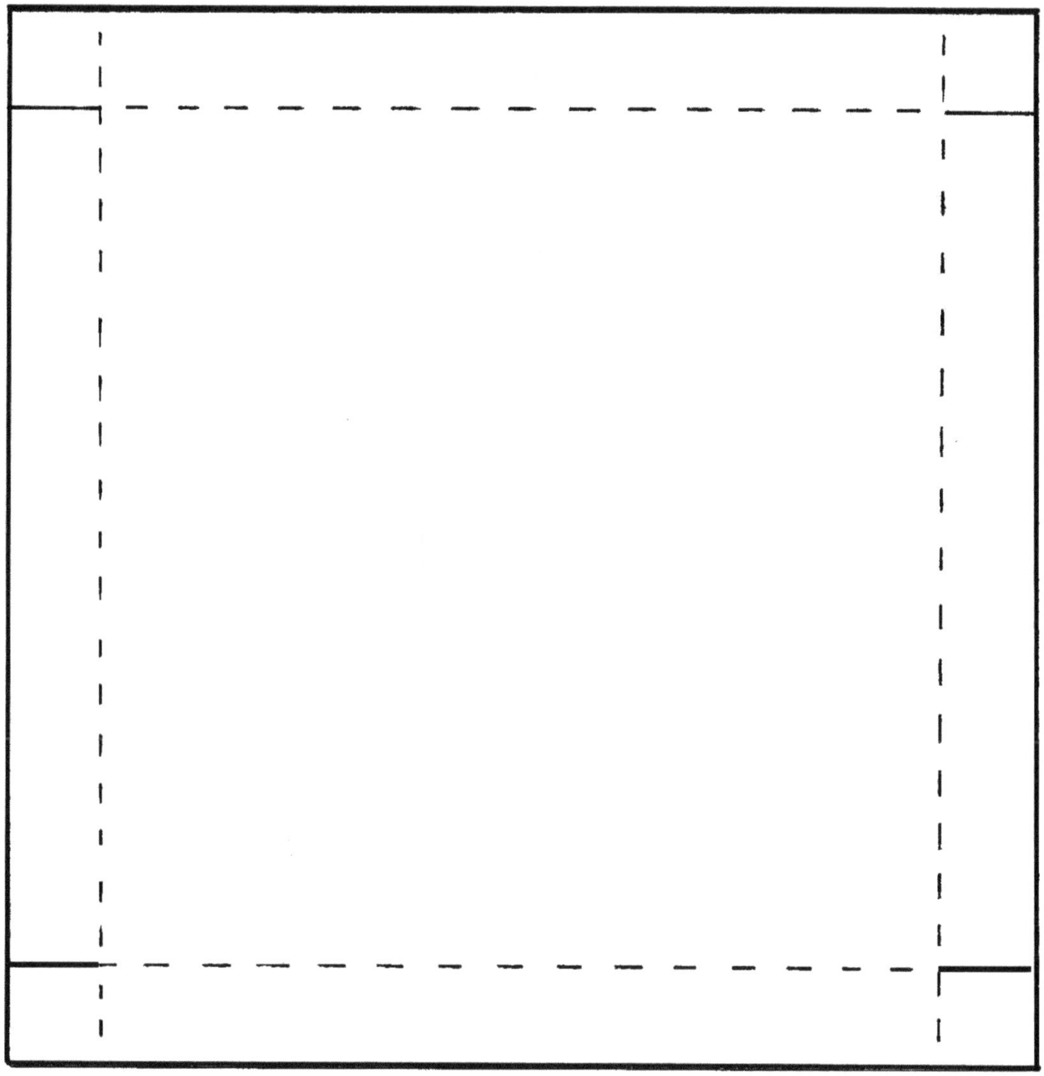

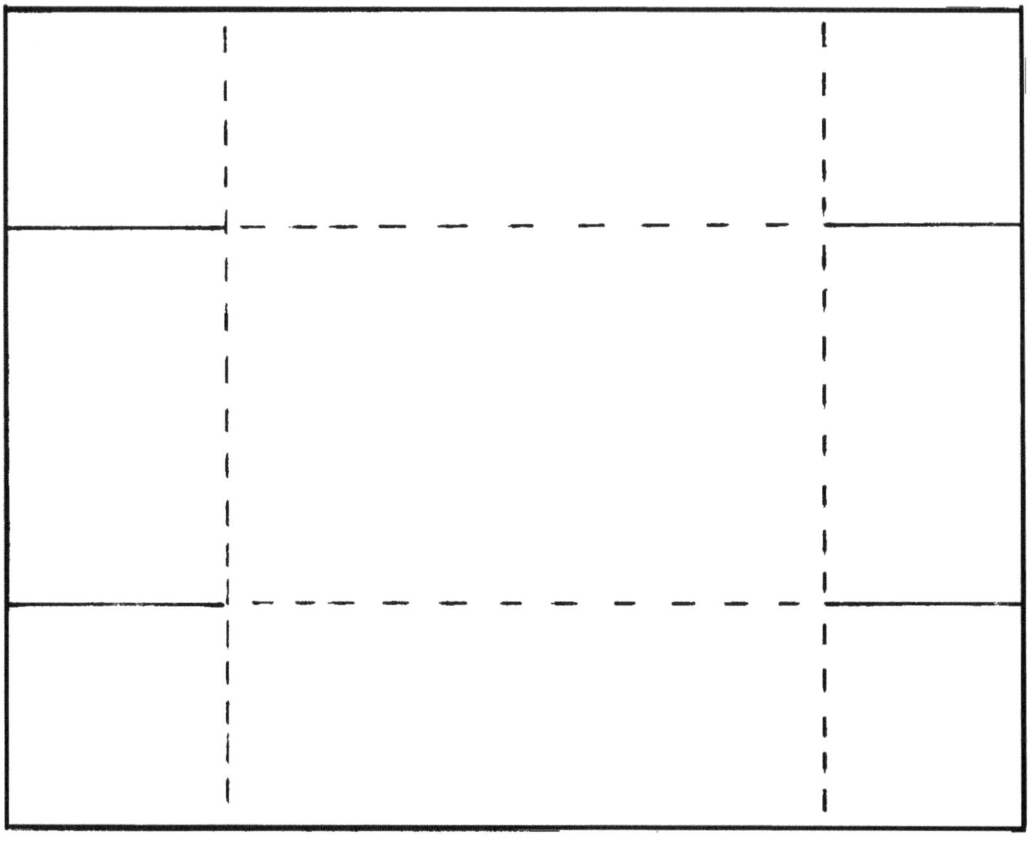

Use these three rectangles for the smaller tubes that insert into the larger tubes. (Roll them lengthwise.)

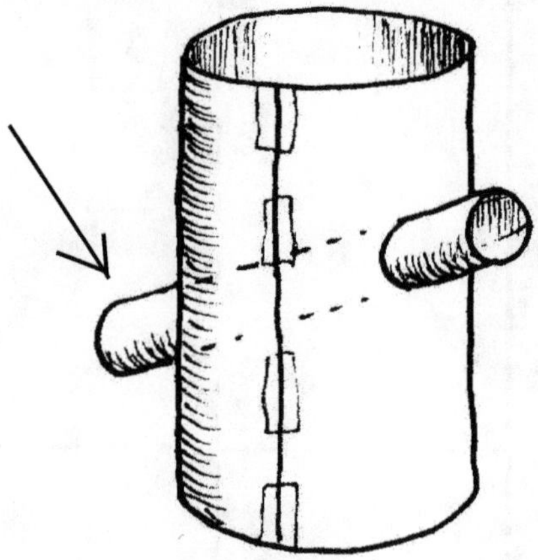

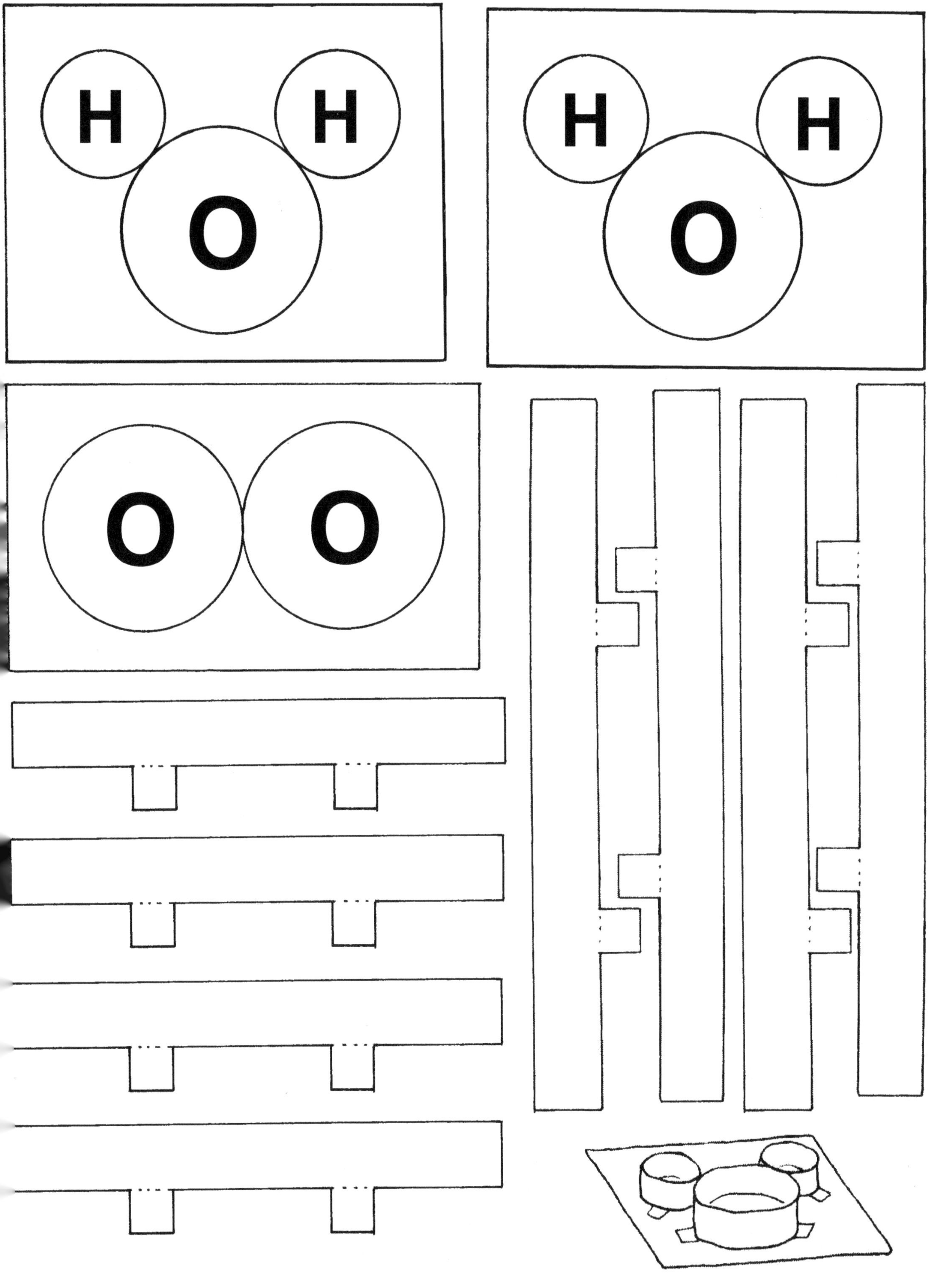
H
H
O
H
H
O
O
O

ACTIVITY IDEA 4B: CRAFT: ATP "POP GUN"

Instructions for this craft can be found at www.ellenjmchenry.com. Click on FREE DOWNLOADS, then HUMAN BODY. (UPDATE NOTE: As of 2021, I have had trouble getting the correct kind of ballpoint pens. The new design for cheap pens has a much wider point and won't work with these instructions. Also, I have been finding that corrugated cardboard has gotten thinner and won't work as well as older boxes.)

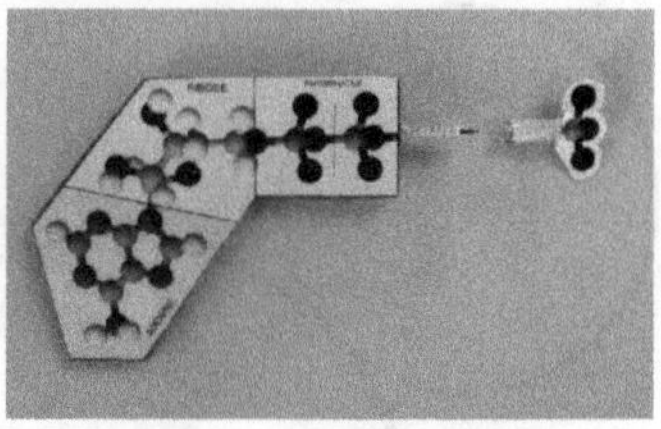

ACTIVITY IDEA 4C: CRAFT: PAPER MODEL OF ATP SYNTHASE

NOTE: Check out the "Cells" YouTube playlist for a video showing how this model works and how to put it together.

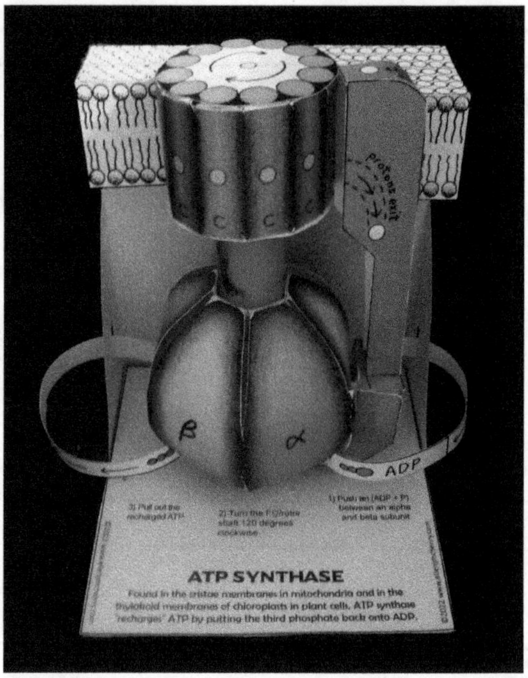

You will need:

- copies of the following pattern pages printed onto card stock
  (If you need a digital copy of the patterns, go to the website www.ellenjmchenry.com, FREE DOWNLOADS, HUMAN BODY, scroll down till you see a link for printable pages for the Cells curriculum.
  NOTE: If you can't print in color, the digital download will give you both color and black and white options.
- really good glue (For white glue use either PVA glue, Elmer's Craft Glue or Aleen's Tacky Glue. For glue stick, use "extra strength" craft bond intended for adult use.)
- good scissors
- small piece of masking tape or any tape that is not too sticky
- sharp craft knife (such as X-Acto knife)
- piece of thick cardboard, either corrugated or the back of a writing tablet (thicker than cereal box cardboard, though in a pinch you could use that) Size needed is approx. 4"x 4" (10 cm)
- something to hold joints while they dry; try either clothespins or paper clips

How to assemble the model:

1) Cut out the F1 spherical subunit (the shape with six wedges) and the little strip below it. Assembling this piece will take several rounds of gluing and drying, so to be time efficient you should go ahead and start gluing these before cutting the rest of your parts.

2) Use your fingers to gently bend the tapered edges of the wedge shapes. (This is similar to curling ribbon with scissors, but here you will be using your fingers instead of scissors.) Stroke the wedges from the center to the outside, bending while stroking. They should curl under quickly and easily.

3) Bend the entire piece into a circle and glue in place using the side glue tab. Bend white tabs inward.

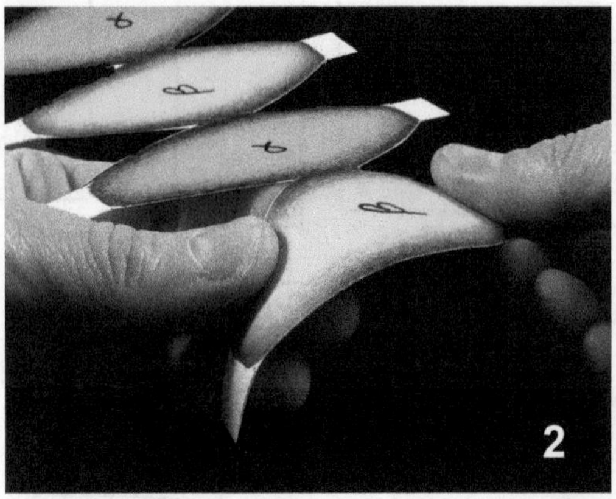

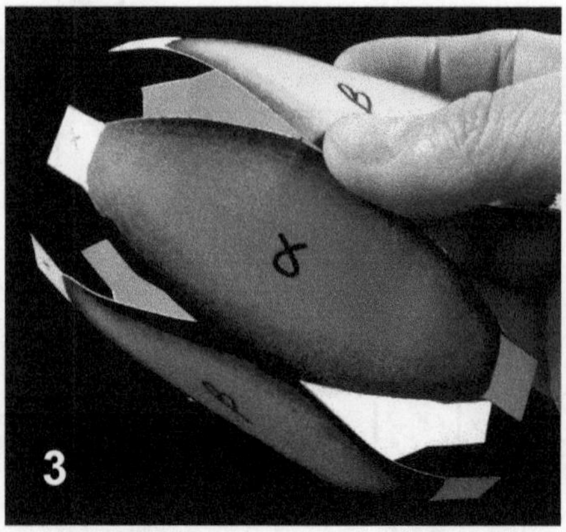

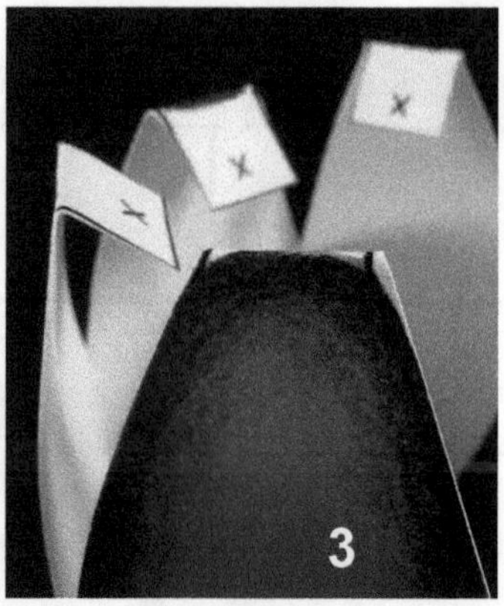

4) Score the dotted lines on the strip, then fold into a hexagon and glue the glue tab.

5) Put a dab of glue on one of the tabs that has an X. Press the glued X tab onto one of the sides of the hexagon ring. Press and hold till secure. Then do the opposite side. (Use clothespins or paperclips if necessary.) You can also reach your finger up inside and press the seam from the inside.

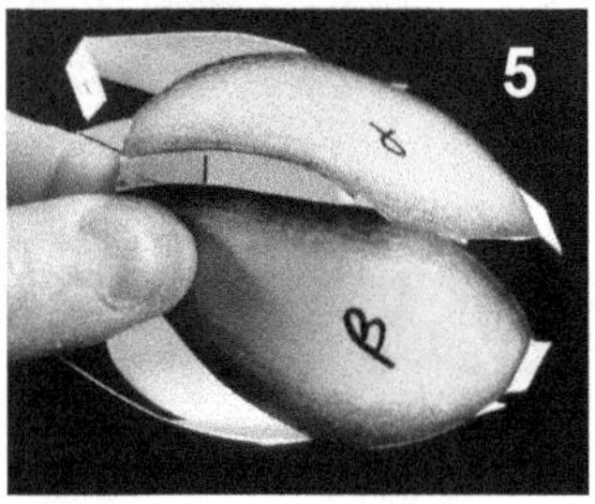

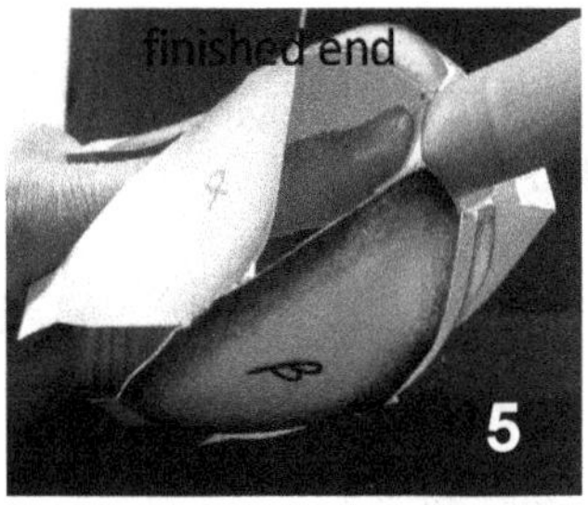

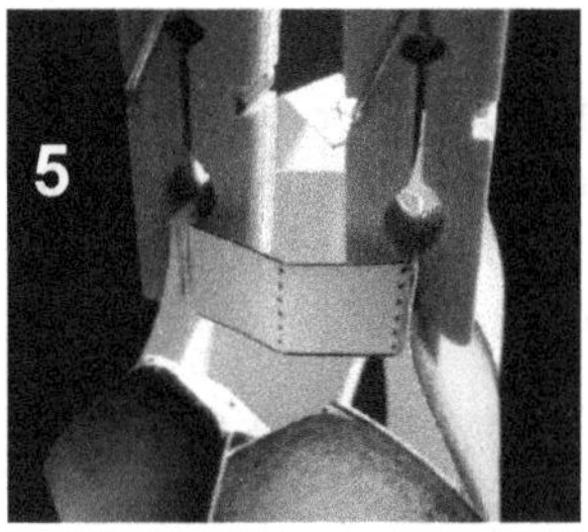

finished end

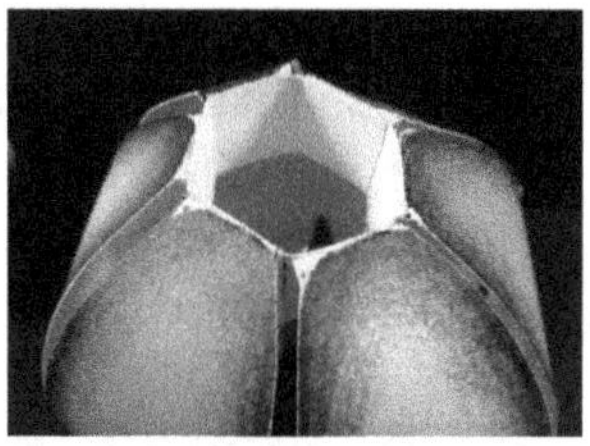

6) Cut out the part labeled ROTOR SHAFT. Carefully roll it into a cylinder shape. Rolling it around something cylindrical can be helpful. Cut 4 pieces of tape (a tape that is not very sticky, just as blue painter's tape, masking tape, or "invisible" tape) and have them ready for the next step.

7) Apply glue to the area marked GLUE, making sure not to get glue past the dotted line. A uniform, thin layer of glue is better than blobs of glue. You can even out the glue with your finger, but make sure to wipe your finger before working with the tape. Overlap the edges to that the non-glued side covers the glue tab right up to the dotted line. Secure the seam with the pieces of tape. Put something cylindrical into the tube you have just made and press the seam firmly. Press up and down the seam for at least half a minute. Set aside to dry.

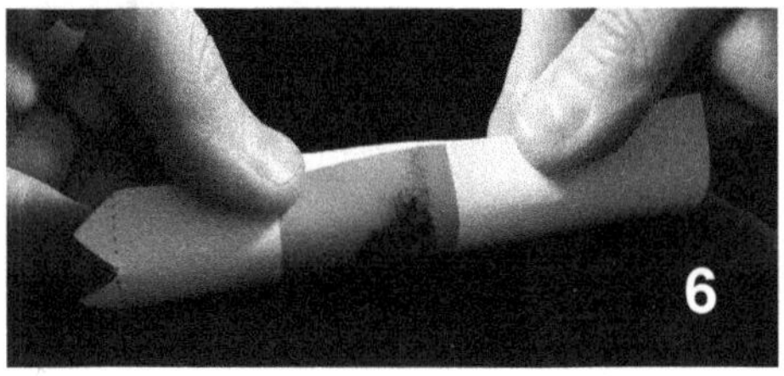

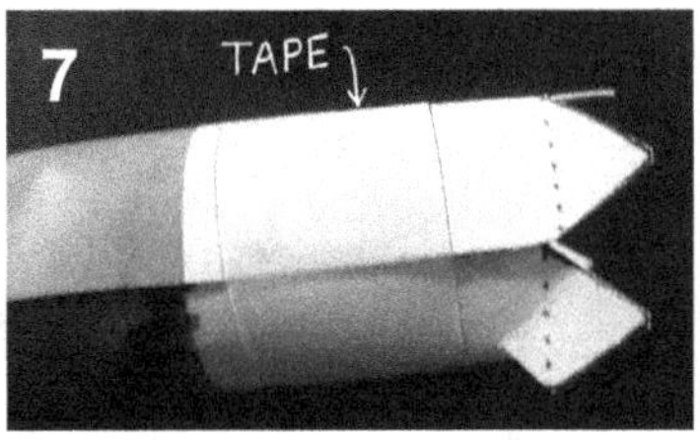

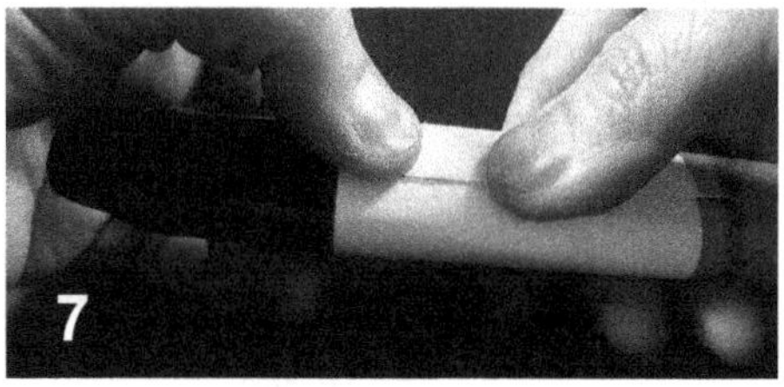

8) Cut out the green parts (the long strip and the two circles). Score along the bottom of the triangular flaps.

9) Curl the long strip into a circle and glue using tab.

10) Cut out the (striped) center of the green circle. Make sure the green cylinder is sitting with the letter C's at the top. Put dots of glue onto the triangular tabs and set the ring on top. After pressing the circle to get it lined up as well as you can, turn the cylinder over and use your finger to press the seams from the inside.

11) Put glue dots on the triangular tabs on the remaining side and set on the piece with the 12 green circles. Press and hold the seam from the top, making sure that the green circles look like they are lined up with the long rectangles beneath. Once it is staying in place, turn it over and use your finger to press the seam from the inside.

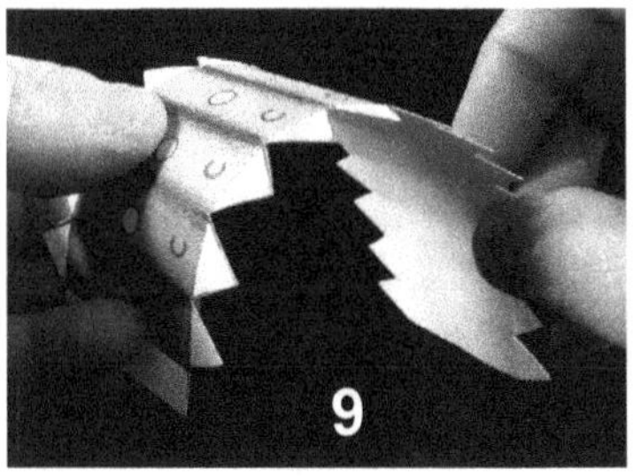

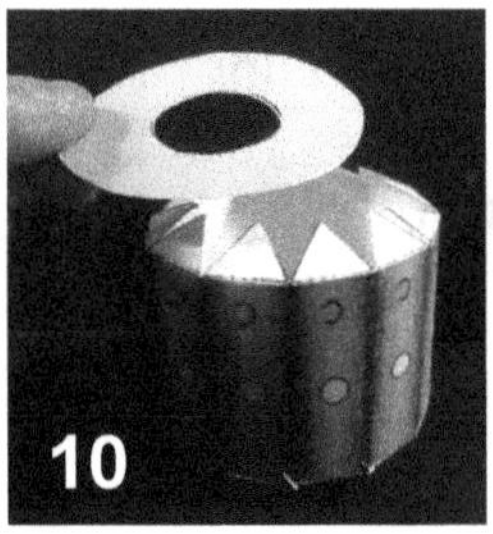

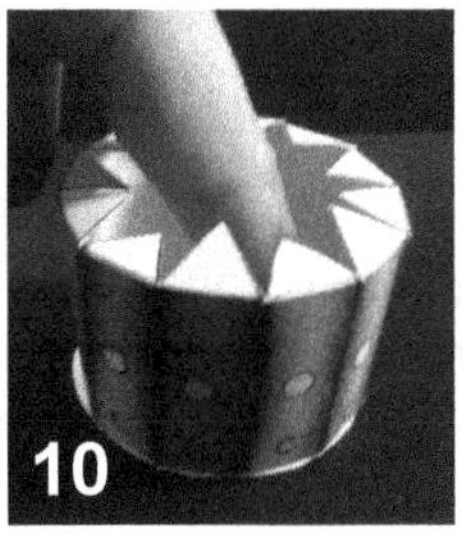

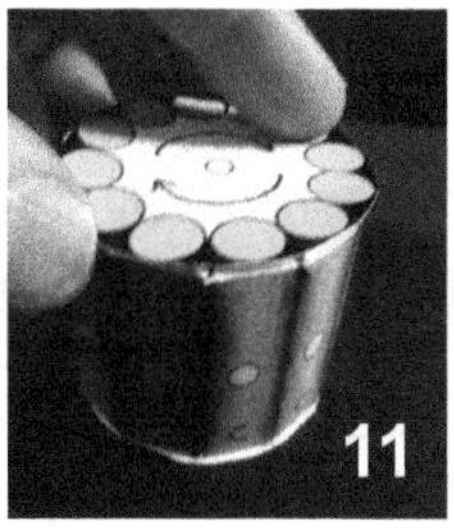

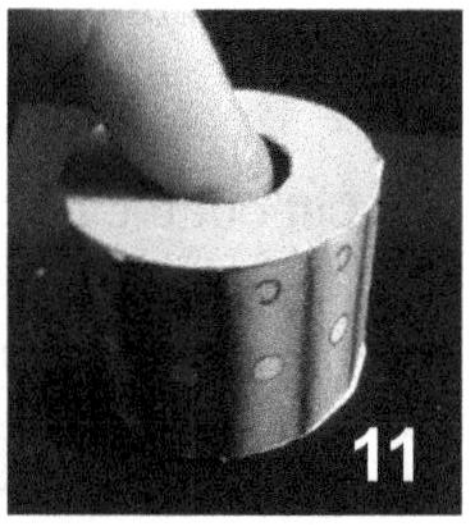

12) Now do the other end of the F1 sphere. Cut out the remaining strip and cut and fold as shown. Secure glue tab.

13) Fold down all the end tabs as shown in picture 13.

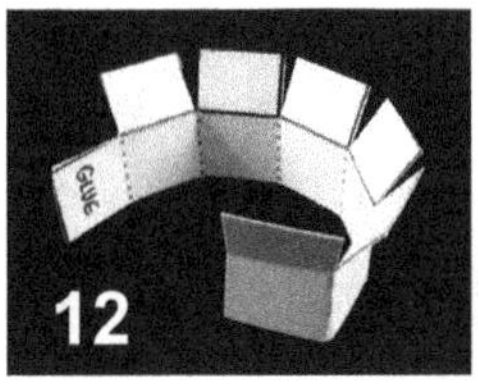

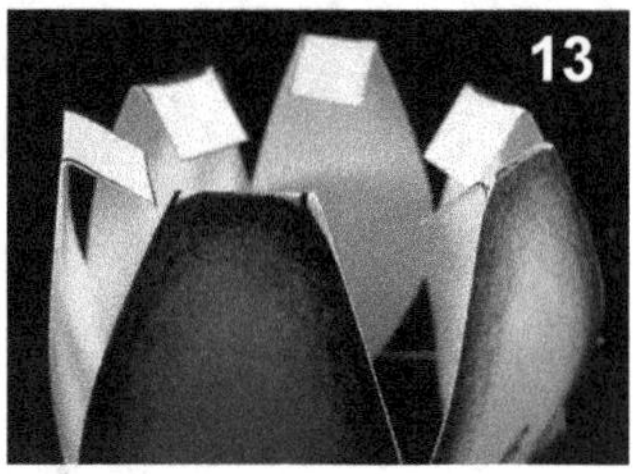

14) Repeat the same procedure you did for the hexagonal ring on the other end, except make sure that the strip's glue tabs are sticking out so your final product looks like the photo. Press and hold the glued seams tightly. Be patient! You can also stick your finger or the end of a scissors up through the sphere to help press the seams from the inside.

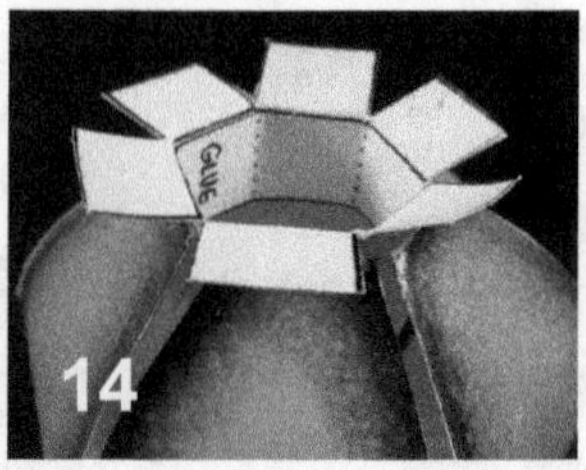

15) Cut out the large piece labeled "BACK WALL AND BILAYER MEMBRANE."

16) Use a sharp knife to cut out the tiny rectangles labeled "CUT OUT."

17) Score on all the lines that will be folded. (Use a ruler or straight edge.)

18) Fold the bilayer membrane section (use picture as guide) and use glue tab to secure in place.

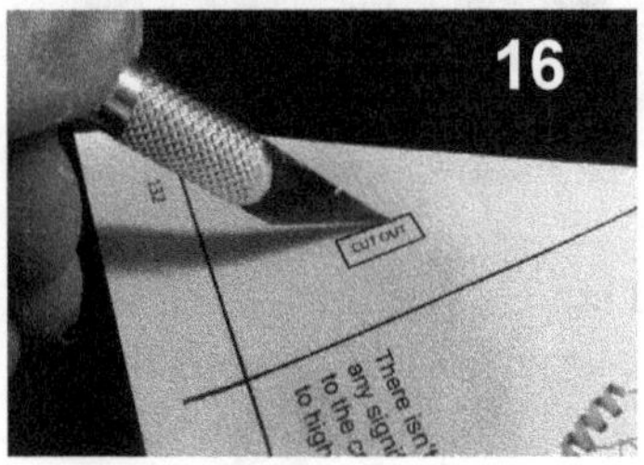

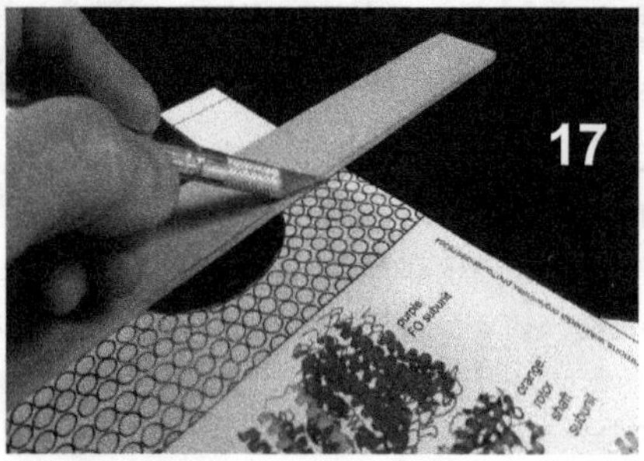

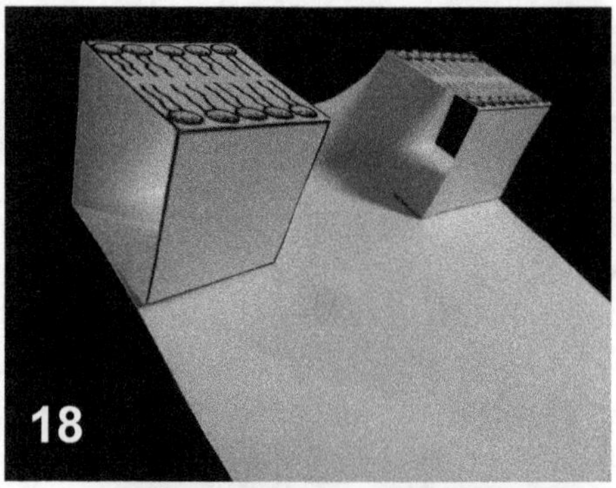

19) Cut the slits as shown in picture 17, and fold along fold lines.

20) Glue the corners as shown in picture 18.

21) Cut out the square with the title ATP SYNTHASE. Cut a piece of cardboard the same size and glue it to the back of the title square.

22) Glue cardboard-backed "floor" piece to backing board as shown in photo.

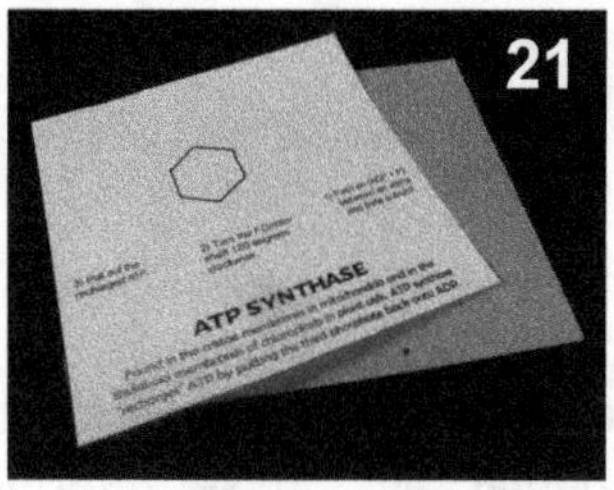

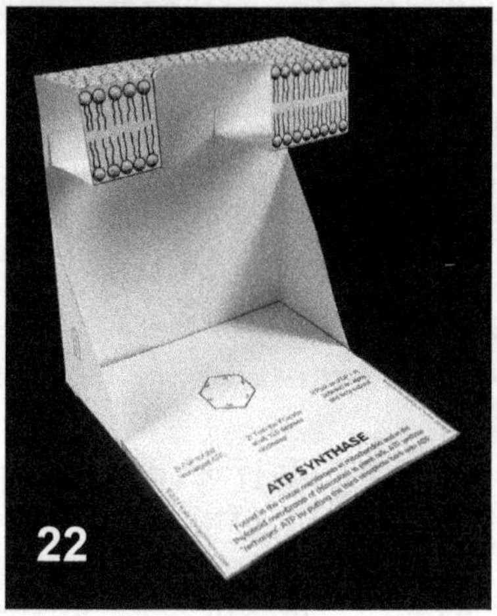

24) The rotor shaft should now be dry enough that you can take the pieces of tape off. After removing tape, put glue on the triangular tabs. Let the triangles point upward a bit. This will ensure that all the tabs come into contact with the surface they are being glued to.

25) Push the glued end of the shaft down into the green cylinder. Let dry.

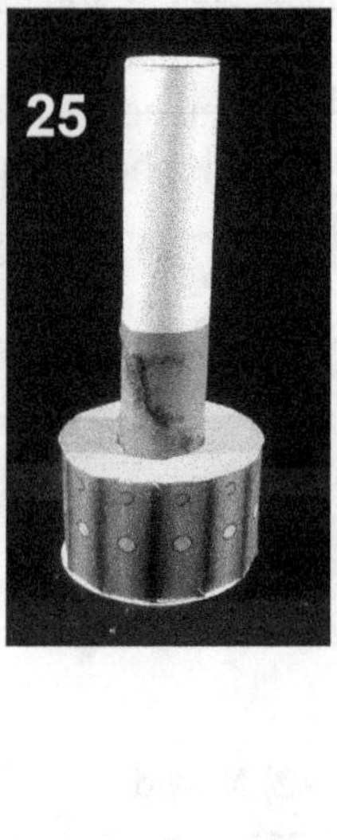

26) Cut out the part marked STATOR SUBUNIT. (In the real protein, this is the stabilizer that keeps the bottom sphere in place, and connected to the bilayer.) Score on the fold lines that run down the length of the stator and include the white tabs. Fold, then glue the white tabs to the curved edge.

27) Cut and fold the tabs as shown in photo 27.

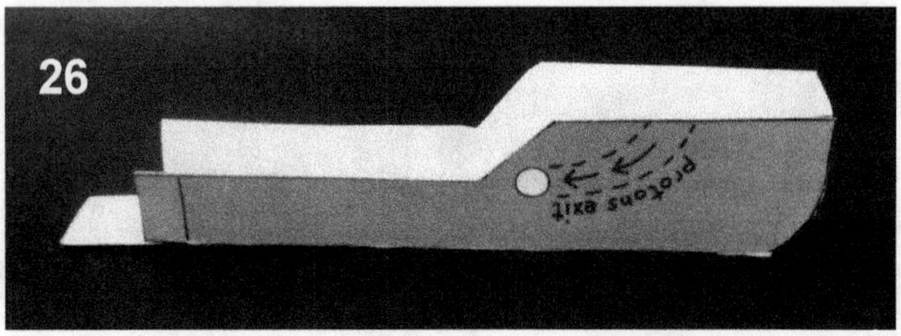

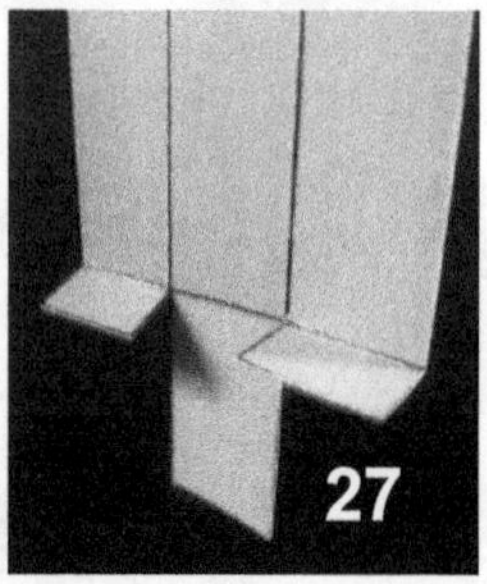

28) Cut out the piece labeled "BOTTOM OF STATOR SUBUNIT."

29) Score along white tabs, then fold as shown in photo 29. **30)** Use glue tabs to make shape shown 30.

31) Use purple tabs on stator to glue it to the purple bottom piece.

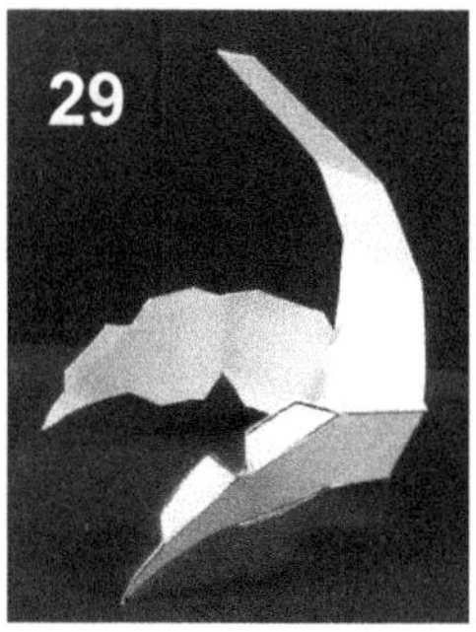

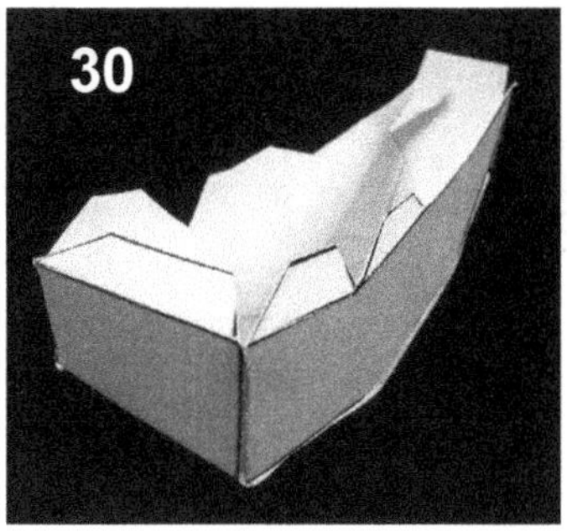

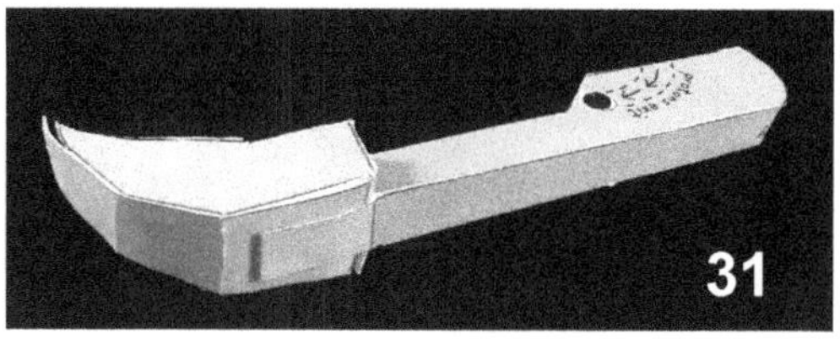

32) Glue the sphere to the floor of the model, placing the hexagonal ring on top of the hexagon on the floor, and making sure to get the alpha and beta wedges placed correctly. Press the tabs firmly. Let dry a minute or two.

33) Cut the ATP and ADP strips while the stator piece is drying.

34) Curl the ADP+P strip and then put it through the space between an alpha and beta subunit, and through the tiny rectangle, as shown. Use glue tab to make strip into continuous circle.

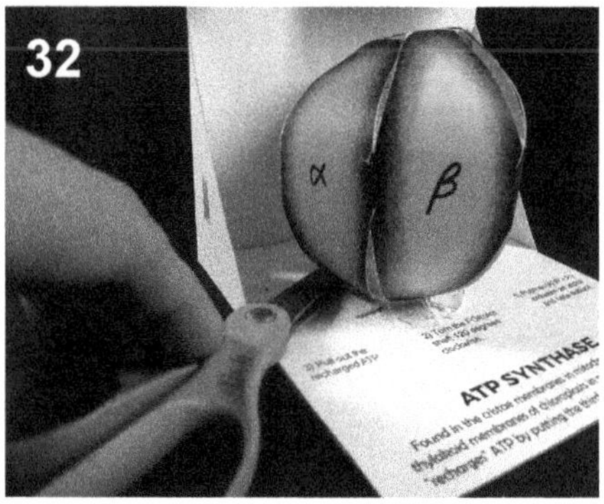

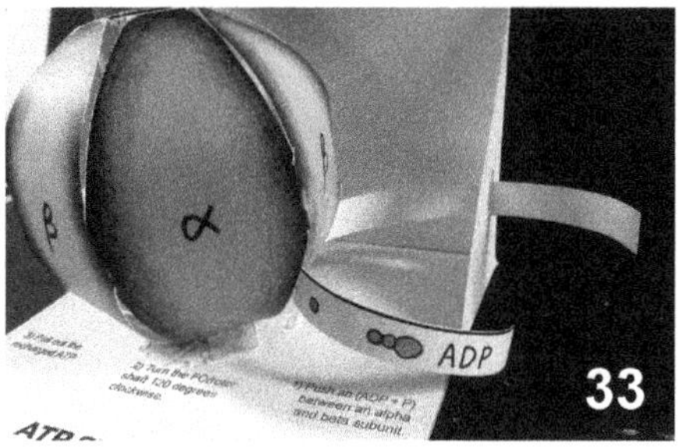

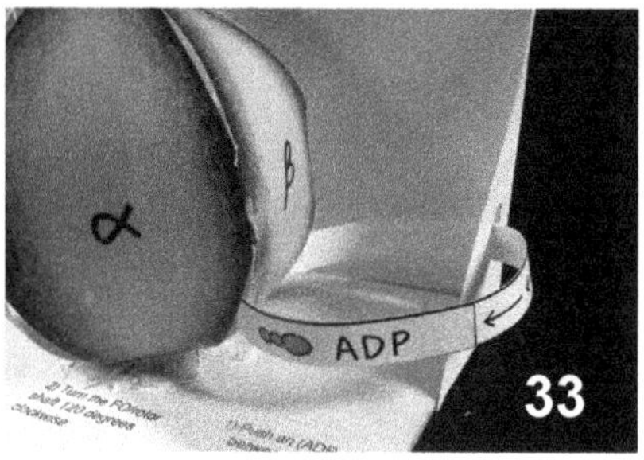

35) Put the ATP strip behind the alpha unit on the other side and glue the tab.

36) Put the rotor shaft down through the center of the sphere and then snug it down into the hexagonal circle at the bottom. Do NOT glue in place! The rotor should be able to turn freely.

37) Glue the (purple) bottom of the stator unit to the side of the sphere. It will attach to a beta unit. Use plenty of glue in this case and press hold it for at least a count to 20. If it isn't sticking, hold it a bit longer.

38) Secure the top of the stator to the bilayer behind it. Just a little drop of glue will hold it. Do NOT glue the stator to the cylinder! The rotor cylinder should turn freely.

39) To "operate" the model: 1) push in the ADP strip, 2) turn the rotor, then 3) pull out the ATP strip.

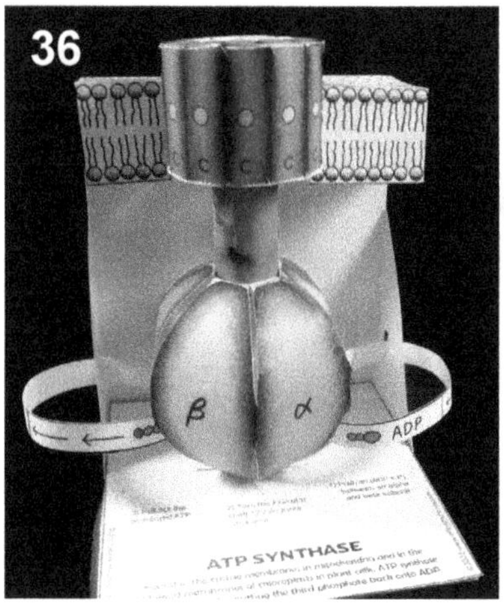

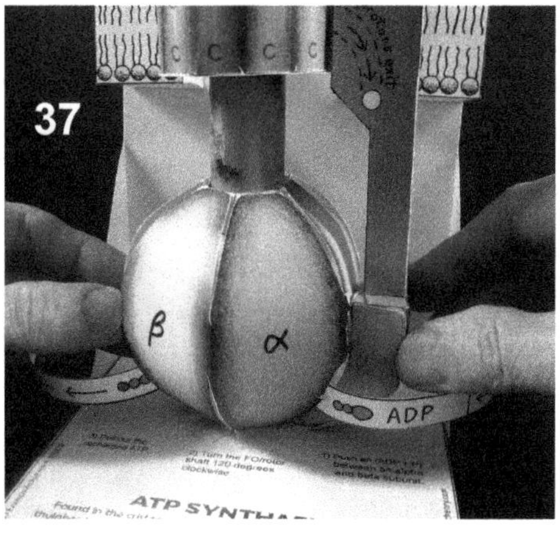

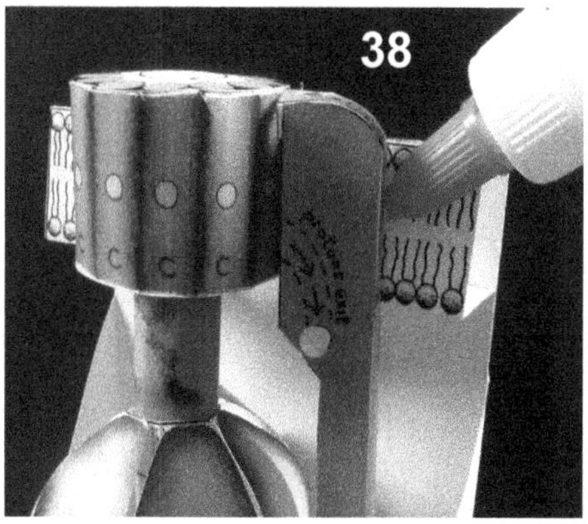

COPY ONTO CARD STOCK

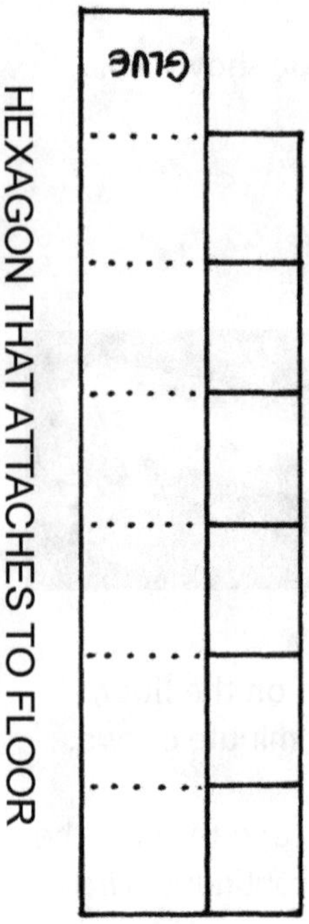

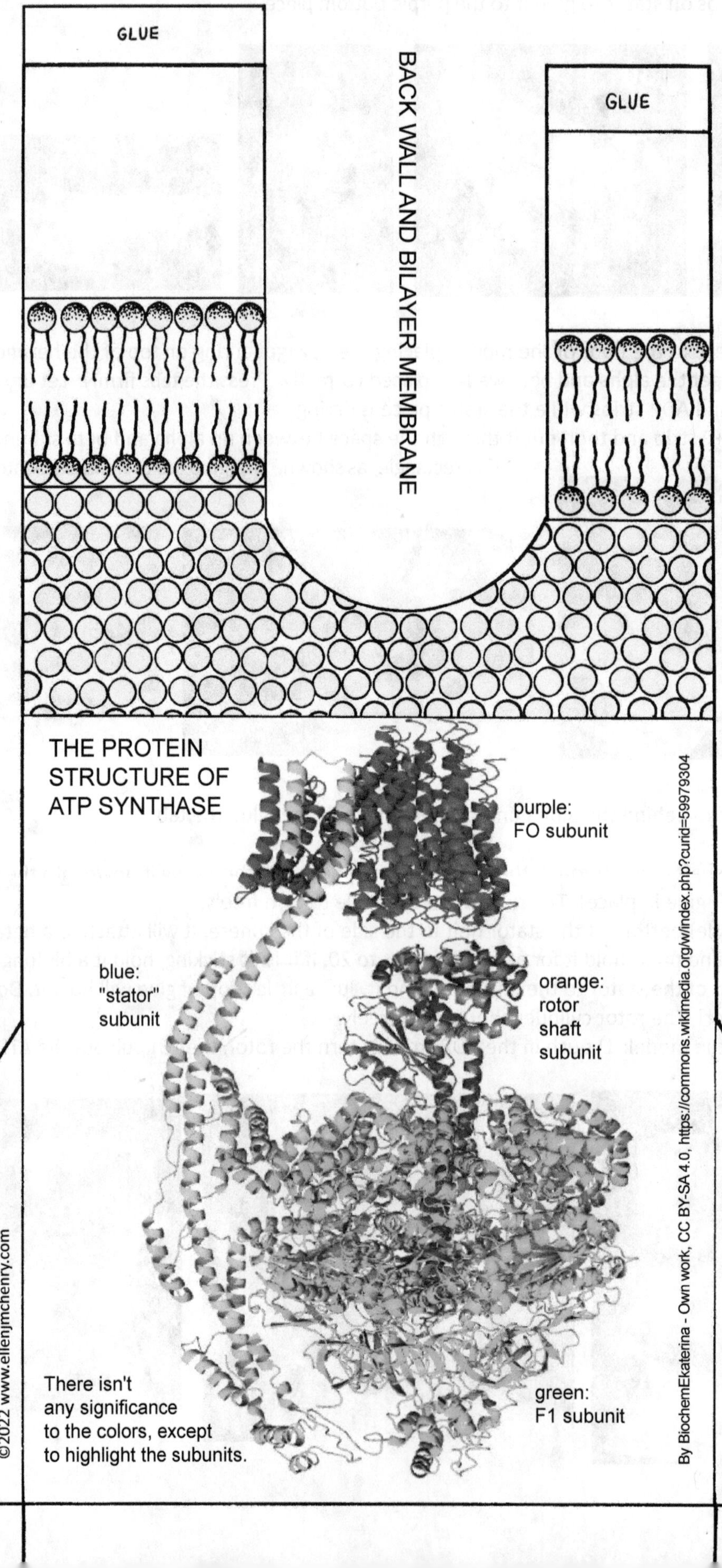

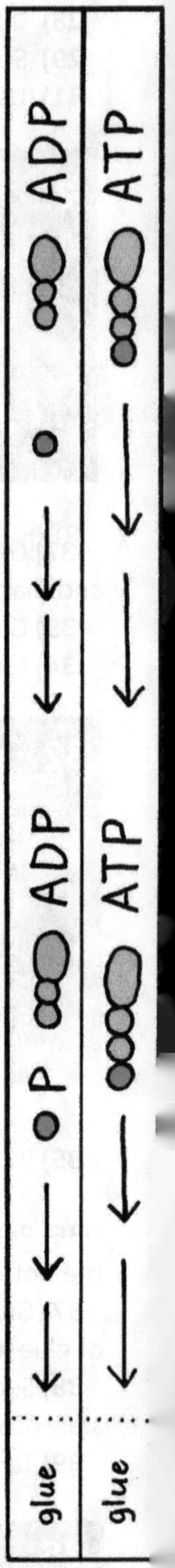

You may want carefully trim off the black lines around the edges of the tabs because they will visible on the floor of the model.

COPY ONTO CARD STOCK

F1 SPHERICAL SUBUNIT

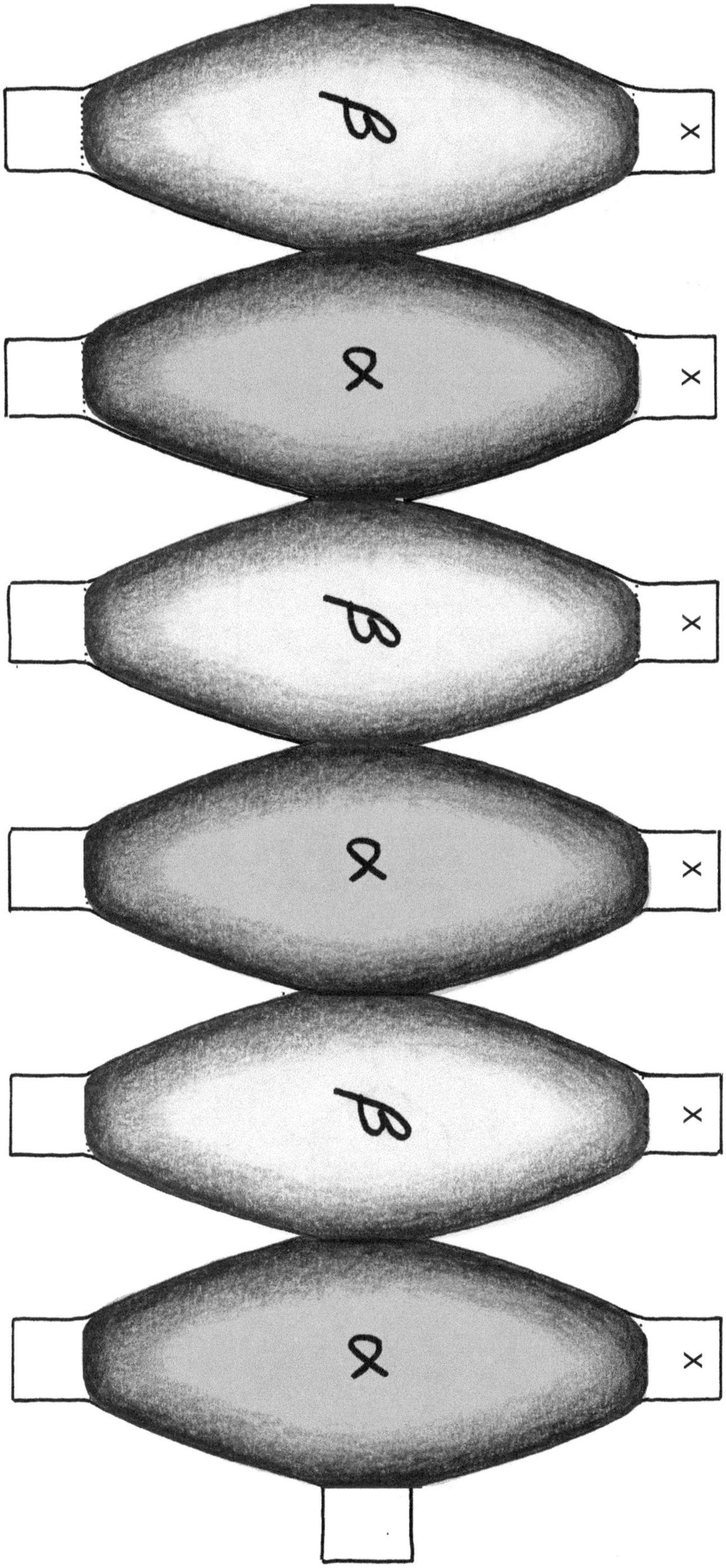

CYLINDER PORTION OF F0 SUBUNIT

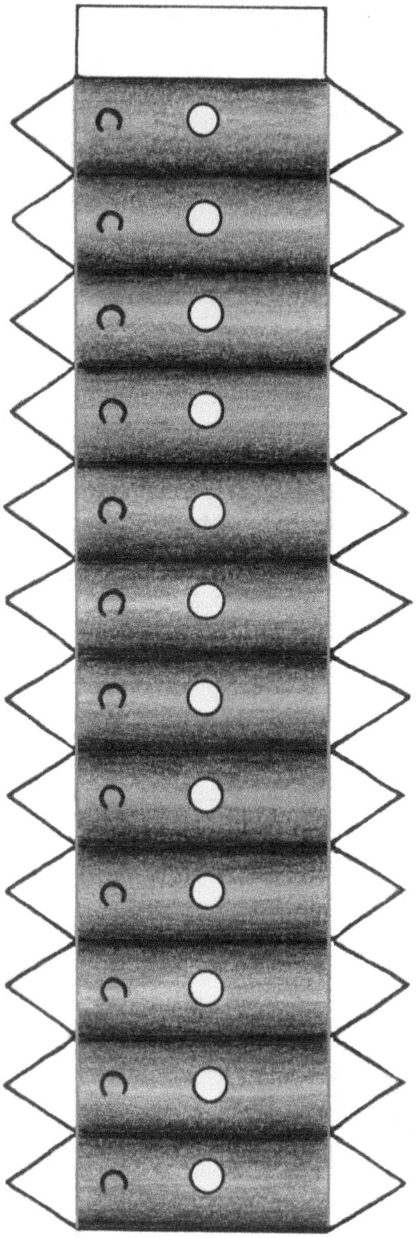

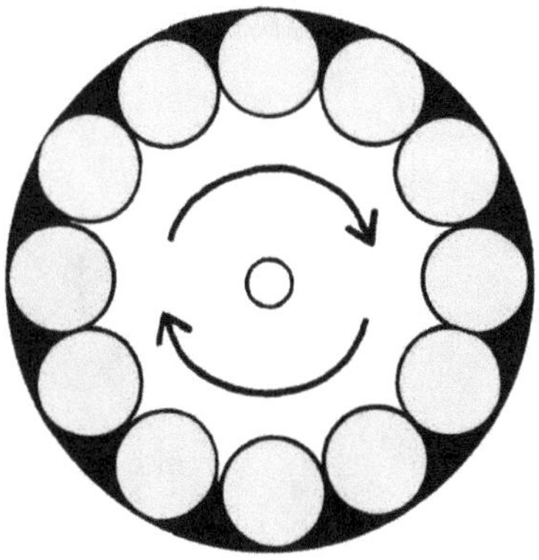

TOP OF F0 SUBUNIT

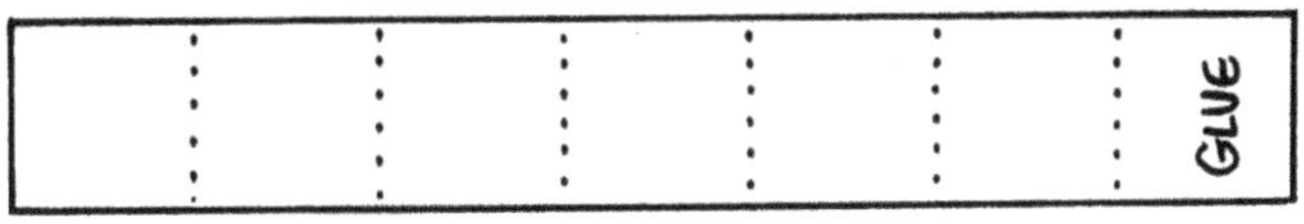

HEXAGONAL RING FOR TOP OF F1 SUBUNIT SPHERE

STATOR SUBUNIT

COPY ONTO CARD STOC

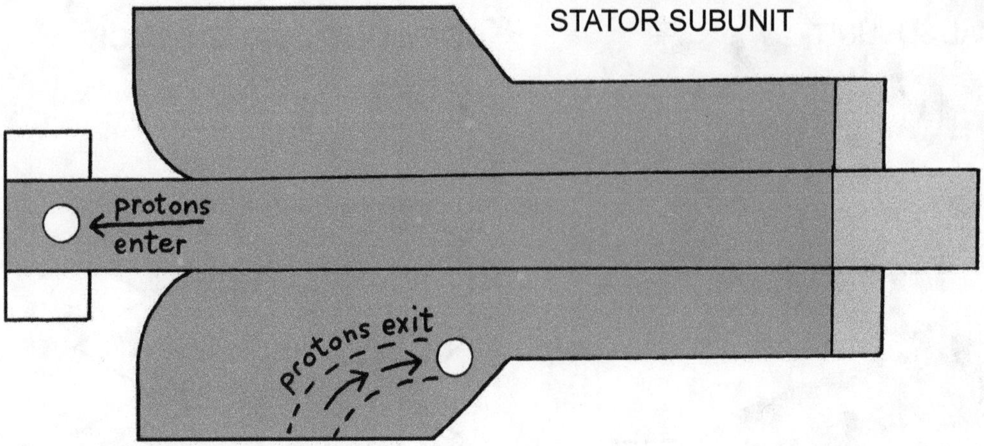

BOTTOM OF
F0 CYLINDER

CUT OUT
CENTER

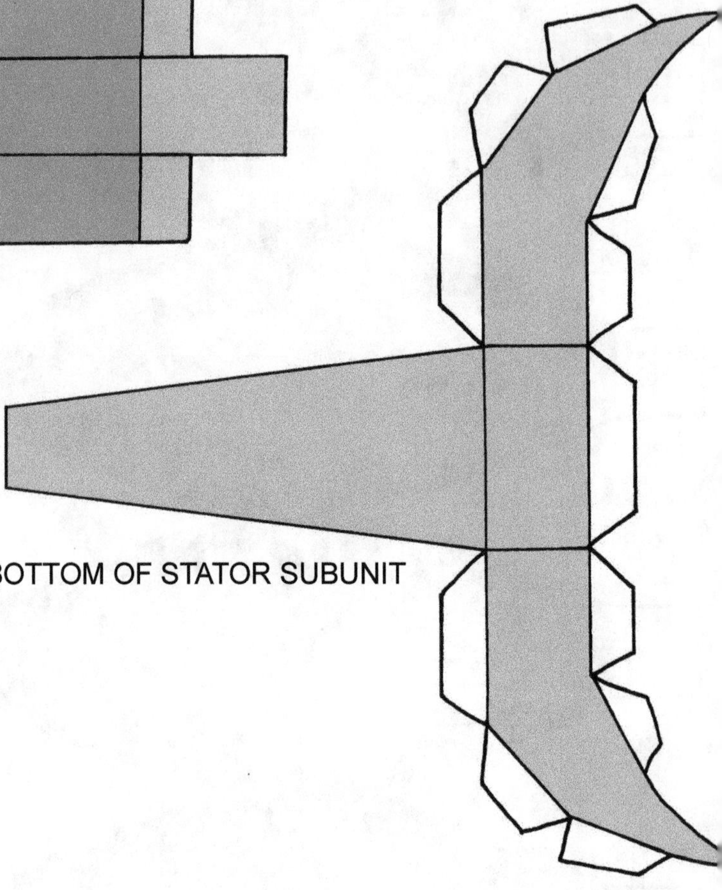

BOTTOM OF STATOR SUBUNIT

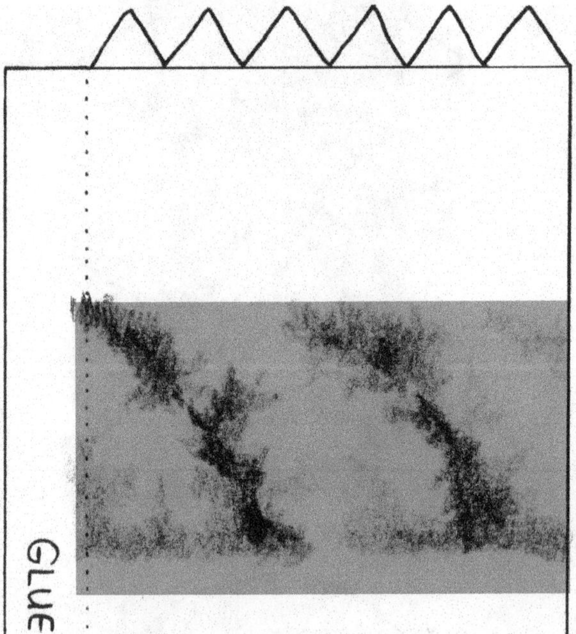

ROTOR SHAFT

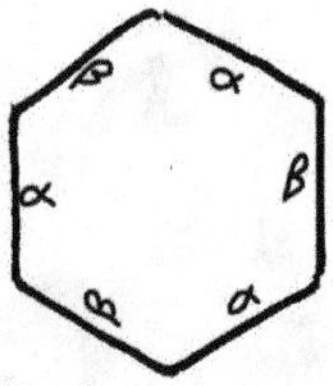

1) Push an (ADP + P) between an alpha and beta subunit.

2) Turn the FO/rotor shaft 120 degrees clockwise.

3) Pull out the recharged ATP.

ATP SYNTHASE

Found in the cristae membranes in mitochondria and in the thylakoid membranes of chloroplasts in plant cells, ATP synthase "recharges" ATP by putting the third phosphate back onto ADP.

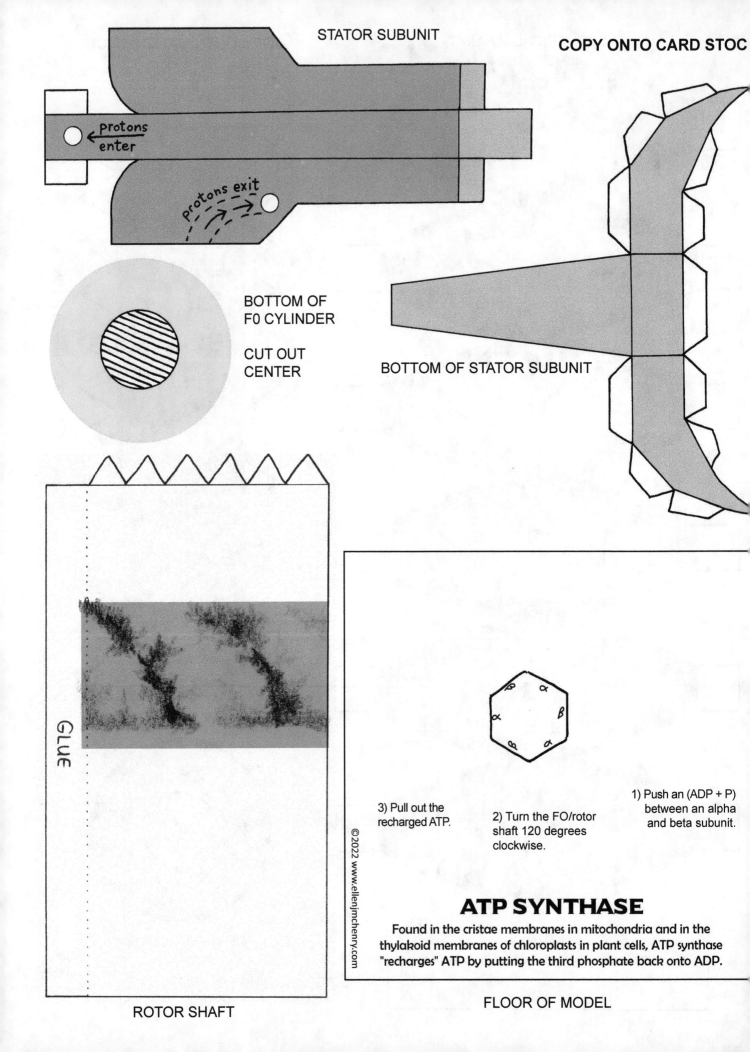

FLOOR OF MODEL

CHAPTER 5

ACTIVITY IDEA 5A: CRAFT: PROTEIN "PENCIL TOPPER"

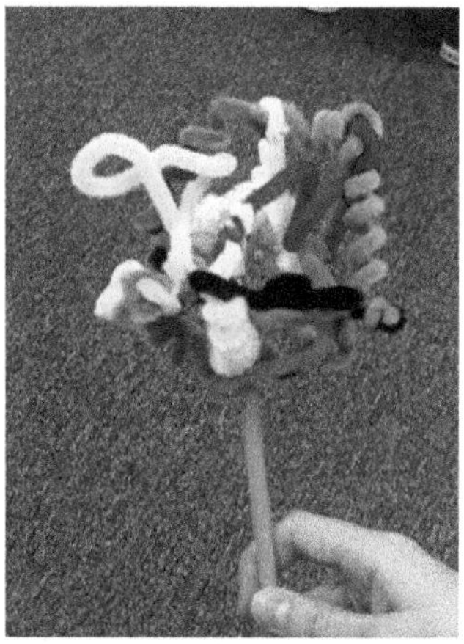

The students will make a "pencil topper" decoration that resembles a computer-generated virtual protein model. This project is a guaranteed success even with students who aren't especially "crafty."

NOTE: The students do not need to try to replicate an actual protein. There are so many different kinds of proteins in the world that whatever they make is likely to have some resemblance to some protein somewhere. The main point of this activity is to become familiar with what virtual protein models look like, be able to recognize the alpha helix and beta sheet structures, and to understand the importance of protein folding in determining what the protein will do in a cell or in a body. You may also want to reinforce the concept that a protein folds into a particular shape because of the interactions of the amino acids it is made of.

What to tell the students:

You will be making a model of a protein using flexible craft materials that you can bend and fold into helix or beta sheet formations. Your protein doesn't have to be patterned after a real protein, though it will be helpful to look at some images of proteins while you are working. There are so many proteins in the world, that whatever you make will probably bear some resemblance to a real protein somewhere in a plant, animal, fungus or bacteria. Your finished model will go on top of a pencil so it will have a dual purpose as both a scientific model and also something you can write with. (You have a motor protein pen, so now you need a protein pencil to go with it!)

You will need:

- a pencil (new and unsharpened is best)
- a pencil-top eraser (shown here on right)

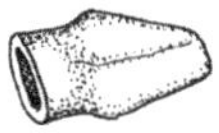

- about a dozen (per student) "chenille stems" of various colors
- optional: medium-size craft beads (with a diameter that will allow them to slide onto a chenille stem)
- access to some pictures of computer-generated virtual protein models (You can use the color picture provided in the color pages supplement at www.ellenjmchenry.com, FREE DOWNLOADS, HUMAN BODY, "Color pictures for "Cells" curriculum," but you can also find your own pictures via the Internet. A good source of these virtual models is the web site listed in chapter four: **http://www.thesgc.org/structures/**)

How to assemble:

1) The students may make any number of alpha helices or beta sheets (also known as ribbons). To make a helix, simply wind the chenille stem around the pencil. To make a beta sheet (a ribbon) fold the chenille stem back and forth in a zigzag, pushing the zigzag together tightly so that there are no gaps. Students may also wish to have some chenille stems be curled into a random shape.

NOTE: Some proteins have, as part of their design, individual atoms, or groups of atoms, stuck on at various places. (Ex: One of the DNA chaperone proteins has "zinc fingers." The spherical subunit of the ATP synthase protein has a magnesium atom at a critical point.) You can represent a metal atom using a bead that you slide onto a chenille stem. If you decide to use beads, put them on before folding the chenille stems.

2) Fasten all the stems together in a long line, making a continuous amino acid chain. Twist the individual stems together at the ends, overlapping them by about an inch. Make sure to twist tightly so they are secure.

3) Finally, bend and fold the chain into a more compact shape, like a real protein bends and folds. Choose where you want to connect your chain to the pencil and add more stem to this end. Curl this end stem around the end of the pencil and wind tightly. (We found that no tape was required. The stem held it adequately.) Slide this curled part down just a bit and put the eraser cap on the end of the pencil, so the curled stem doesn't slide off the end of the pencil. Voila—a protein pencil topper!

ACTIVITY IDEA 5B: CRAFT: BUILD A MODEL OF DNA

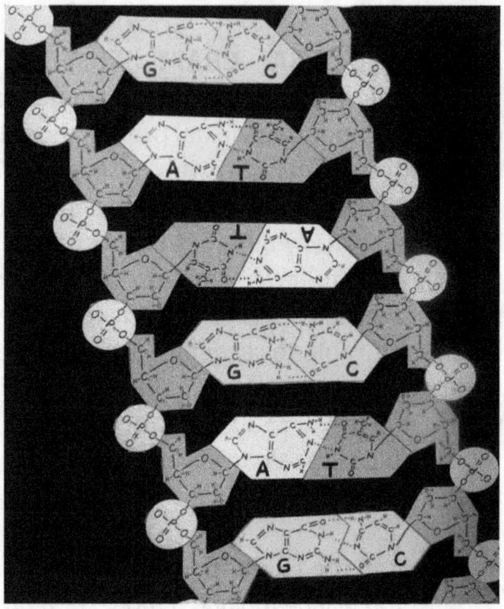

Notice that the ladder is not perfectly straight up and down. This is scientifically accurate.

This activity can be done as a cooperative project with students working to make one very long classroom model, or you can have each student make their own shorter model. If you are doing a cooperative model, you might want to assign twice as many students to cutting the phosphates and sugars compared to bases because the circles and hexagons are more time consuming and you need more of them.

For individual projects, you can choose to make the model a little smaller. Most copiers and scanners have settings where you can choose to print at greater than, or less than, 100 percent.

Another option you will need to think about is how to connect the bases. **If you want to use this model to demonstrate transcription, you will want to use paperclips to fasten C to G and A to T.** This will allow you to "unzip" your DNA and have your student use the ribose sugar (and Us instead of Ts) to make mRNA. If you would rather have a permanent display model, use glue and/or tape to secure all joints.

What to tell the students:

This paper model of DNA is more detailed than most. The problem with making a model of DNA is that it is very hard to show both the 3-dimensional shape while also showing all the atoms. Models that show both are very complicated to look at, let alone to construct! This model will favor showing all the atoms in great detail, rather than worrying about trying to make the helix shape. You are welcome to twist your model when it is done, but its main purpose will be to let you see exactly what DNA is made of and how all the atoms are connected.

You will need:

- copies of the following pattern pages (If you want to make your model smaller or larger, any copier or scanner will have settings where you choose a scale larger or smaller than 100 percent.)

I put each item on a single page so that you can copy them onto different colors. I recommend using pastel colors. Pastel paper is easy to find in any stationery department. They often sell packets containing a selection of pastel colors, you probably won't need to buy a ream of each.

- scissors
- glue stick, or white glue
- stapler or tape (instead of glue if you want to be less messy)
- lots of paperclips if you want to make the center of you DNA able to "unzip" to demonstrate transcription

How to assemble:

1) Copy each page onto a different color of paper. (If this is not possible, the model will still work.)

2) Cut and assemble many nucleotides, as shown on right. Notice that the A and G have that blank area on the end. Be careful not to cut this off. If you want to make RNA, make some U nucleotides, also, and be sure to use ribose sugar.

3) Put the nucleotides together to form DNA (or RNA). It does not matter if the letters are upside down. If you don't want to unzip your DNA, use glue or tape to join bases. If you want your model to be able to demonstrate transcription, use paper clips to join the bases.

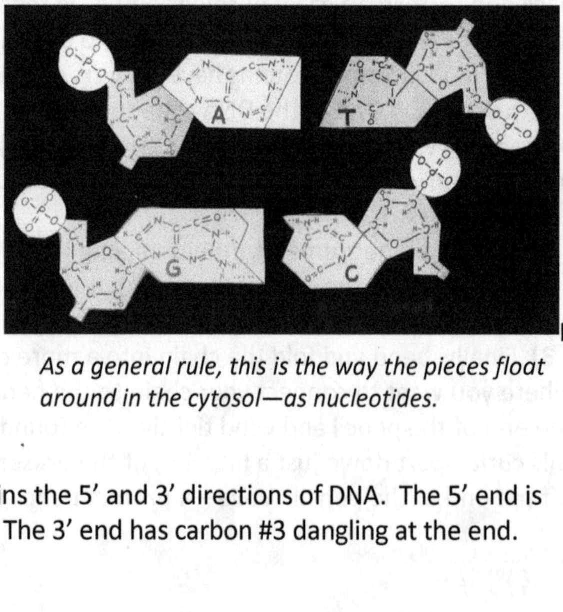

As a general rule, this is the way the pieces float around in the cytosol—as nucleotides.

5) Notice that the carbons are labeled with numbers. This explains the 5' and 3' directions of DNA. The 5' end is the end that has carbon #5 not connected to another nucleotide. The 3' end has carbon #3 dangling at the end.

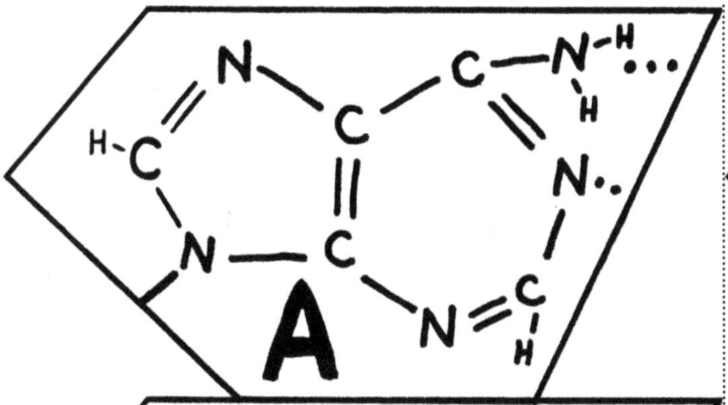

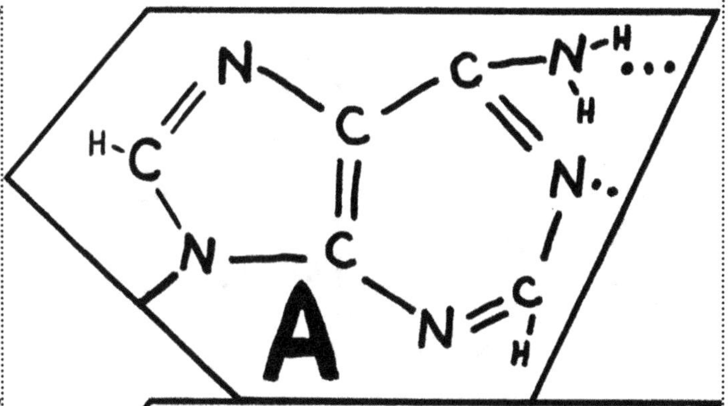

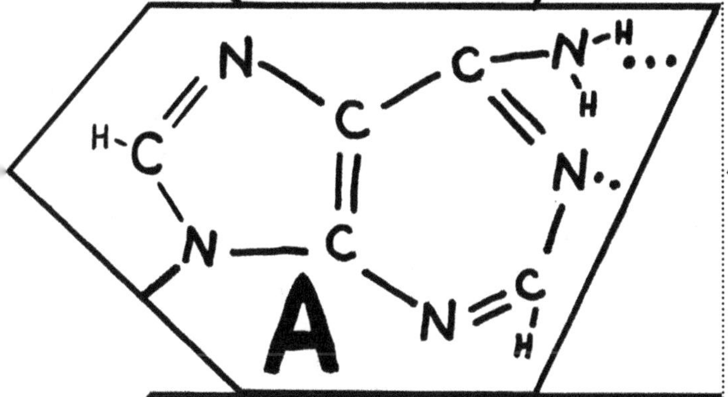

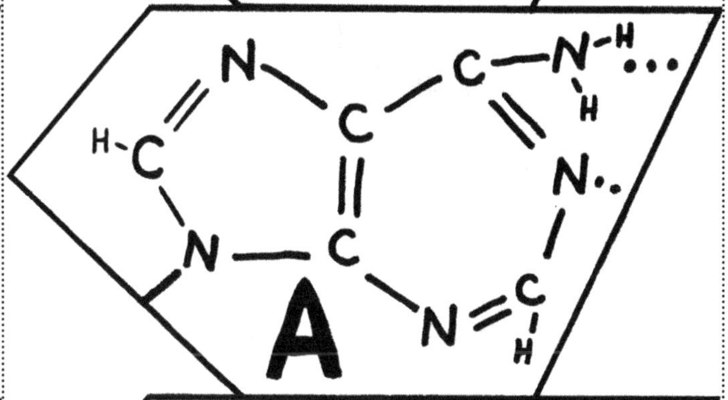

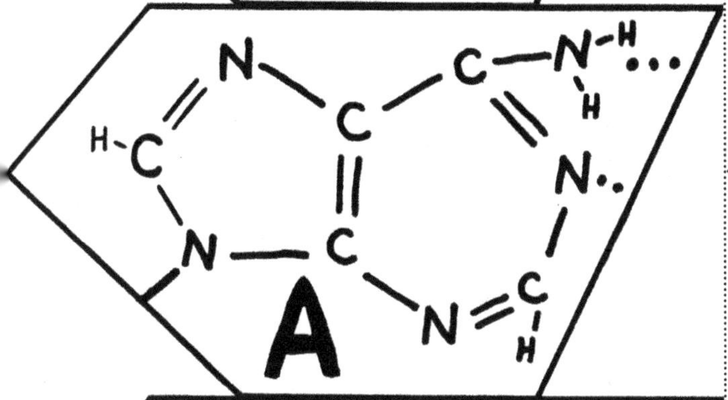

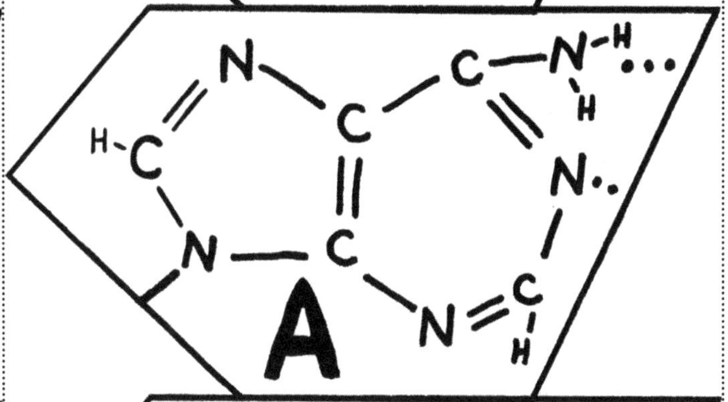

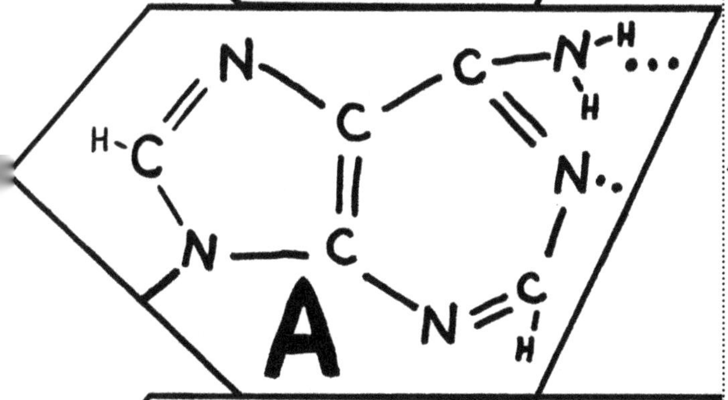

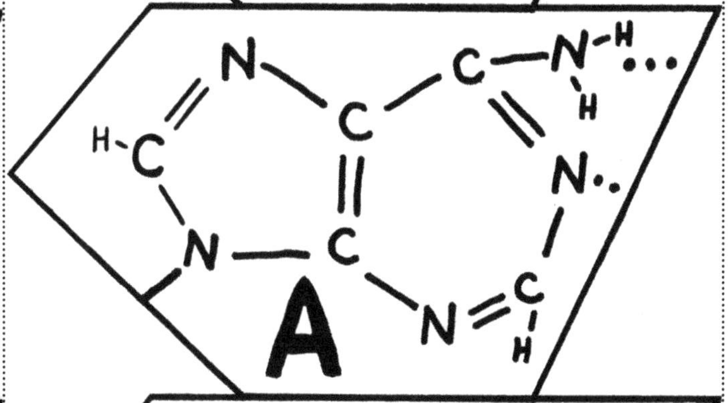

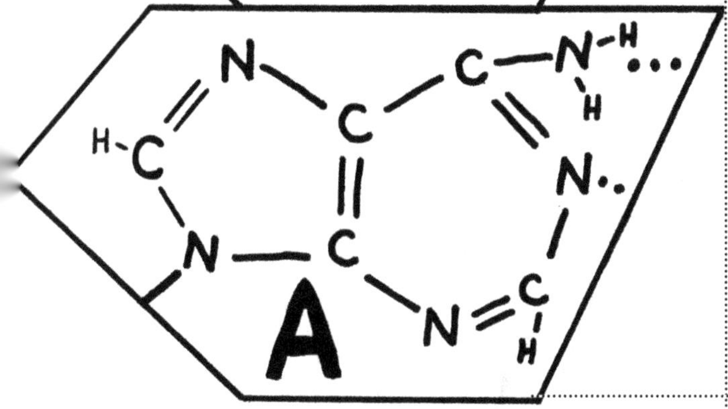

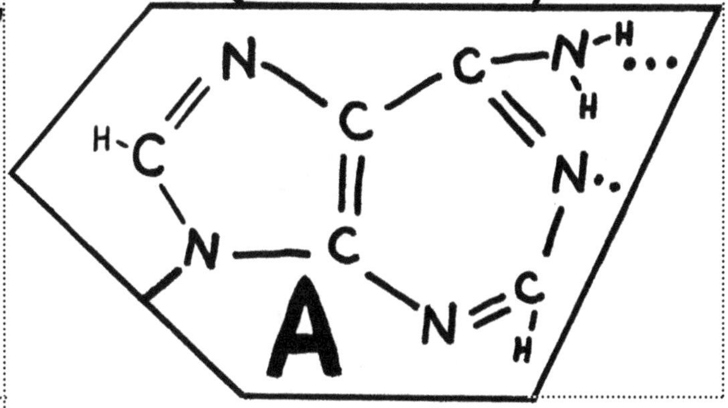

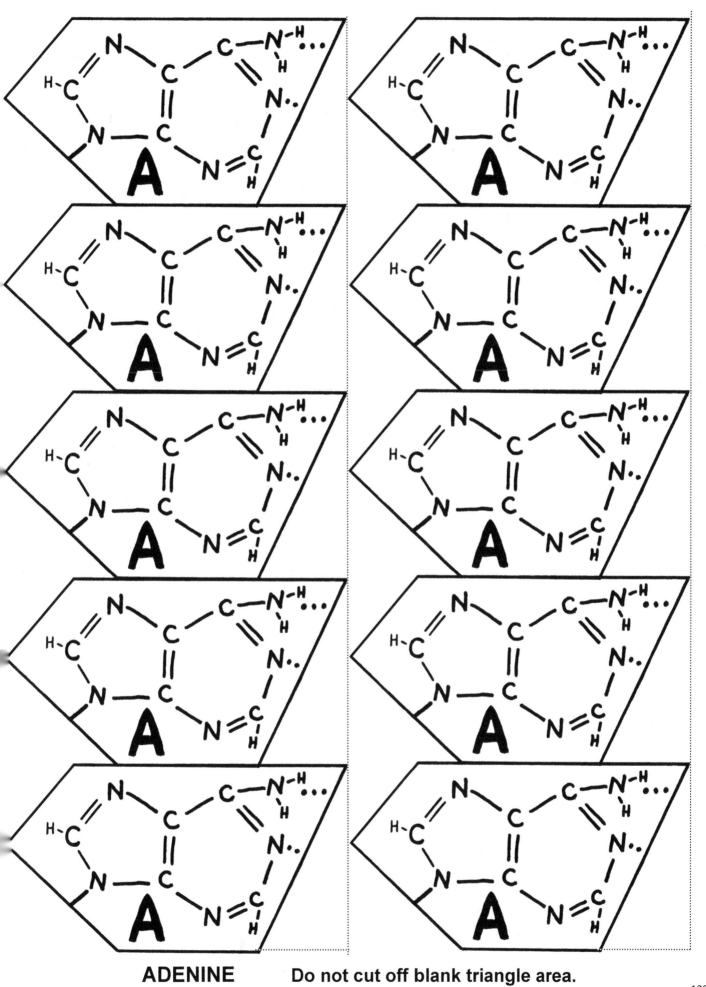

ADENINE **Do not cut off blank triangle area.**

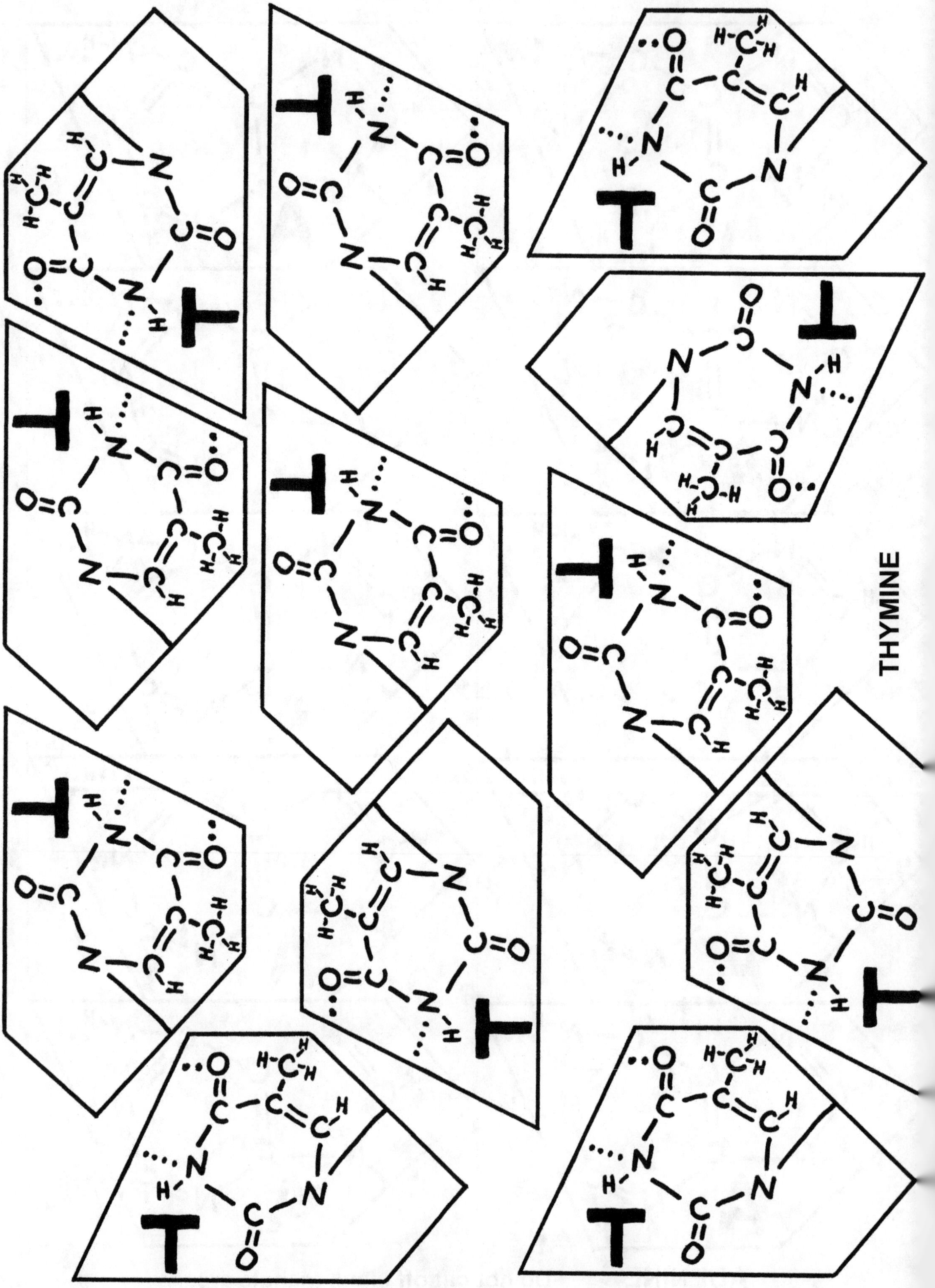

THYMINE

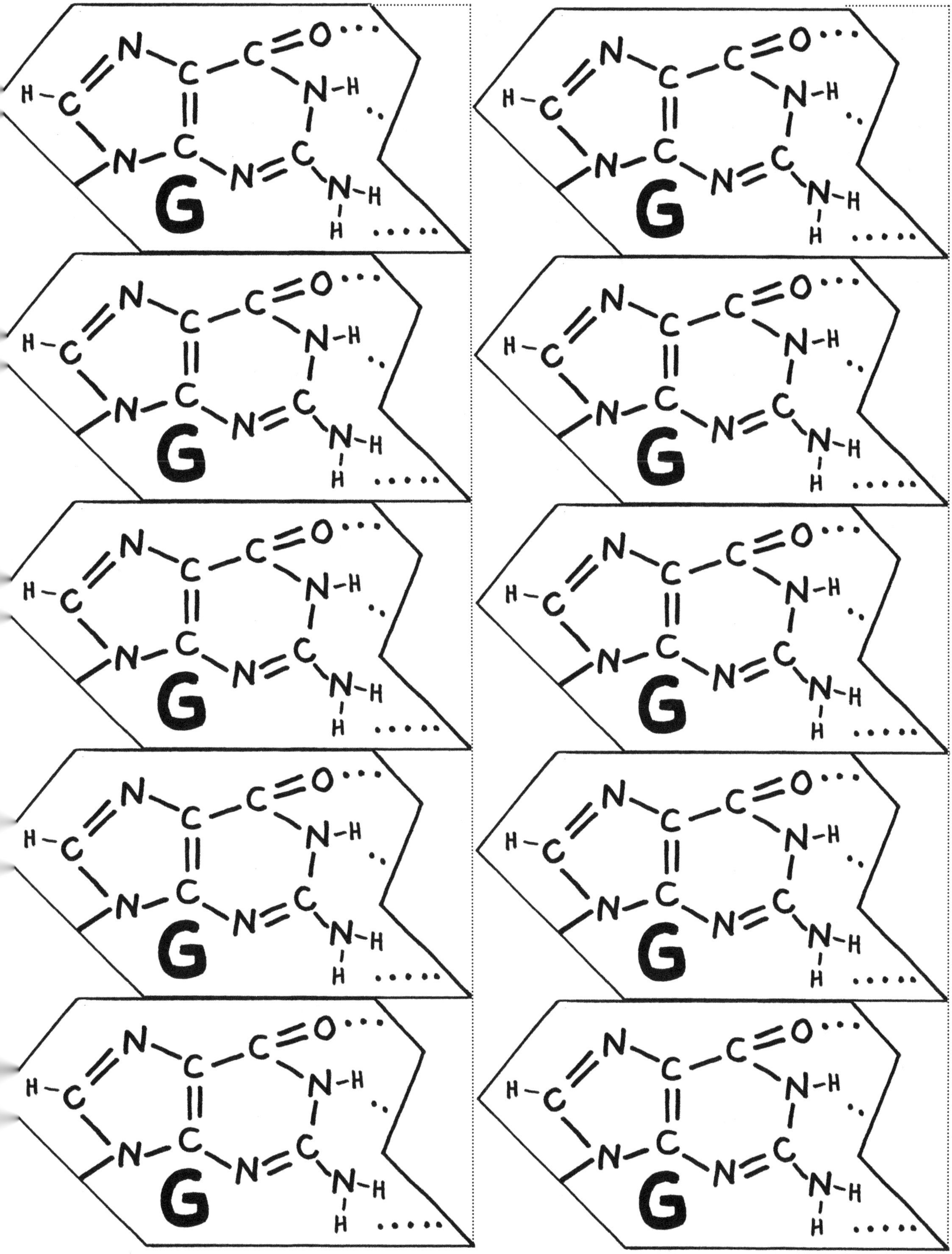

GUANINE **Include blank area inside dotted line.**

CYTOSINE

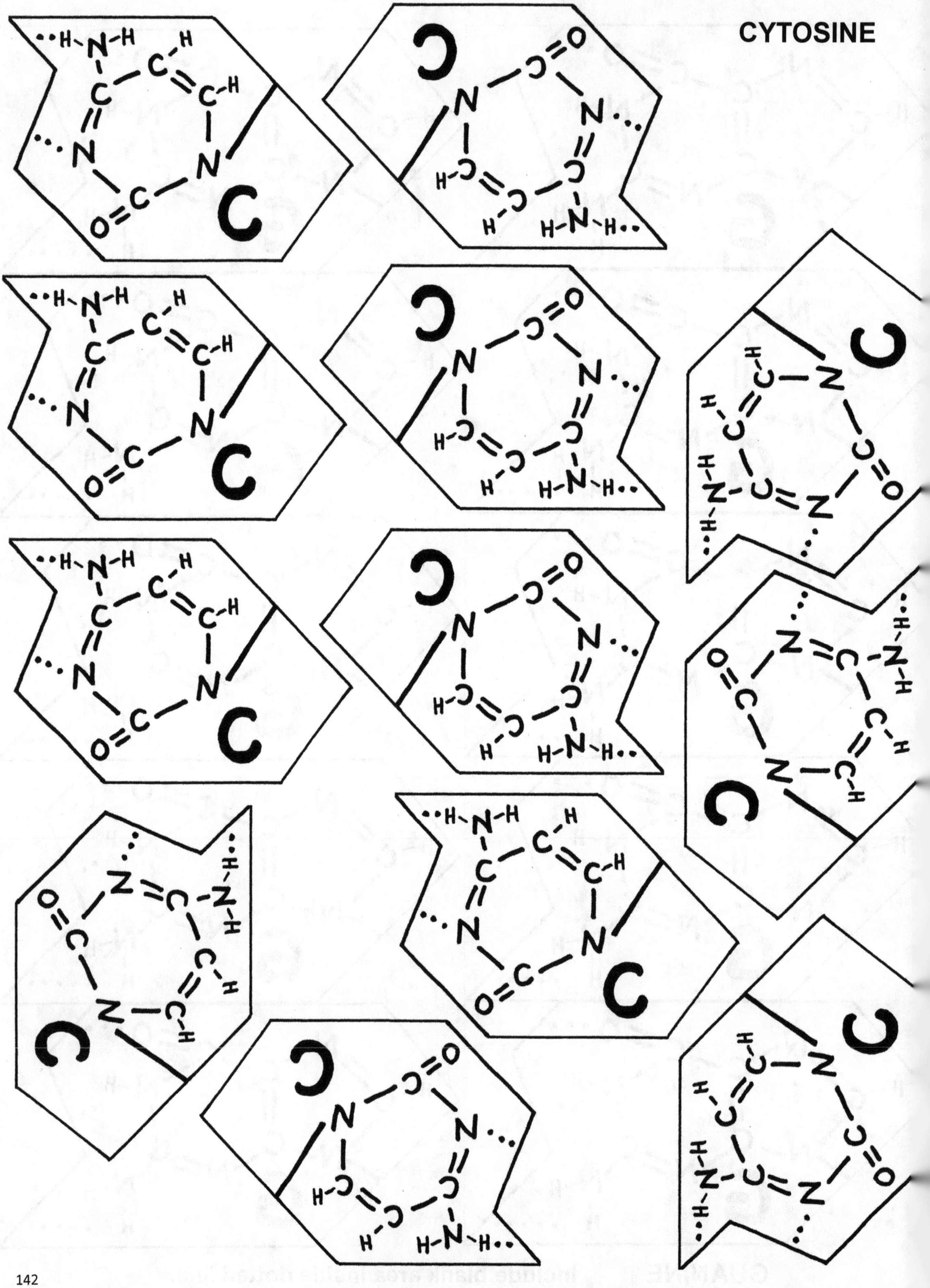

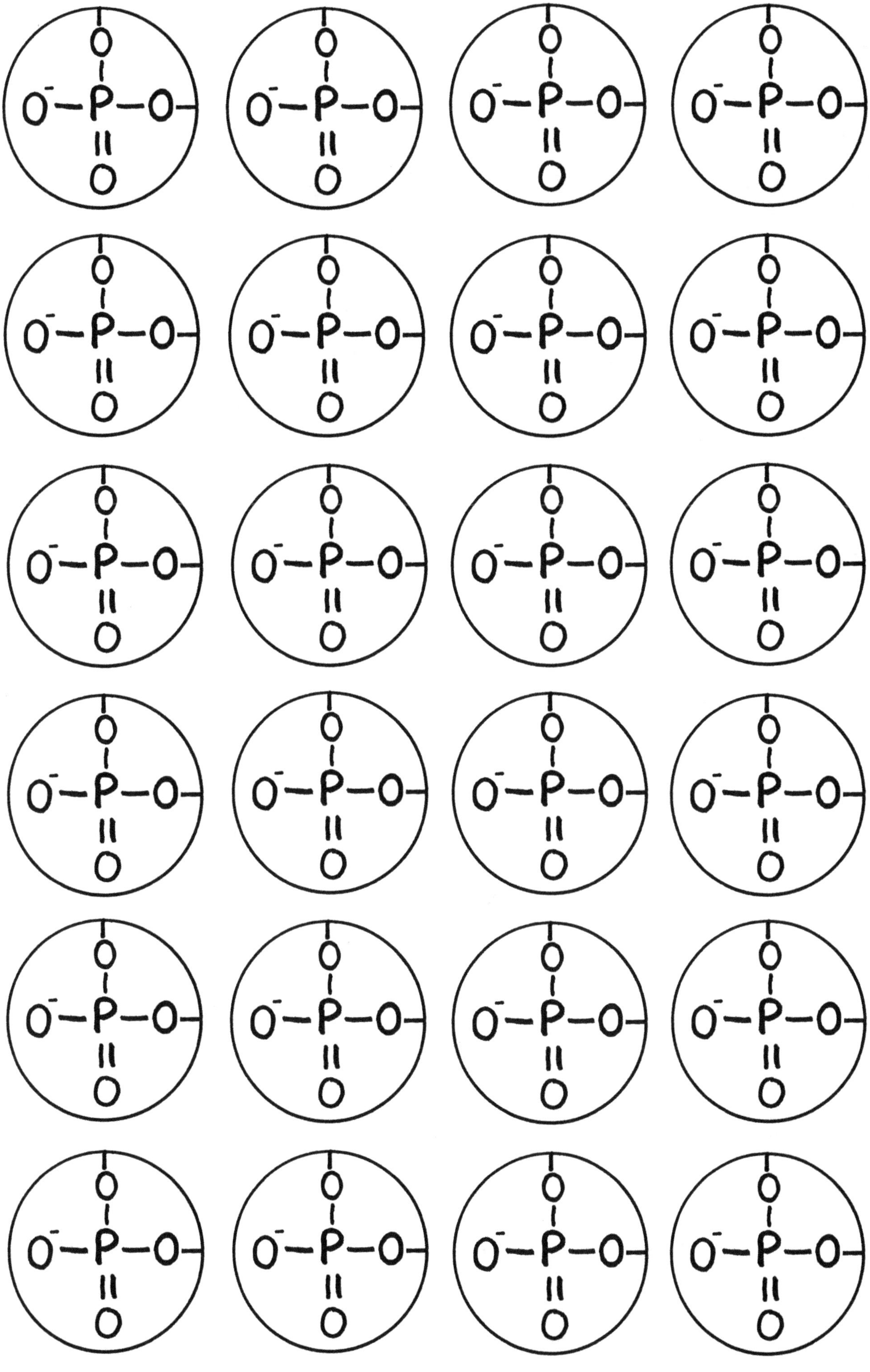

PHOSPHATES

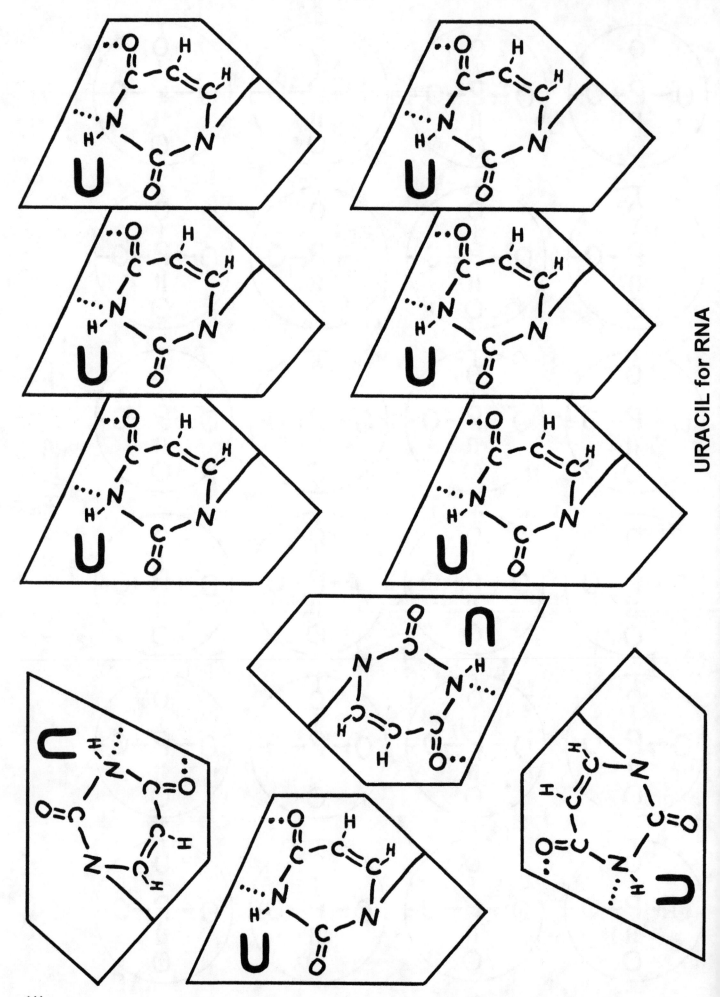

URACIL for RNA

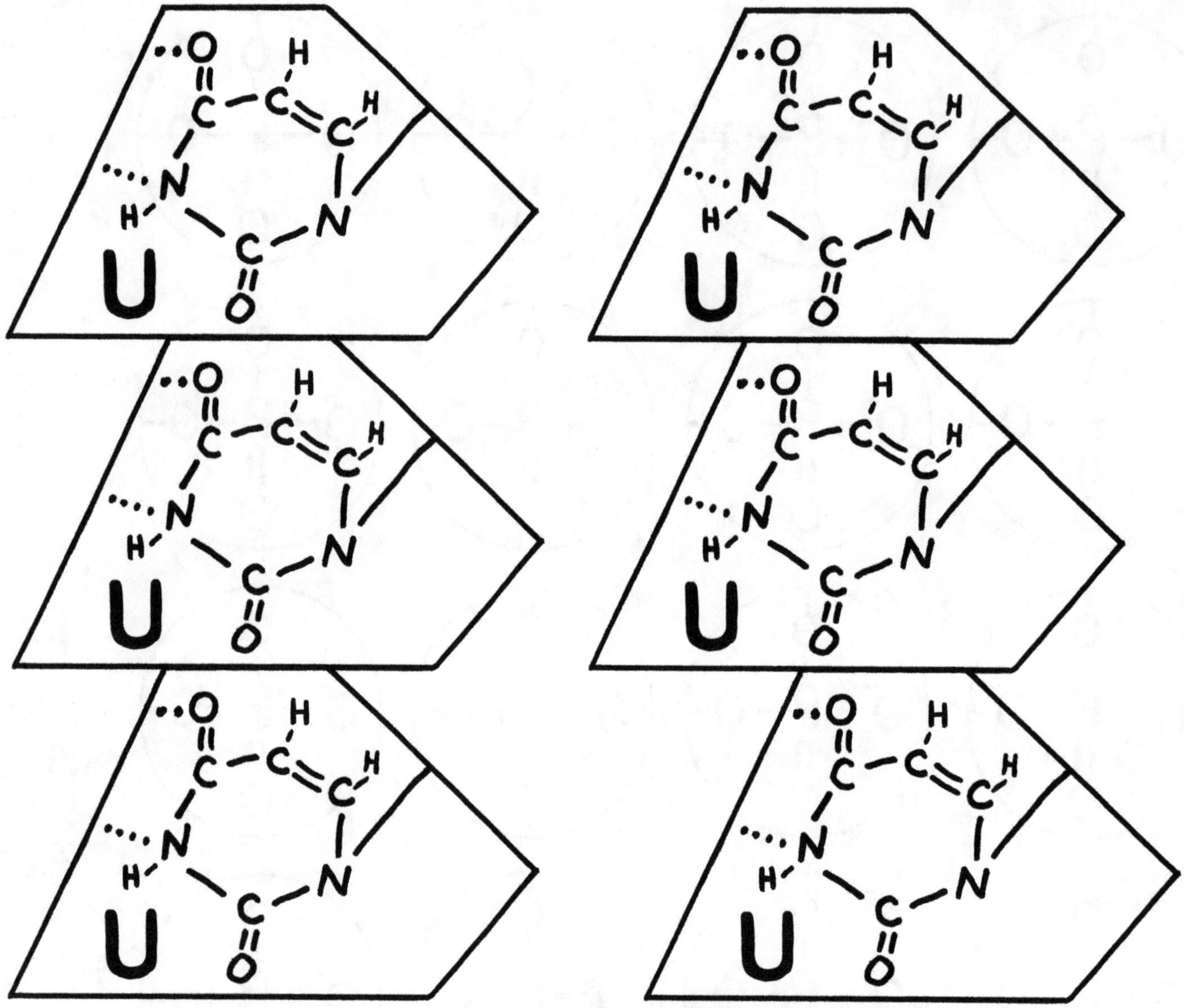

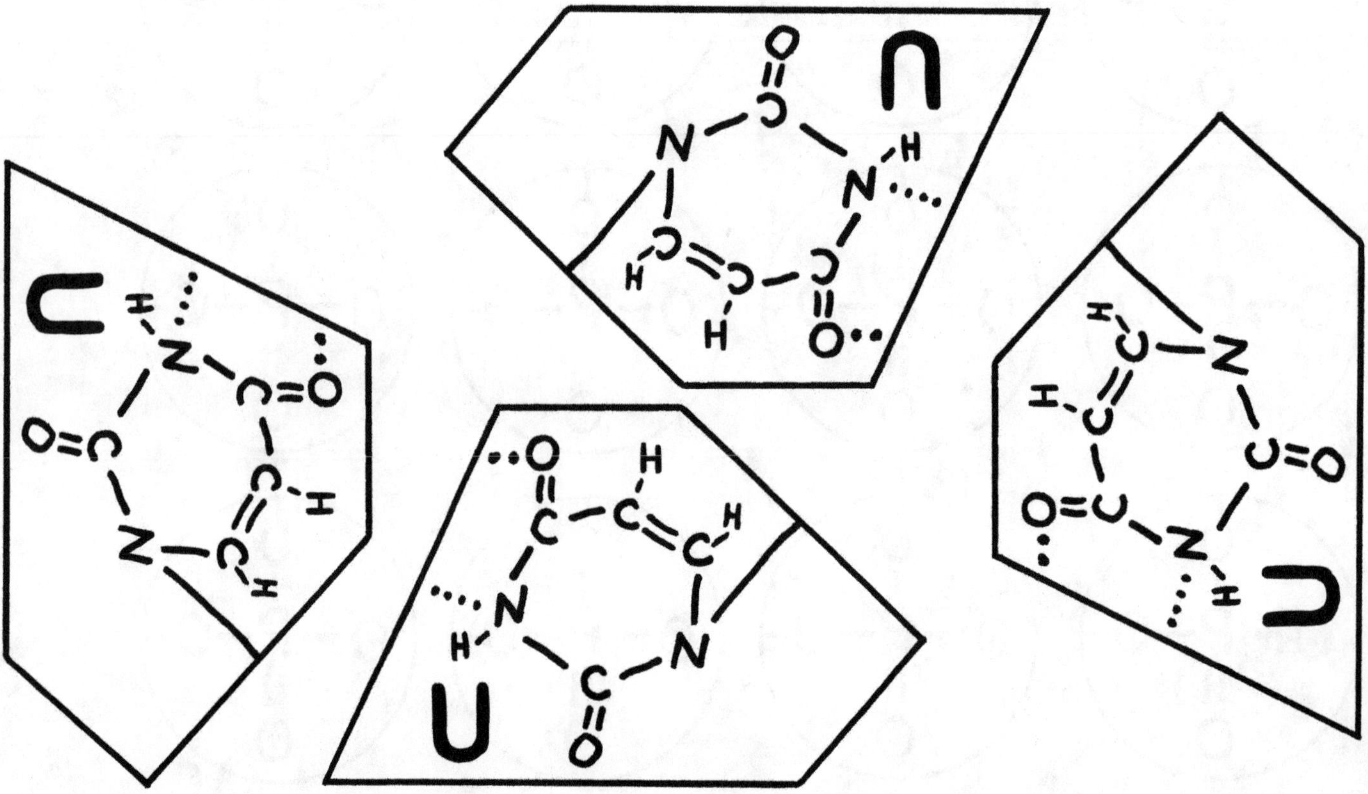

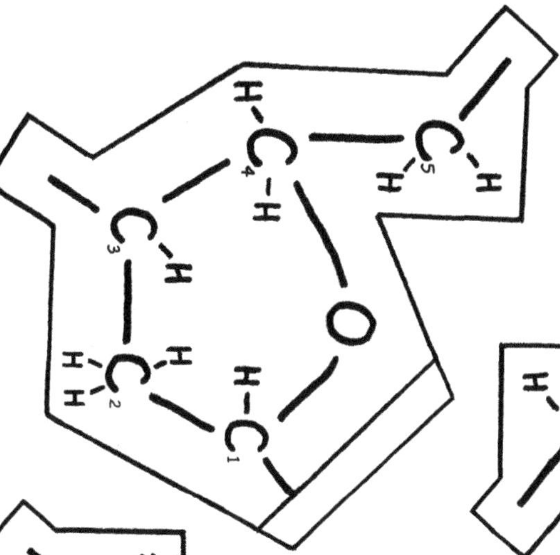
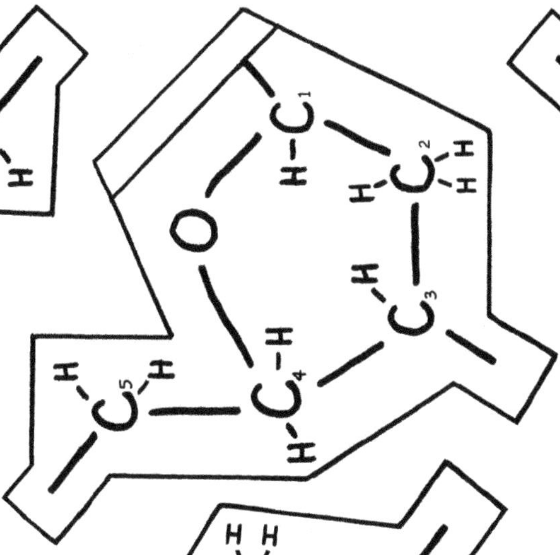
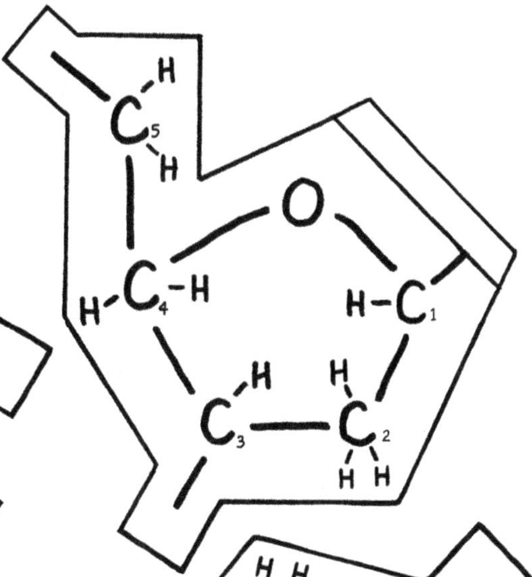
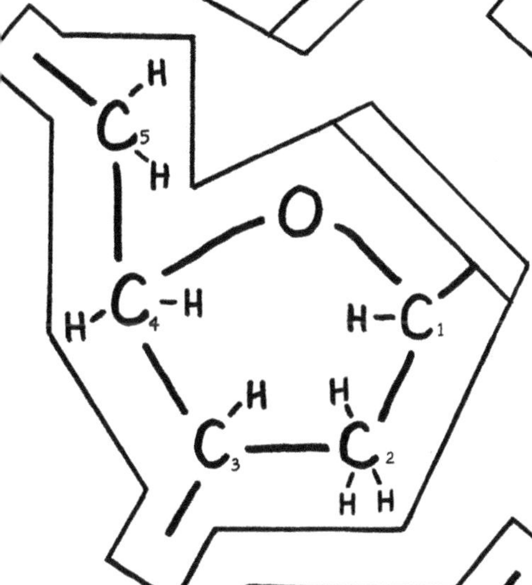
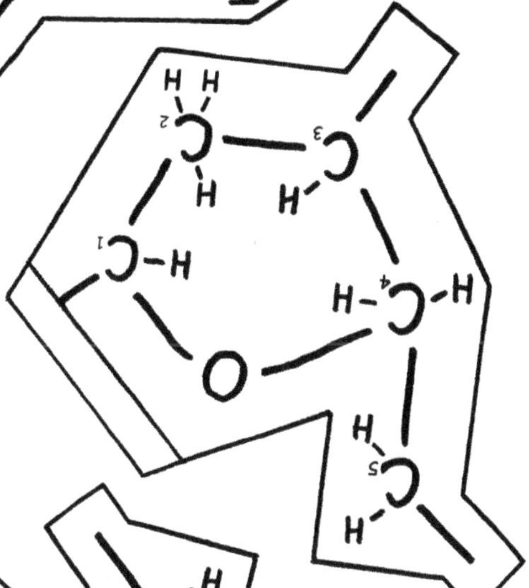
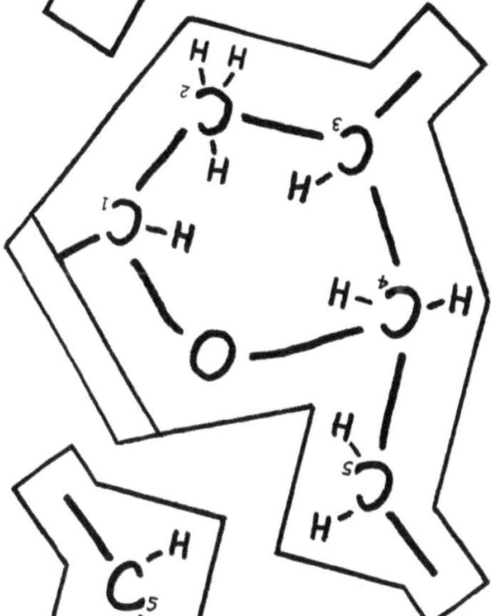
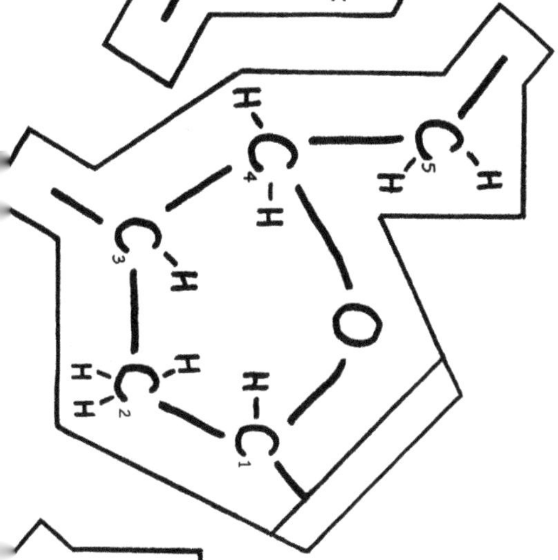
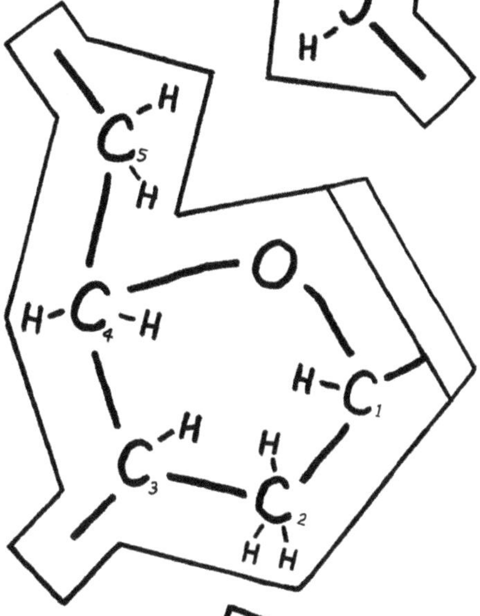
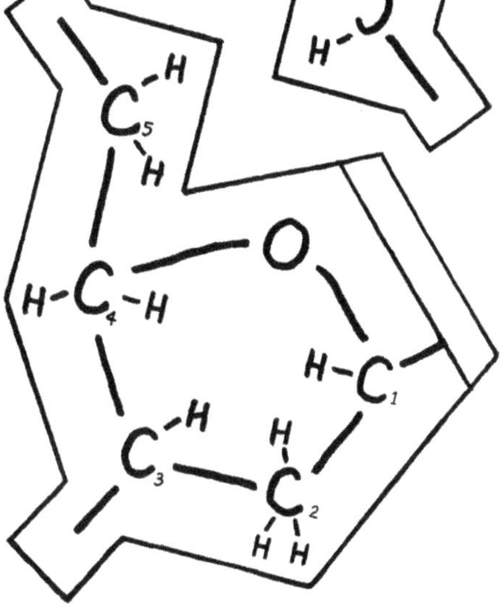
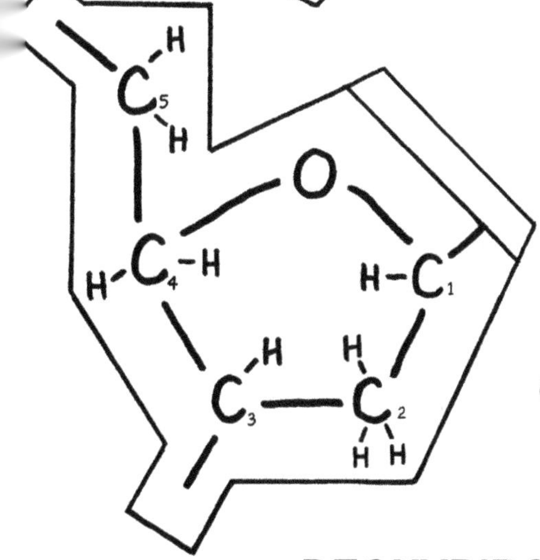
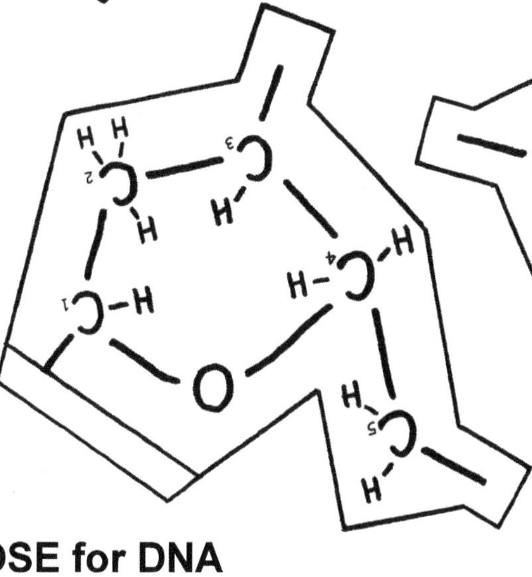
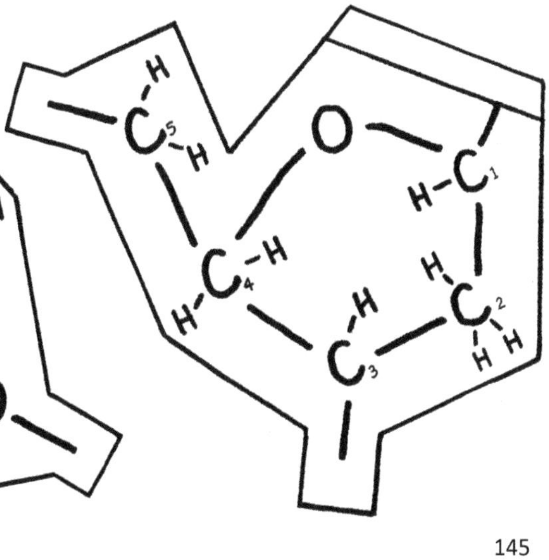

DEOXYRIBOSE for DNA

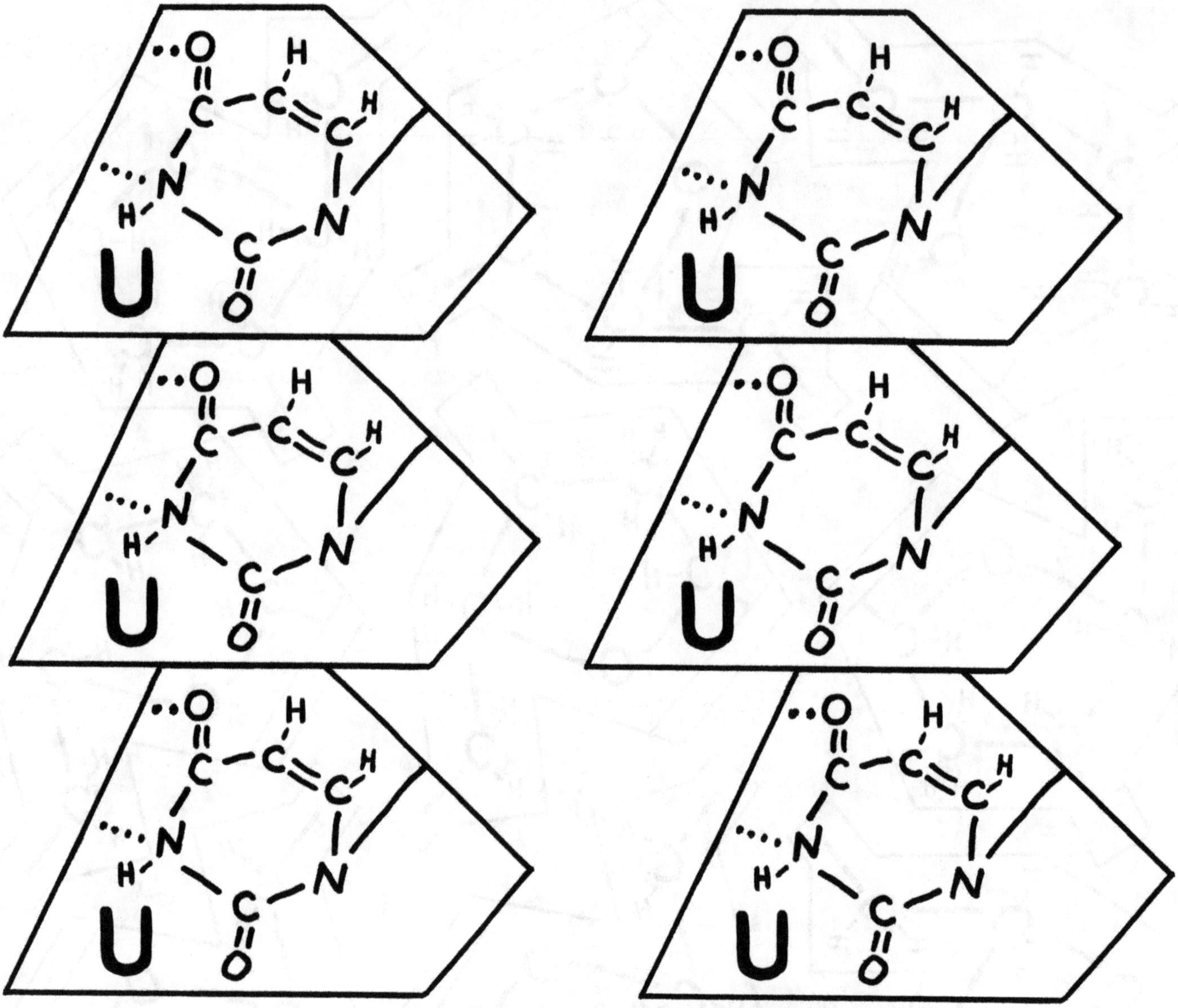

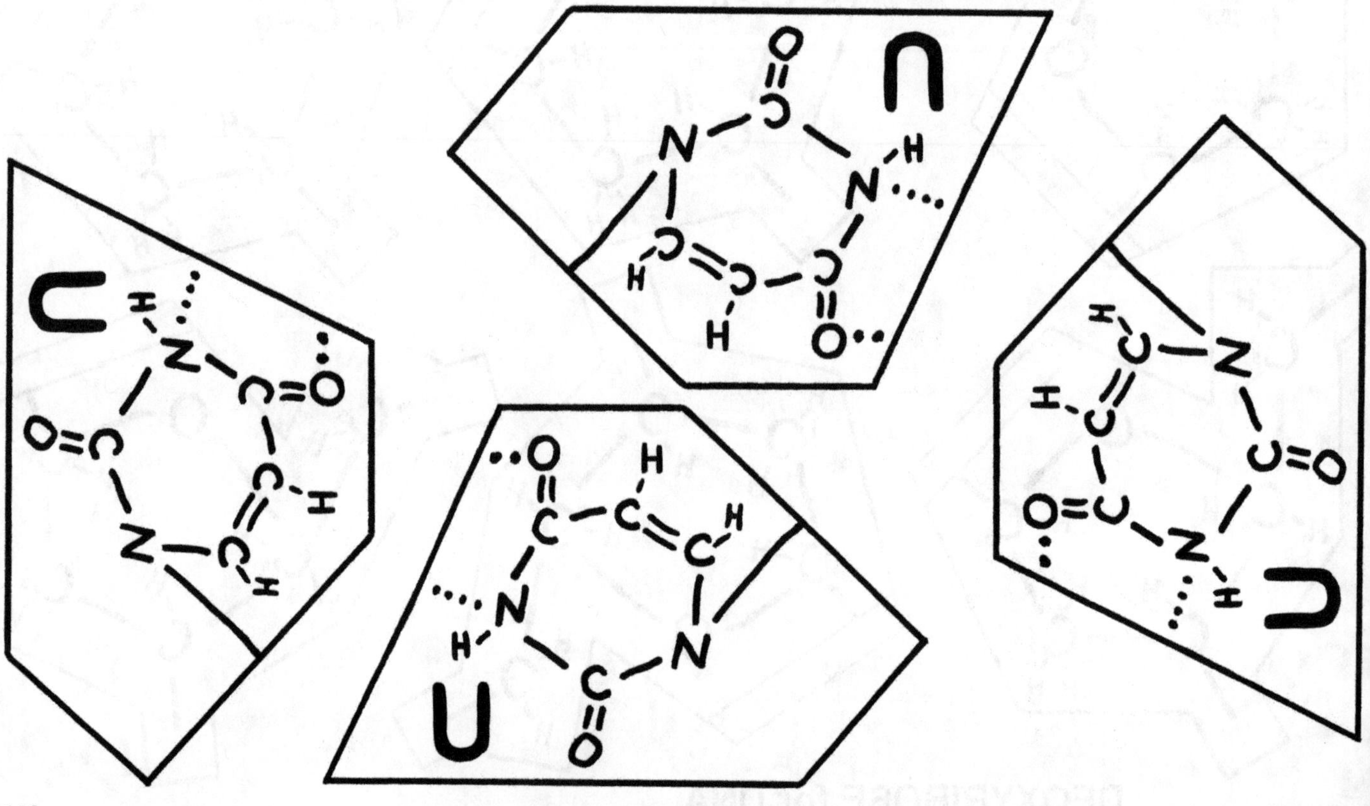

URACIL for RNA

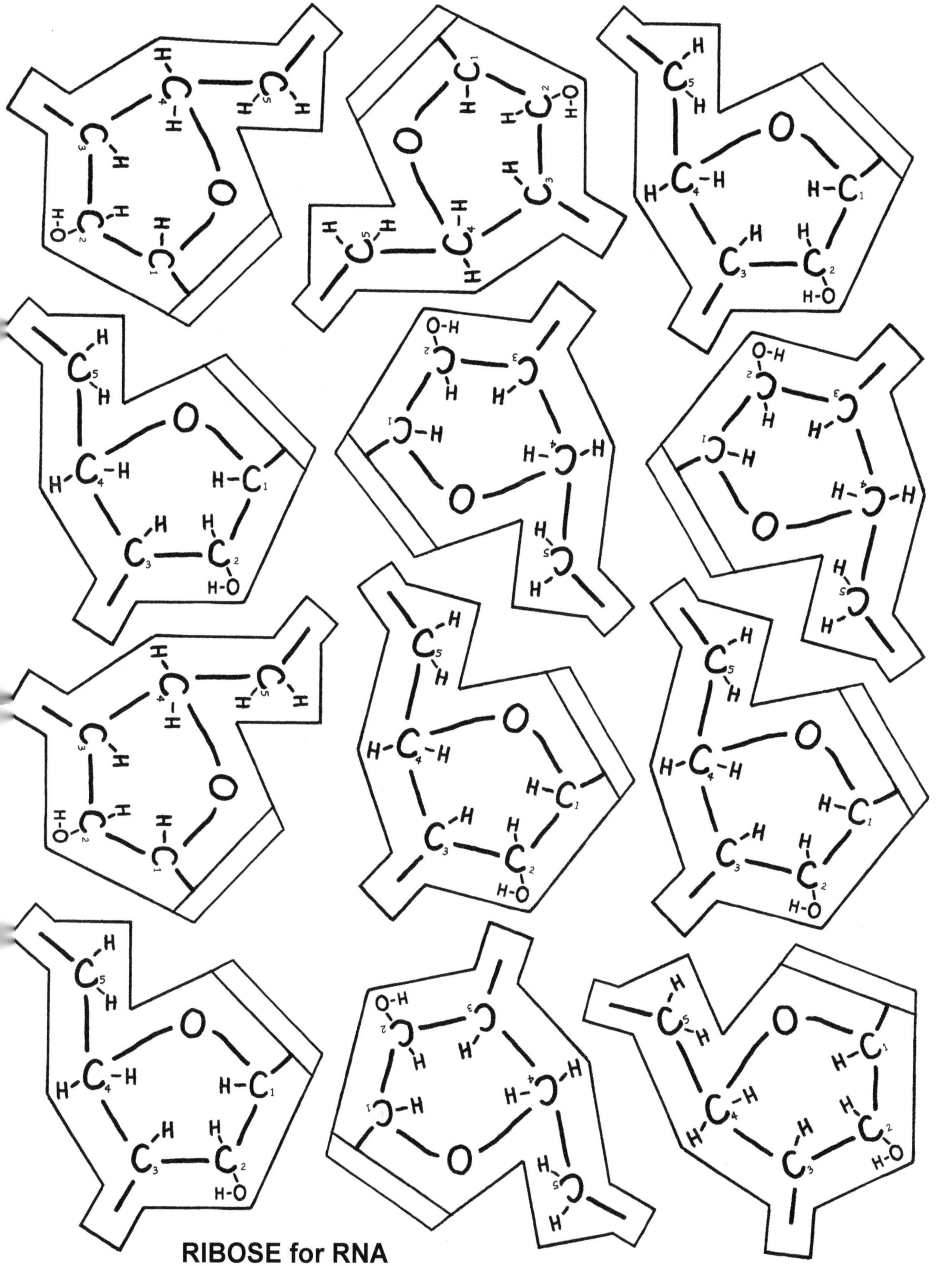

RIBOSE for RNA

ACTIVITY IDEA 5C: EDIBLE CRAFT: TRANSFER RNA COOKIES

This craft can be adapted for special nutritional needs (gluten-free, sugar-free, etc.). You could even make an inedible version that is just for decoration.

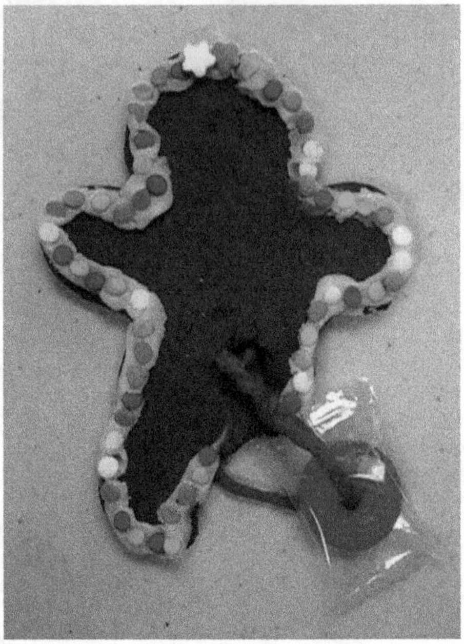

You will need:

- a cookie recipe and the ingredients it calls for (rolled gingerbread is best because the finished cookies won't break easily, but a rolled "sugar cookie" will do)
- a gingerbread man cookie cutter (but can manage without one)
- some very small cake decorations to represent amino acids
- white icing
- some small plastic bags to make disposable "piping" bags for the icing (TIP: We found that quart size baggies were stronger and less likely to split open under pressure than sandwich size.)
- small bits of yarn or string
- some individually wrapped circular candies ("Lifesavers®")
- a table knife if you are cutting the cookies "freehand" or an adapted gingerbread man cutter

How to prepare:

Mix up enough dough so that each student will be able to make several cookies. Provide a gingerbread man cookie cutters of the desired size. You can adapt the cookie cutters by bending the shape, or you can just cut off one of the legs after cutting the shape. You could also provide table knives and just cut them out freehand. Since each student will only be making a few cookies, it won't take too long.

Also prepare the icing and put it into some plastic bags, pushing the icing into one corner. Seal the bag shut, and cut a TINY snip off the corner of the bag. By squeezing the icing out of the snipped hole, you can make the icing come out in a thin stream, making it very easy to apply it to the edges of the cookies.

A simple icing recipe is to use a spoon of shortening, a cup of powdered sugar (confectioner's sugar), half a teaspoon of vanilla and just a dribble of water. Chop the shortening into the sugar, then add the vanilla and just a tiny amount of water and stir briskly. Keep adding a few drops of water at a time until the mixture becomes the right consistency for spreading—not too thin, not too thick. (You can always add more sugar or water if you find that the icing is coming out of the bags too fast or too slow.)

How to make the cookies:

1) Cut out (with or without cutter) the tRNA shapes.

2) Poke a small hole in the cookie right over where the missing leg should be. (The end of a drinking straw works really well for this.) This will be where the yarn goes through after baking.

3) Bake and let cool.

4) Using your icing baggies, squeeze a bead of icing around the outside of the cookie. Immediately stick on the little colored decorations that represent the amino acids A, C, G and U.

5) Cut a piece of yarn and feed it through the hole in the cookie. Poke a hole in the individually wrapped candies and put the yarn through. Tie the yarn. The candy represents an amino acid. (We threaded our yarn through very large darning needles.)

ACTIVITY IDEA 5D: GAME: "TRANSLATION TAXI"

NOTE: This game also appears in my "Mapping the Body with Art" video ecourse. If you will be doing that course in the future you might want to save this game until then. (But playing it twice is fine, too!)

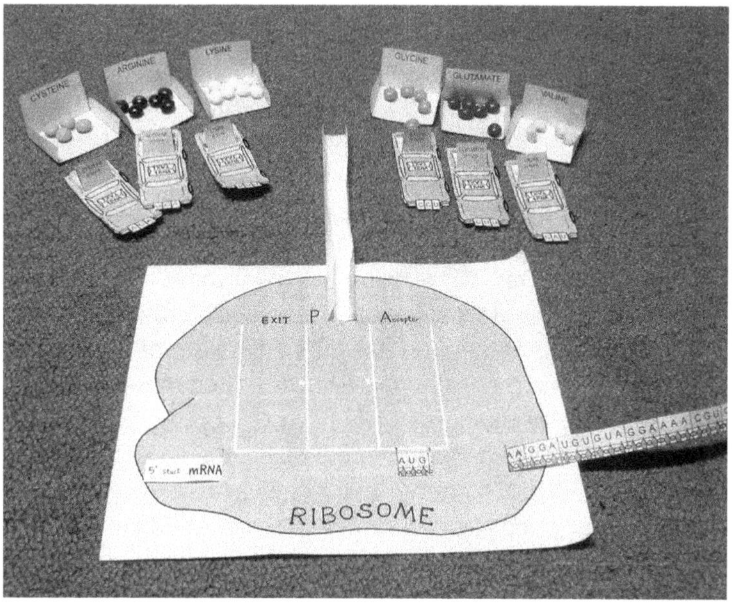

What to tell the students:

This game shows you the inside of a ribosome and lets you simulate the process of using a messenger RNA pattern to make a chain of amino acids. You will use paper taxi cars to represent transfer RNA; their license plates will display their anti-codons and their trunks will hold their amino acid. You will see that ribosomes have three "parking places" for tRNAs.

You will need:

- copies of the pattern pages
- a blank piece of paper
- glue stick and a roll of tape
- scissors
- X-Acto knife (or razor blade)
- Six different "tokens" to represent amino acids (candies, cereal bits, nuts, dried fruits--whatever you want to use) ALTERNATIVE: Use paper circles to represent amino acids, and staple them together.

How to make the game:

Make enough copies of the ribosome page so that each student has a copy. Copy the taxi page and the amino acid boxes once for every two students. (If you are really pressed for time/materials, three players could possibly share a set of taxis.) Copy the mRNA page enough times so that each student will have one strip. As you can see, one page will give you five strips.

Ribosome page: Cut along the four vertical lines below the parking places, making them into slits. Also, cut a strip of paper from the blank sheet and make a long, thin paper tray extending out from above parking place P. The tray has to be just wide enough to hold your largest tokens. Don't make the tray too wide. Secure the tray with glue if you want it permanently stuck on or with tape if you want to take the tray back off so you can store the pieces for later use.

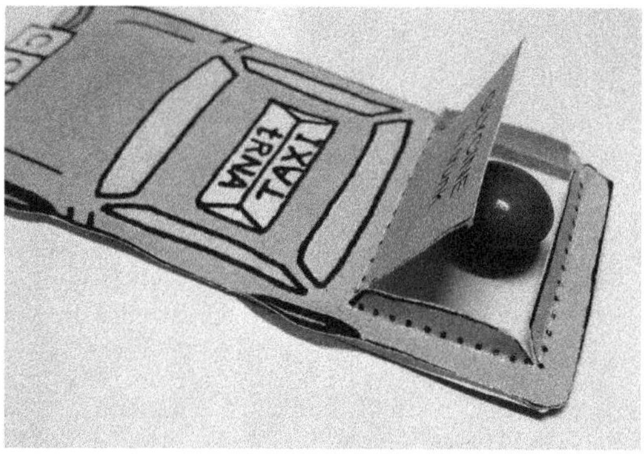

Taxis: Cut out the tRNA taxis. Fold the gray side so it becomes the bottom of the car. Unfold it temporarily and cut the trunk slits. Fold up the rims around the trunk and pull the trunk hood up. (see picture below) TIP: Don't apply the glue to the gray side! You'll end up getting glue in the trunk area. Apply the glue to the back of the yellow side, but being careful not to apply it to the headlight or license plate. Press the front and backs together, pressing carefully around the rim of the trunk so that the paper rims do not get folded down. The rims will help to stop the tokens from slipping out as the car is moved.

mRNA page: You can use either tape or a glue stick to put the two halves of the mRNA strips together. If you want to use tape, just put the strips end to end and apply tape around the joint. Make sure the joint is smooth. Do trimming if necessary. If you want to join them with glue stick, leave a small paper tab on the left side of each bottom strip. Apply glue to the tab and join to the top half. If the joint isn't smooth, trim with scissors so that it is. (Make sure there is nothing sticking out that will catch as the strip slides through the slots.) Notice that the strips are labeled 1-5, right above the "mRNA." They are all different, so if you play a second or third time you can switch to a different strip if you want to.

How to play the game:

NOTE: Video instructions can be found on the Cells playlist at www.Youtube.com/TheBasementWorkshop.

Slide the mRNA tape to AUG, which means "START." The codon AUG codes for the amino acid methionine. Methionine's only job is to be the "starter" as the first amino on the string. It is not used for anything else. You can use a special token for methionine, or you can decide to skip methionine or you can simply tell the students that AUG means start and then move on to the next codon on the strip.

Look at the first codon and find its matching anti-codon on the license plate of a taxi. Bring the taxi and park it in the A site. Move the mRNA over to the next codon and find the matching taxi. When the second taxi is brought to the ribosome, the taxi in the A site may then move to the P site. Now you have two taxis, one in the A site and one in the P site. Take the amino out of the trunk of the taxi in the P site and put it in the long paper tray. Then move the taxi to the Exit site. Then move the taxi in the A site over to the P site. Slide the next codon over and look for the correct taxi to come occupy the now empty A site. Bring the taxi over and fill the A site. Then you take the amino out of the taxi in the P site and add it below the others. Move all taxis one space to the left, whatever that might be. You'll have a taxi exiting, a taxi moving to the exit and a taxi moving to the P site.

Keep going like this, remembering that an amino can NOT be added to the string unless there is a taxi sitting in the A site. Also, don't forget that the new amino must be added at the bottom, not the top of the growing string. Yes, it is a bit cumbersome to slide all the amino ups every time, but you must do it in order to get the science right. No point in learning wrong science.

Players will have to be constantly refilling the trunks of the taxis. This simulates the reality that the tRNAs go out of the ribosome and replenish themselves by finding another amino acid to transport.

Here are my students playing the game using mini M&Ms for amino acids. I had not prepared enough boxes to hold the supplies of candies (not realizing that many had been destroyed and were not longer in my Cells unit bin), so I quickly improvised with paper plates. The cars couldn't back up to the plates like they could with the squares boxes, but the players seemed not to mind and game was still a great success.

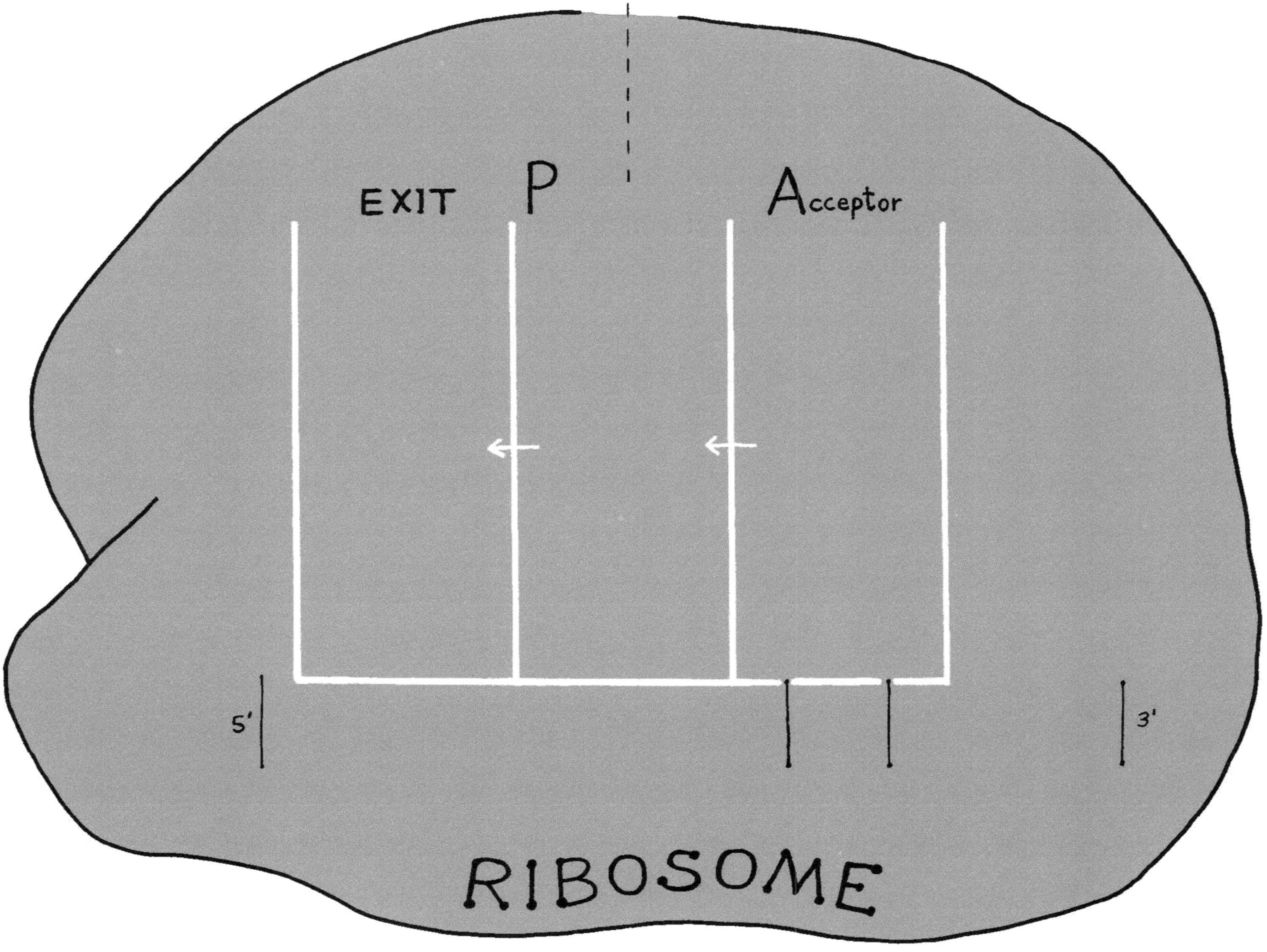
EXIT
P
Acceptor
5'
3'
RIBOSOME

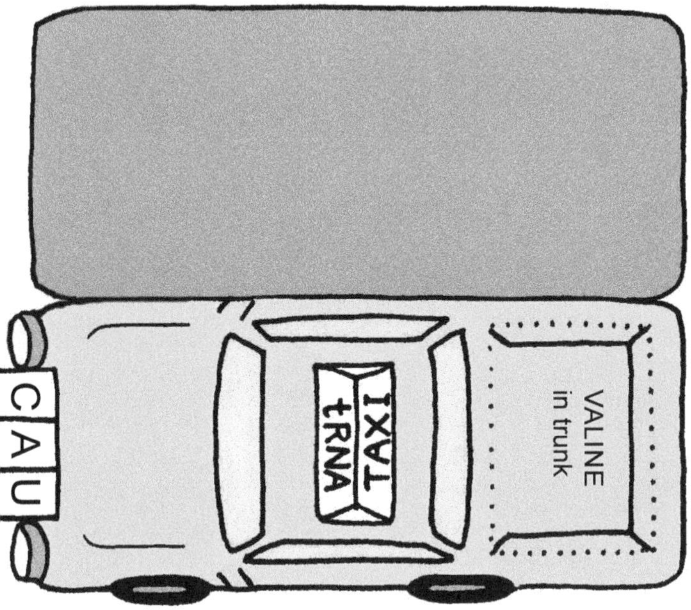
VALINE
in trunk
TAXI
tRNA
C A U

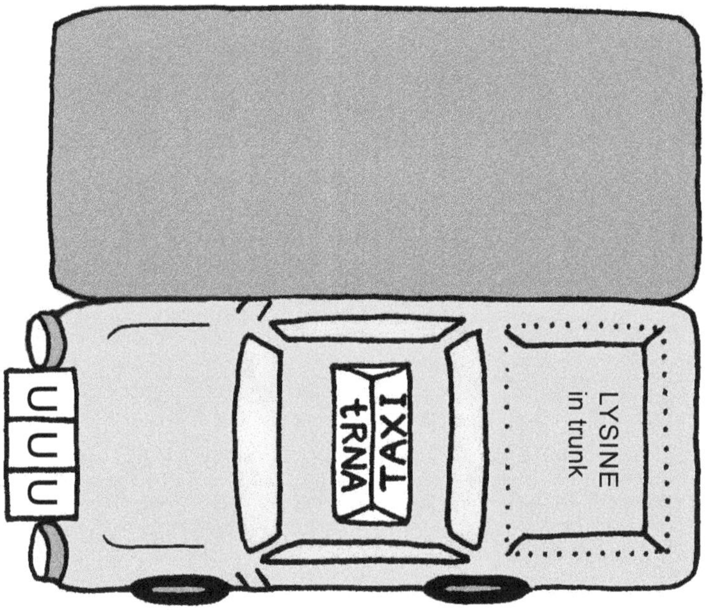
LYSINE
in trunk
TAXI
tRNA
U U U

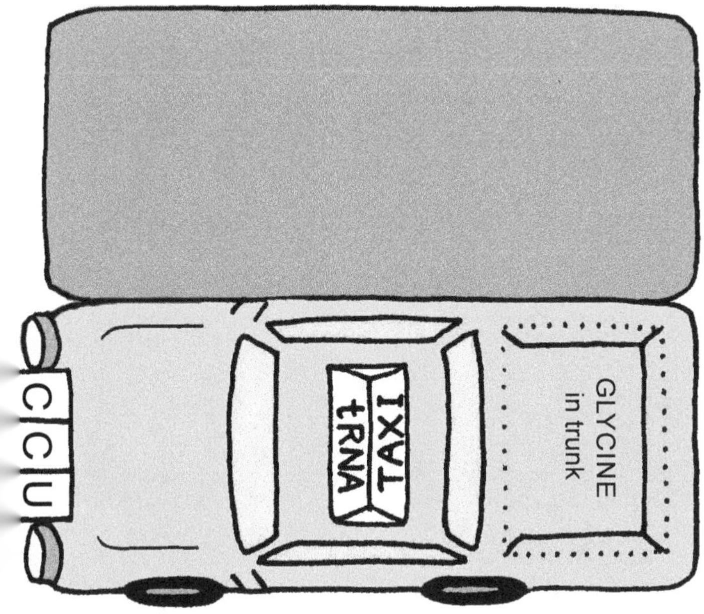
GLYCINE
in trunk
TAXI
tRNA
C C U

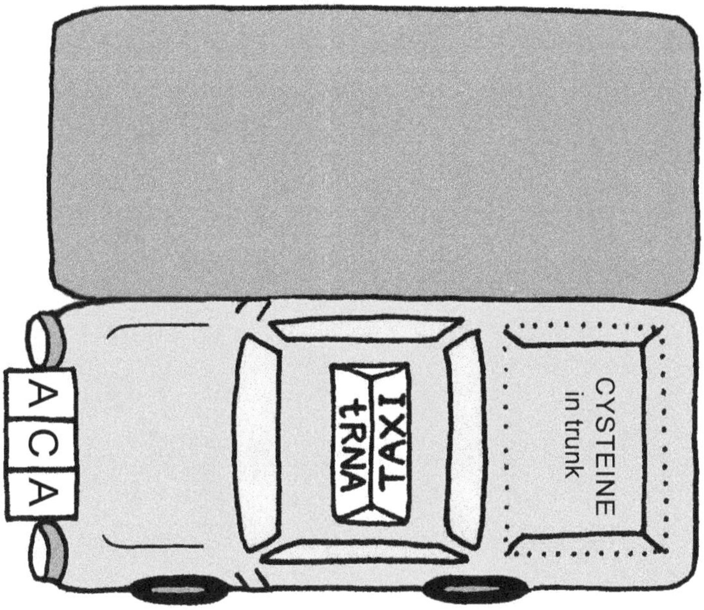
CYSTEINE
in trunk
TAXI
tRNA
A C A

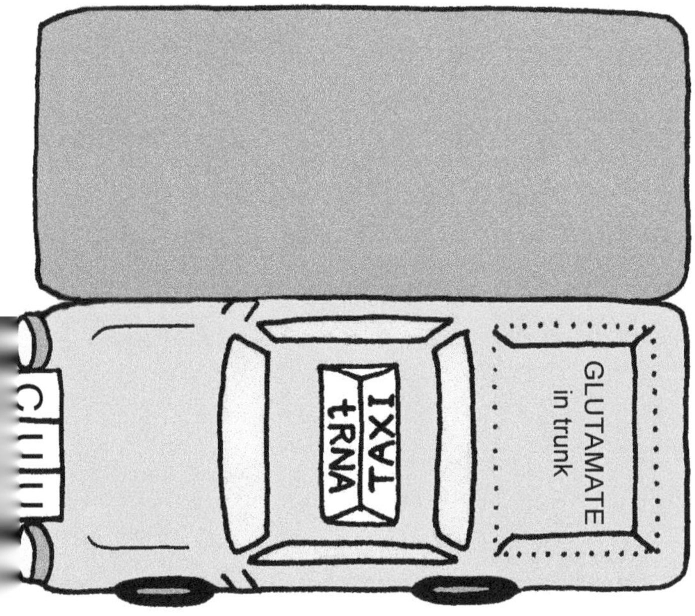
GLUTAMATE
in trunk
TAXI
tRNA

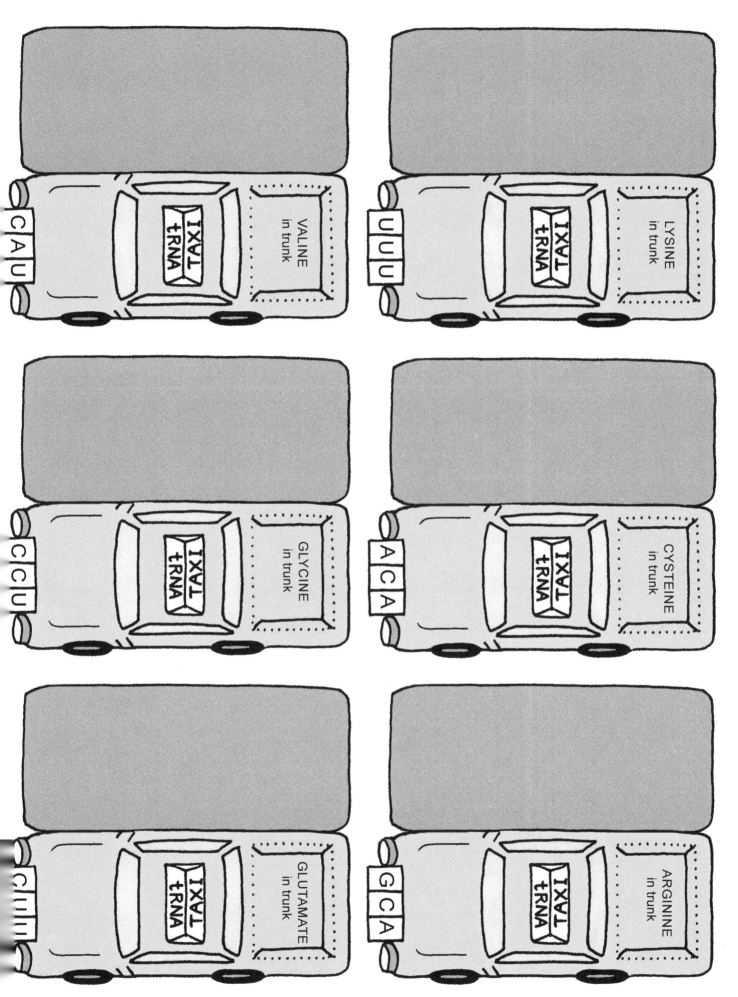

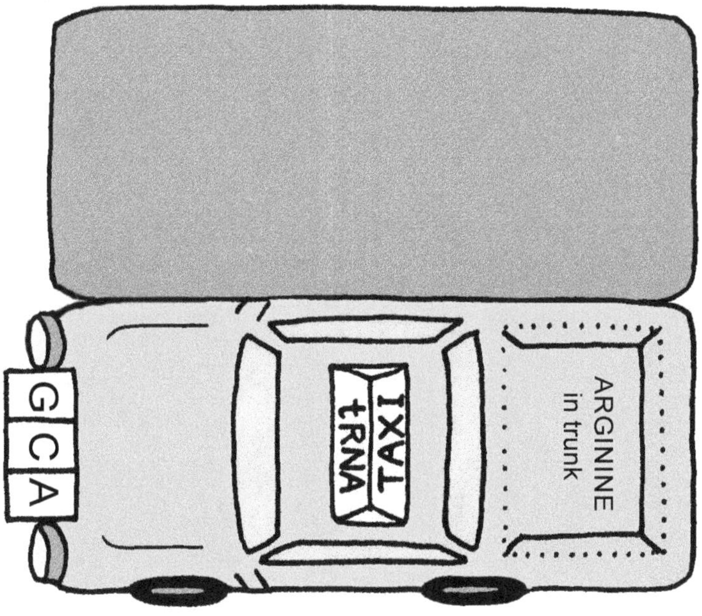
ARGININE
in trunk
TAXI
tRNA
G C A

5′ start **mRNA** [1] AUG GUA AAA GGA UGU

GAA CGU GUA CGU UGU GAA AAA end 3′

5′ start **mRNA** [2] AUG AAA GUA GAA GGA

UGU GUA GGA AAA CGU GAA UGU end 3′

5′ start **mRNA** [3] AUG GGA UGU GUA AAA

GUA AAA UGU GAA GGA CGU GGA end 3′

5′ start **mRNA** [4] AUG UGU GGA GAA GUA

AAA GGA AAA GUA GAA GGA UGU end 3′

5′ start **mRNA** [5] AUG GGA UGU AAA GAA

GUA GGA GUA UGU GGA GAA AAA end 3′

CYSTEINE

VALINE

LYSINE

GLUTAMATE

GLYCINE

ARGININE

ACTIVITY IDEA 5E: GAME: "ROLLER COASTER REVIEW"

NOTE: You can also use this activity after chapter 6.

You will need:

- a piece of poster board (or other heavyweight paper)
- scissors and clear tape
- colored markers (permanent markers work best)
- one paper clip for each team (a large clip, if possible)
- copies of the following 4 pages of quiz cards, printed onto heavy paper

(If you will be making more than one game, print each set of cards on a different color, or mark the cards in some way so that it will be obvious which set they belong to. This will make clean up easier—you won't have to re-sort the cards every time!)

What to tell the students:

This game is called "Roller Coaster Review." The board for this game will be a long strip of paper that will represent a protein. It will end up looking more like a race track or a roller coaster than a board for a table game. Since proteins are folded into all kinds of weird shapes, your protein race track will also be twisted or folded into any shape you want, even one that looks like a corkscrew roller coaster! Since the playing surface will be going upside down at some points you will have to keep your tokens from falling off the track by using a paperclip. To advance along the track, you will answer questions about cells and proteins. If you give the right answer on your first guess, you advance three spaces on the track. If it takes you two guesses, you advance two spaces. If if takes you three guesses, you advance one space. And since there are only three answer options, this means that you will always advance at least one space. And the best part is that if you don't know the right answer, you'll learn it by the end of your turn! The first team to reach the end of their proteins wins the game.

How to prepare:

1) Cut some long strips of poster board. You can make the track as long or as short as you want. I used four strips cut from a sheet of poster board, each about 2 inches wide and 20 inches long.

2) Divide the students into teams of two, three or four players. (The more players there are on the team, the less frequently each one will get a turn to answer questions.) Distribute several strips to each team (one per player is ideal) and hand out markers, as well.

3) Have the students mark off squares on the strips. (Or, you might want to pre-mark the squares with pencil ahead of time to make sure all strips have the same number of squares, and have the students trace over the pencil lines with their markers.) Each square will represent an amino acid. They can choose to write the name, or use the 1-letter or 3-letter symbol. Page 39 will be very helpful for this. It doesn't matter which aminos they choose since this will be an imaginary protein. The point is to have them look at the chart and become familiar with the names.

4) Tape the strips together, end to end, to make one long strip. Tape one end down to the table, then let the students bend or twist the strip into any shape. Then secure the other end to the table.

5) Give each team a paper clip to keep track of which square they are on. Start the paper clip on the first square of one end.

6) You will need a set of cards for each group of teams that are playing against each other. For example, in our class we had four teams with three on each team. We had two teams play each other, so we needed two sets of cards. (If you are playing with only a few students, they can each make their own track and just use one set of cards.)

7) If you happen to run out of cards, just start them over again. This time they should really know the answers if they were paying attention the first time around.

Photos I took when my class did this unit in the late 1990s:

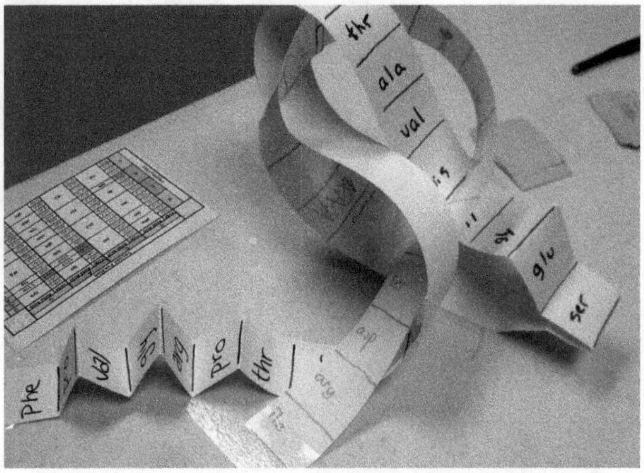

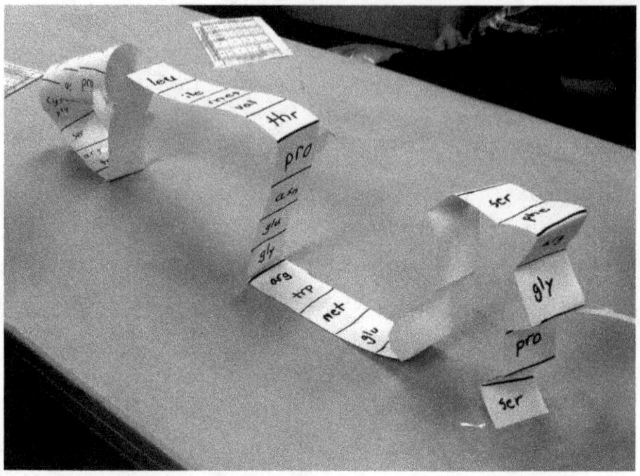

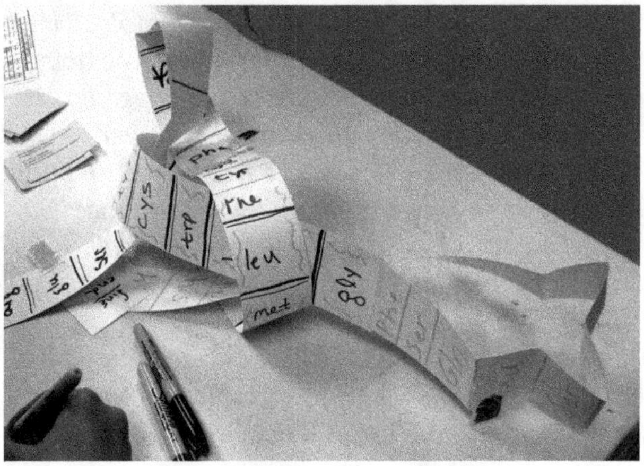

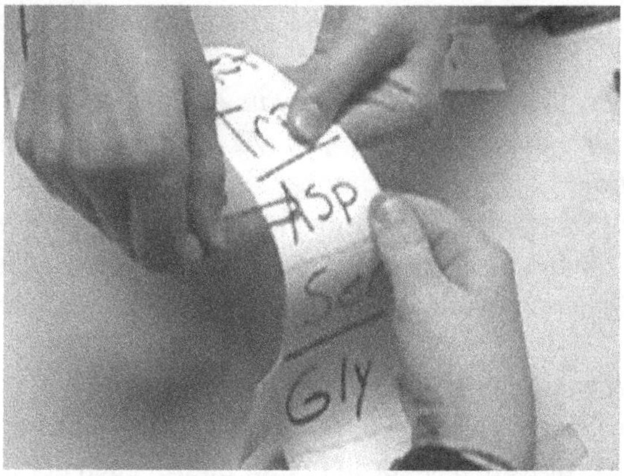

Use a paper clip to keep track which square you are on.

My 2021 class playing their game:

If you join an electron and a proton, what do you get? a) a water molecule *b) a hydrogen atom c) an acid	What does "hydrophobic" mean? a) made from water b) "loves" water *c) "fears" water	If amyloplasts are organelles in plant cells that store starch, what do you think the protein enzyme “amylase” might do in your body? a) store extra starch in fat cells *b) break down starches as part of digestion c) use light from the sun to make sugar
What happens when electrons pass through the pumps in the ETC? a) ATP is formed b) water is formed *c) two protons are pumped upward	What is the inside of the mitochondria called? a) membrane *b) matrix c) cytoplasm	A protein called "scramblase" is found in plasma membranes. Can you guess what it does? a) acts as a gateway or portal b) runs the proton pumps *c) mixes up phospholipid molecules by movingt hem from one side to the other
How many layers are in a cell’s plasma membrane? a) 1 *b) 2 c) 4	What happens when the third phosphate is popped off ATP? *a) energy is released b) a molecule of water is formed c) a proton is released	What do you think the proteins called “immunoglobulins” help your body do? *a) fight infections b) digest your food c) carry oxygen in the blood

If the Greek word for glue is “kolla,” where do you think the protein “collagen” might be found? a) in the brain *b) in connective tissues such as tendons and ligaments c) in heart muscles	What does the protein “transferrin” do? a) helps the body to fight viruses *b) carries iron in the bloodstream c) nothing	Where do you think you would find the protein called “pepsin”? *a) in the stomach, digesting food b) in the brain, transmitting signals c) in the eye, gathering light
What does the protein “insulin” do? a) regulates your heart beat b) regulates your body temperature *c) decreases the amount of glucose in your blood	What does the protein called “hemo-globin” do? a) acts as an identification flag on the plasma membrane *b) carries oxygen through the blood c) copies DNA to make mRNA	What do you think the protein called “elastin” does? *a) gives skin its flexibility b) help blood to clot c) help the liver to produce bile for digestion
What do cells make with the protein called “tubulin”? *a) microtubules b) cell membranes c) proton pumps	Take a guess as to what the protein “fibrin” does: a) makes the muscles contract *b) makes protein fibers that allow the blood to clot and form a scab c) acts as a messenger to other cells	What do you think the protein called “porin” does? a) digests lipids and sugars in the intestines b) starts the process of transcription *c) acts as a portal or gateway in the outer membrane of cells

<table>
<tr>
<td>What energy source was used to dis-cover the shape of DNA?

a) electrons
*b) x-rays
c) light</td>
<td>What is the most natural shape for a bunch of phospholipid molecules to form?

a) a flat surface
*b) a ball
c) a long chain</td>
<td>Lipid rafts are made of phospholipids and...

a) water
*b) cholesterol
c) microfilaments</td>
</tr>
<tr>
<td>What does the centrisome do?

a) acts as a gathering point for pro-teins that are floating around the cell
b) sends and receives messages
*c) acts as a central point of organiza-tion for the cytoskeleton</td>
<td>What kind of atom (meaning what element on the Periodic Table) marks an organic molecule as a protein?

*a) nitrogen
b) oxygen
c) carbon</td>
<td>What is the fluid inside a cell called?

a) water
b) matrix gel
*c) cytosol</td>
</tr>
<tr>
<td>What do you call a group of three amino acids?

*a) a codon
b) a secret code
c) a nucleic acid</td>
<td>Glycine, alanine, lysine, proline and tyrosine are examples of _____.

a) nucleic acids
b) hydorchloric acids
*c) amino acids</td>
<td>What do you call a phophorus atom with 4 oxygen atoms attached to it?

a) phospholipid
*b) phospate
c) glycerol</td>
</tr>
</table>

What is it called when a lot of some-thing goes to a place where there is less of it? a) transcription b) combustion *c) diffusion	What does glycerol do in the phospho-lipid molecule? *a) keep the phosphate and lipid to-gether b) keep the phosphoate and lipid apart c) push the phosphate toward water	What is the most basic unit of energy used by living things? a) an amino acid *b) ATP c) sugar
How many proton pumps are in the electron transport chain? a) 1 b) 2 *c) 3	What did Watson and Crick discover? a) ATP synthase *b) the shape of DNA c) protein folding	What part of the cell helps it to keep its shape and also provides a network for transportation? a) cytoplasm *b) cytoskeleton c) plasma membrane
How many amino acids are there? a) 4 *b) 20 c) hundreds	What is the name of the shape that DNA forms? a) coil b) sheet *c) helix	What does the Greek word “soma” mean? *a) body b) cell c) center

CHAPTER 6

ACTIVITY IDEA 6A: GROUP GAME: LYSOSOME SIMULATION GAME

The goal of this game is to reinforce the fact that the lysosome's job is to disassemble molecules. You will be using a block of Legos™ to simulate a large organic molecule. The players will pretend to be enzymes. Each enzyme can only do one thing, so each player will be assigned to blocks of one particular color. Players will have to work together to disassemble their block.

What to tell the students:

In this game you will pretend to be an enzyme inside. a lysosome. Each team will represent a lysosome. It's a race to see which lysosome can disassemble a molecule in the least amount of time. Since an enzyme can only do one job, your team of enzymes must work together to disassemble your molecule.

You will need:

- enough Lego® bricks to make a block that is 2-3 inches (5-8 cm) on a side for each small group (2-6 players).

(The blocks don't have to be Lego® brand—any type of building block will do.)

How to prepare:

Make a block for each small group that will be playing the game. Ideally, the number of colors will equal the number of players in a group so that each player can be assigned to one color. For example, for a group of 4 players, you choose just four colors. However, it is possible to have a player play more than one color if need be. For each block count out the same number of each color. You can use various sizes as long as the number of colored pieces is about equal. For example, if you are building a block for a group of 5 players, you might use 12 reds, 12 blues, 12 whites, 12 greens and 12 yellows. Build the block so that the colors are evenly distributed throughout. As you can see from the picture, the block does not need to be perfectly square. In fact, have it being a little odd-shaped makes it look more like an organic molecule.

How to play the game:

Divide the players into teams. Give a block to each team.

Assign a color to each member of each team. In other words, each team should have a person doing red, a person doing blue, a person doing white, etc. If you have only a few players on each team, some, or all, of the players will be in charge of more than one color. This means that they will be playing the part of more than one enzyme. Remind the students that an enzyme is very specific and can only disassemble (or assemble) one particular type of molecule.

The object of the game is to disassemble the molecule down to a pile of individual blocks. (In a real cell, these "building blocks" would be amino acids, glucose or other small sugars, nucleotides, and fatty acids.)

Players can remove only outer blocks. In other words, if you are the red remover, you can't pull off a white block to get to a red one. Therefore, the block must be passed around to various players according to which blocks are on the outside and available for removal. When all the players have removed at least one block and have it sitting in front of them, it will be easy to keep track of who is in charge of what color. Players will naturally start handing the block to the person whose color is on top. I always see lots of cooperation in my groups.

The first team to completely disassemble their molecule wins.

If you want to play a second round, the blocks will have to be reassembled. The players will cooperate to reconstruct the block. It doesn't have to look anything like its original shape. Tell the teams that they will be switching blocks with another team, so their goal is to make the deconstruction take as long as possible.

Bonus idea: To reinforce the concept that a lysosome is being simulated, have each team sit on the floor inside a large oval of string or rope that represents the lysosome's plasma membrane.

ACTIVITY IDEA 6B: LAB: SIMULATION OF MERGING VESICLES USING OIL DROPLETS

What to tell the students:

In this activity, you will be observing tiny oil droplets merging with larger ones. This is very similar to what happens in a cell when vesicles merge with Golgi bodies or with the plasma membrane.

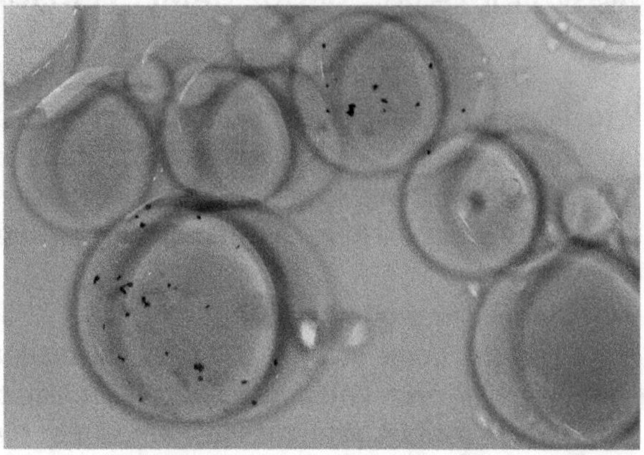

The double rings are from shadows cast by the oil drops.

You will need:

- vegetable oil of any kind (TIP: darker oils, like virgin olive oil, are easier to see in the water)
- water
- pepper
- eye droppers (though you can still do the lab without; fingertips dipped in oil might also work)
- small paper cups (or other small containers to hold a few tablespoons of oil)
- plates or shallow trays to hold a thin layer of water (1/2 inch or 1 cm deep)
- paper towels
- Optional: food coloring (Coloring the water can help the oil to be more visible, as shown in the photo.)

How to prepare:

If possible, each student should have their own plate or tray. However, if the trays are bigger than a plate, students might work in pairs. Each student should have an eye dropper, a small cup of oil, and a few paper towels.

What to do:

Pour a little water in each student's plate or tray. You only need the water to be deep enough so that oil will float on it. Even 1/4 or 1/2 half inch (.5 or 1 cm) is enough.

Tell the students to create some relatively large (no larger than a golf ball) areas of oil to represent either a lysosome or a Golgi body. Then have them add some smaller droplets, away from the larger ones. Have them guide (gently push or pull using the eye dropper or a finger) the smaller droplets over to the larger ones. Watch what happens as they touch. Usually, the tiny droplet will instantly disappear and become part of the larger one.

Have them try squeezing out a line of oil. Watch what happens. Does the line stay there or does its shape change? Why the change in shape? (Remember what the most efficient shape is for hydrophobic substances.)

Challenge them to see if they can put a water drop inside an oil drop. What about oil inside water inside oil? What eventually happens to the water bubbles inside the oil drops?

Have them put a few grains of pepper on top of an oil droplet. The pepper will represent some kind of particles inside a vesicle. Push the vesicle over to a larger area of oil and merge with it. What happens to the pepper? It should suddenly find itself "inside" that larger oil circle. When a vesicle merges with a lysosome or Golgi body, its contents end up inside that organelle.

Do vesicles ever sit right next to an oil droplet and NOT merge with it? Students will likely observe this. Ask the students to lightly touch the spot where the two oil droplets are touching. Does this cause them to merge? This type of thing also happens in cells. For example, vesicles in nerve cells sit right near the plasma membrane, ready to dump their contents into the space outside the cell. The contents are neurotransmitters that will cross a tiny gap and bump into a neighboring cell, causing it to start an electrical signal. Those vesicles full of neurotransmitter have to wait until the right time to release their cargo. The trigger for release is an influx of calcium atoms. When calcium ions come across the plasma membrane, they cause the vesicle to suddenly merge with the plasma membrane and thus release their cargo of molecules into the space outside the cell.

Put one drop of food coloring onto an oil droplet and also onto the water. Tell them not to touch the coloring drops but to simply observe what happens over the course of a minute. Is the food coloring hydrophobic or hydrophilic?

NOTE: There is a short demo video on the "Cells" playlist, though I don't cover every suggestion listed here.

<u>ACTIVITY IDEA 6C:</u> ACTIVE GAME: TRANSLATION RELAY RACE (also includes Golgi packaging)

NOTE: I recommend watching the video about this relay, posted on the "Cells" playlist on www.YouTube.com/TheBasementWorkshop. Look for: "How to set up the Translation Relay Race"
If you watch the video first, the materials list and the preparation instructions will make more sense.

The goal of this activity is to review the process of translation (how tRNA brings the amino acids and joins them together to form a protein chain) and the role of the Golgi bodies (folding and packing the proteins). Each team will make a protein, then fold it, package it, and ship it.

This game idea is very flexible. It can be adapted to suit any number of players. You can also adapt the game to fit your space restrictions. We used the full length of a gym and made the kids run half the length of the gym to deliver their amino acids, but you don't have to do this. You can scale down the size of the parts and make it into a table top activity if necessary. This activity could even be done by a single student, just as a learning activity, not as a race. It will be an effective review activity no matter how it is done.

You will need:

- copies of the following tRNA pattern page printed onto card stock (3 copies is enough for a dozen players)
- envelopes (at least one per player)
- colored paper (one color for each amino acid you will be using)
- white paper (for sugar strings and Golgi mailing labels)
- a long strip of white paper (cash register tape is ideal)
- 4 rolls of clear tape (one for each ribosome station and one for each Golgi body station)
- two cardboard boxes (represent vesicles)
- black marker of some kind
- pencils
- paper clips
- two very large rubber bands (if you happen to have them, otherwise use tape)

Time needed in class to do this activity:

30-60 minutes, depending now how many players you have and how many times you repeat the relay. I recommend doing the relay at least twice—the first time for practice and the second as a race.

Preparation:

1) Decide how many players will be on each team. This is the number of amino acids you will need to prepare. You will need a sheet of colored paper and two envelopes for each amino acid.

2) Cut rectangles from the colored paper that are about the size shown here. Divide the pile of colored paper slips in half so you have two piles of each color. Put each pile into an envelope.

3) Cut apart the tRNA cards and write the names of the amino acids above the oval that says "amino acid." Then choose two of the associated anti-codons and write them at the top of the card' in large letters. (These cards are visible in photo below.)

3) Assign the name of an amino acid to each color and write the name of the amino acid on the front of the envelope. Each team will have an identical set of envelopes. It doesn't matter which amino acids you choose, or which colors you assign to them. Use the following chart to choose two codons for each amino. Write the **<u>anti-codons</u>** on top of the tRNA cards and put the cards into the envelopes.

I prepared for teams of six players, so I had two sets that looked like this:

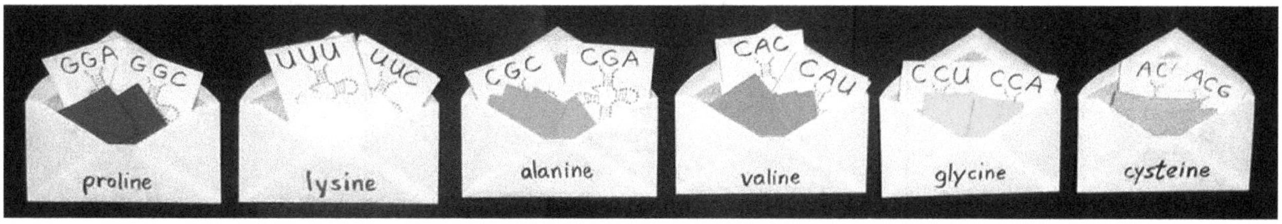

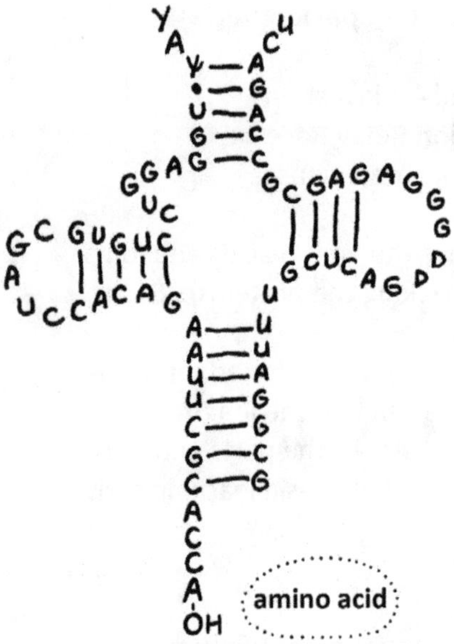
amino acid
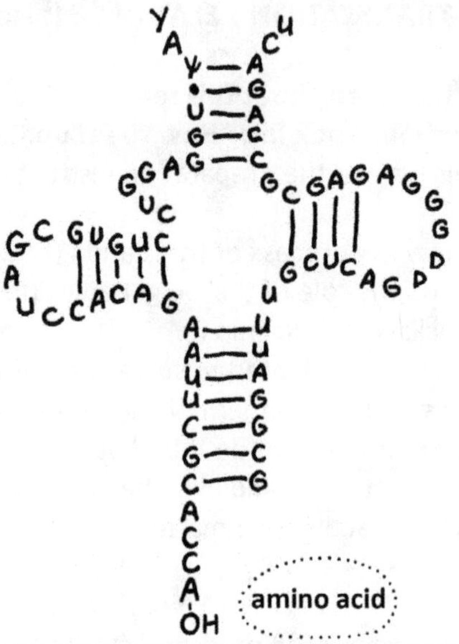
amino acid
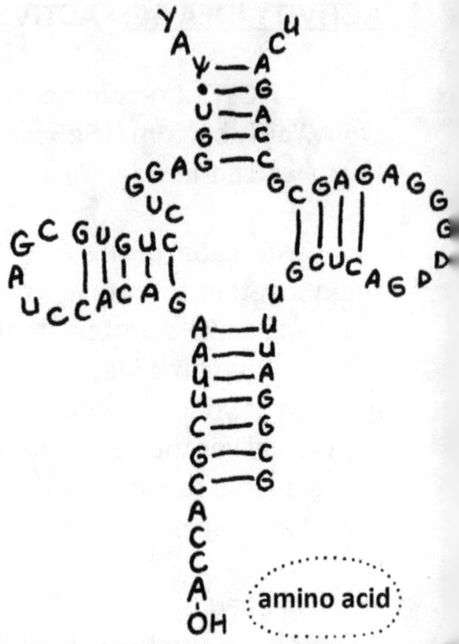
amino acid
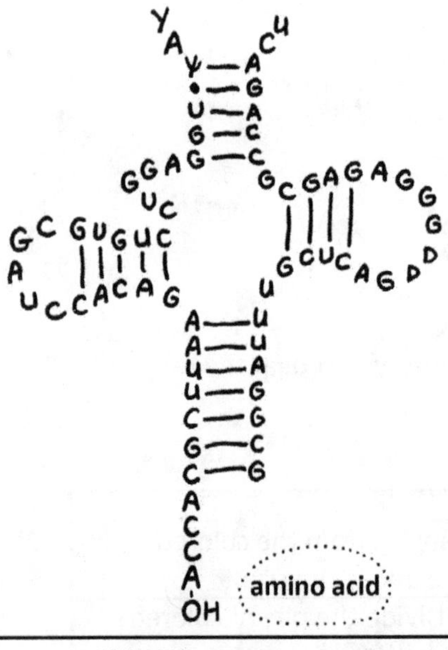
amino acid
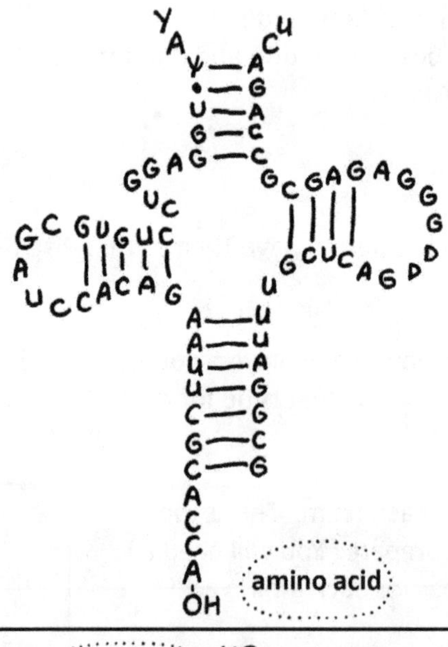
amino acid
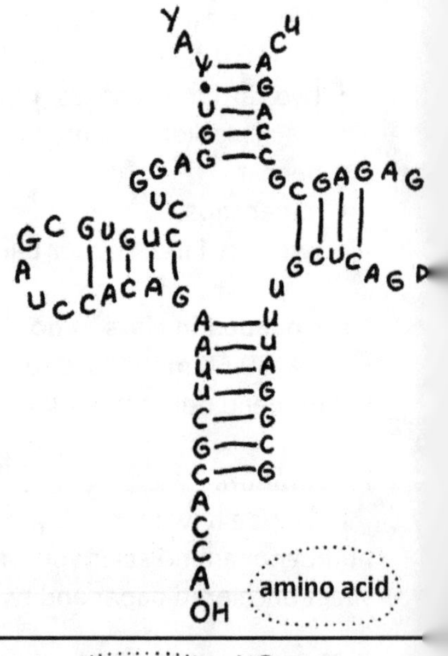
amino acid
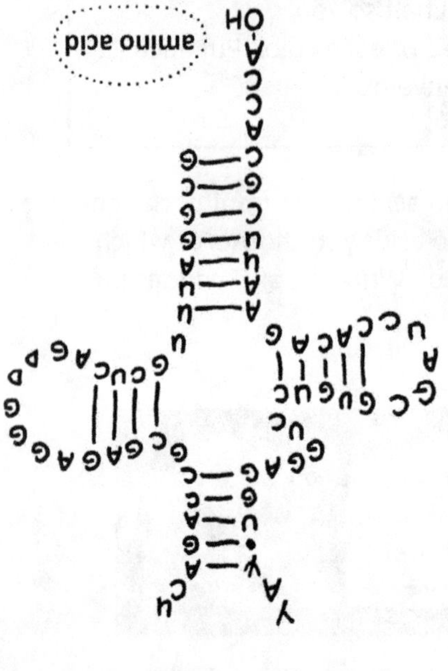
amino acid
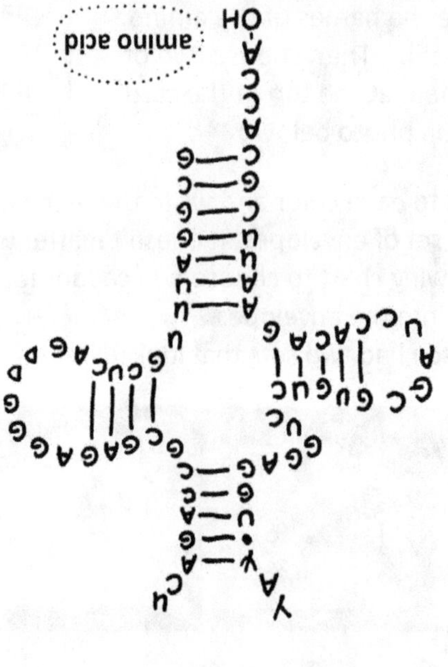
amino acid
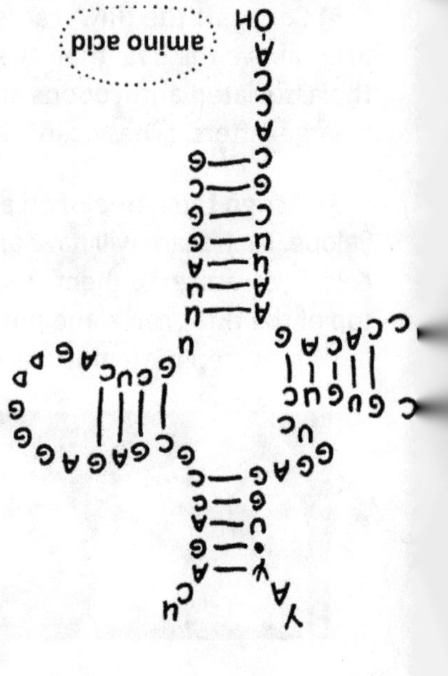
amino acid

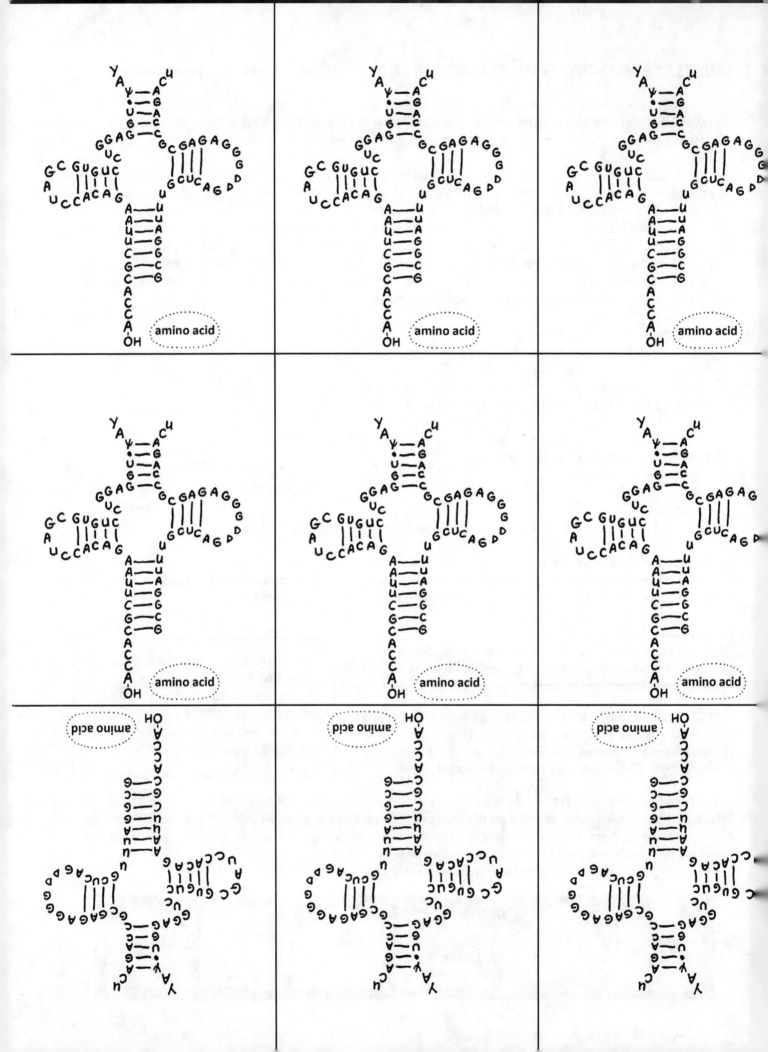

LIST OF CODONS:

FIRST LETTER	SECOND LETTER	THIRD LETTER	AMINO ACID
A	A	A or G	Lysine
		C or U	Asparagine
	G	A or G	Arginine
		C or U	Serine
	C	A, G, C, or U	Threonine
	U	A	Isoleucine
		G	Methionine (START)
		C or U	Isoleucine
G	A	A or G	Glutamic acid
		C or U	Aspartic acid
	G	A, G, C or U	Glycine
	C	A, G, C or U	Alanine
	U	A, G, C or U	Valine
C	A	A or G	Glutamine
		C or U	Histadine
	G	A, G, C or U	Arginine
	C	A, G, C or U	Proline
	U	A, G, C or U	Leucine
U	A	A or G	STOP
		C or U	Tyrosine
	G	A	STOP
		G	Tryptophan
		C or U	Cysteine
	C	A, G, C or U	Serine
	U	A or G	Leucine
		C or U	Phenylalanine

5) Decide how long you want to make your protein chain. I recommend making a chain that has about 20-30 amino acids (pieces of colored paper) so the total length would be about 7 feet (2 m) long. Make sure each player gets to contribute a link at least two times. If you are working with a small number of players, they might contribute a link five or six times.

6) To make the messenger RNA, roll out a long strip of paper that is just a bit longer than the length you want to make your protein chain. (If you don't have cash register tape, you can make a long strip by cutting strips of paper and taping them end to end.) This strip of messenger RNA has just come from the nucleus and is ready for translation in a ribosome.

Write "START" on one end of the strip, then begin marking off sections that are exactly the width of one of the tRNA cards. (If you are adapting this game to a smaller scale, adjust accordingly.) Count off the number of aminos you want in your chain then write "END" after the last one. (For my group of a dozen players I made the mRNA 24 aminos long.) Then go back and write a codon in each of the sections. You will need to use tRNA cards while you do this, making sure to use each codon at least once. I laid out the dozen tRNA cards along the first half of strip, arranged in random order, then used them as a guide to write the mRNA codons. Remember, the codons are the MATCHES for the anti-codons, and vice verse. If an anti-codon is UCG, then the codon is AGC.

long strip of paper (only part of it seen here)

I have temporarily set the tRNA cards here, in order to write the codons of the mRNA

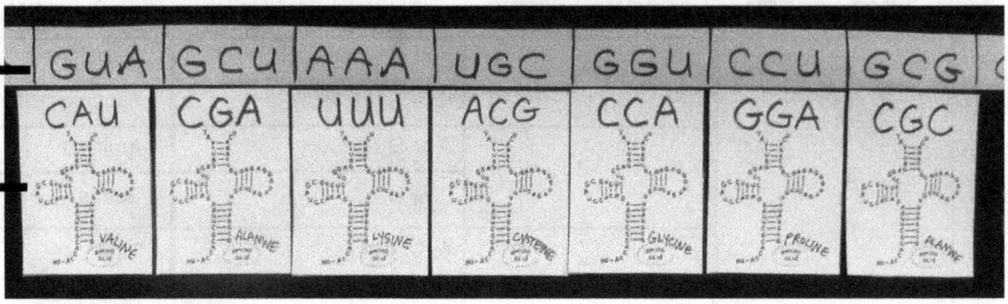

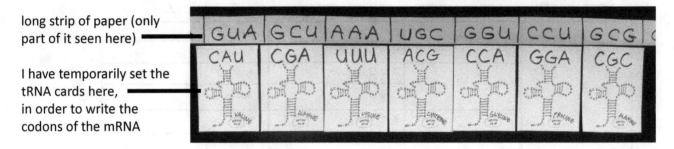

7) Make another mRNA strip for the other team. Mark the backs: Team A — strip 1/Team B — strip 1.

8) Repeat this process again, making a second pair of mRNA strips. The first set will be for your "practice round," and the second set will be for your race.

Set up needed right before class: (HELPFUL TIP: the demo video on the playlist shows my set up)

1) I recommend using four long tables, two for ribosomes and two for Golgi bodies. (If you are playing in a smaller space or with just a few players, you'll have to adapt the game to your situation).

For the ribosome tables, roll out the first set of mRNA strips and tape one to each table. Also make sure there is a roll of tape at each table. You may also want to label each table "RIBOSOME."

For the Golgi tables, put on each table: a box, some paper clips (at least two per player), a roll of tape, a pencil, some slips of white paper, some paper sugar strings (they don't have to look like mine–they can be just plain strips), You might also want to label the table "GOLGI BODY," and perhaps even label the steps that need to be performed: folding, glycosylating (adding sugars), packaging and mailing.

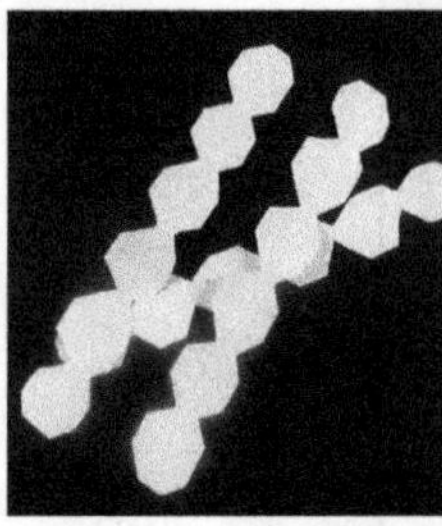

I made sugar strings out of hexagons but yours don't have to look like this.

2) You will need a place that is designated as the place to which the protein will be shipped. In our game, we wrote "Outside the Cell" and had the players end the race by throwing the box out a door. You might want to choose to have the protein shipped to a mitochondrion or somewhere else. Your protein chain could be any type of protein, shipping to just about anywhere. Choose what is best for you.

How to play:

In the first stage of this relay, the players will pretend to be transfer RNA (tRNA) and will assemble the protein at the ribosome station. In the second stage, the players will all go over to the Golgi body area and each player will do something to prepare the protein for packing and shipping. The game ends when the packaged protein is delivered to its destination.

1) Divide players into two teams. Give one set of envelopes to each team, and have each player take one envelope. (If the numbers of players and envelopes are not equal, assign some players to two aminos instead of one.) Tell the players to look inside their envelopes and explain that the colored slips of paper are amino acids. They each should also have two "codon cards." When one of their codons is called, they must bring that tRNA card and one of the colored papers to the ribosome area.

Have the players come over to a ribosome table and demonstrate how they will lay the anti-codon on their tRNA card down next to the codon on the mRNA to make sure they match. Show them how to tape their "amino acid" (colored paper) next to the previous one and thus make the chain grow longer. Remind them that they will be listening for the ANTI-CODON that matches their codons. They will have to listen carefully!

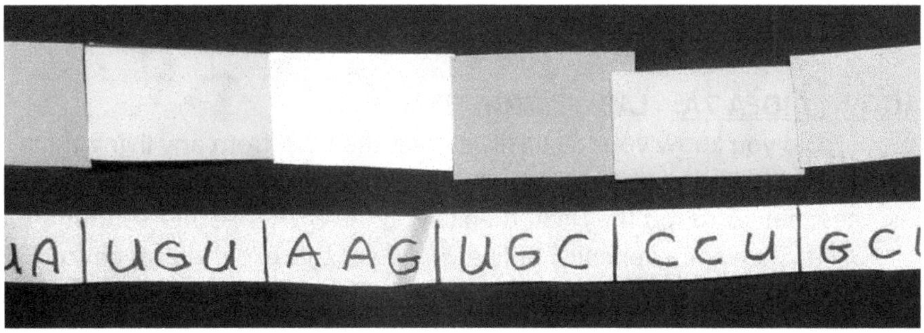

On left, player is checking to see if their tRNA card matches the codon on the mRNA strip that was called out.

Above, you can see the pieces of paper (amino acids) joined together above the mRNA strip, to form the polypeptide chain.

NOTE: If you are short on time and/or space, you can end the relay after the amino acid chain is finished.

2) Next, explain the second stage of the relay. After the chain of amino acids is finished, the players will assume a different role: they will be chaperone proteins inside a Golgi body. After taking their protein chain to the Golgi table each player will put a fold into the paper chain and clip it with a paper clip. (Or assign one player to this task.) It does not matter what the twist or fold looks like. Just twist or fold it, and clip it. The next step is to add some sugar tags. We don't know exactly what this protein is, so we can't say for sure what the paper sugars represent. We just know that often sugars are added (which is called glycosylation). Each player will clip or tape on a paper sugar. (Or assign one player to do the sugars.)

Then the folded and glycosylated protein will go into a vesicle (the box). One player should be assigned to shut the box and either put a large rubber band around it or tape it shut. Then one final step: have a player assigned to put a "mailing label" on the box so that the motor proteins will know where to take it. You can have the labeler write the label or have the label written and ready to tape on. In our game I had a pre-written label that read "Outside the Cell" (meaning the vesicle will merge with the plasma membrane and the protein will be dumped out. Show them the designated area where their vesicle should be taken.

If you are playing competitively, first team to deliver their vesicle wins the game.

<u>ACTIVITY IDEA 6D:</u> EDIBLE CRAFT: GOLGI BODY COOKIES

You will need:
- cookie dough (and baking sheets and an oven)
- icing (something fairly stiff and not too gooey, so it doesn't drip)
- table knives for spreading icing
- optional: aluminum foil and a permanent marker

Tip: If you are working with a large class, have the students put their unbaked cookies on a piece of aluminum foil and then write their name on the foil. Put the foil sheets onto the baking sheets. When the cookies come out of the oven there won't be any arguing over which cookies belong to whom.

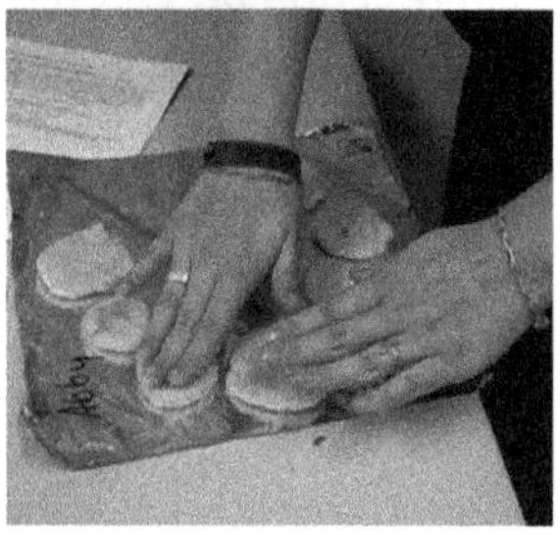

What to do:

Make the pancake shapes and put budding vesicles on two of them (these will be the ends). Bake the cookies, then use icing to stick the "pancakes" together, making sure that the vesicles are on the ends (as shown in photo).

CHAPTER 7

ACTIVITY IDEA 7A: LAB: EXTRACTING DNA

Did you know you can easily extract the DNA from any living thing? All you have to do is break down the barriers (the phospholipid membranes) that hold it inside the cell, remove the chaperone proteins and histones that surround the DNA, then clump the DNA together so that it forms a blob big enough for you to see.

If you want a very nice online version of this experiment, go to:

http://learn.genetics.utah.edu/content/labs/extraction/howto/

You will need:

- a plant or animal product you can put in the blender (liver, peas, or strawberries are often recommended, but you can use anything that contains DNA)
- a blender
- a small strainer
- a bowl, a spoon, and a few small, clear glasses (or beakers or test tubes)
- salt
- cold water
- liquid dish detergent
- meat tenderizer (if you can't get this, try pineapple juice or contact lens cleaning solution)
- isopropyl ("rubbing") alcohol (90% is better, but 70% will probably work)
- optional: some cotton swabs for pulling DNA out of final goop

What to tell the students:

If you wanted to extract DNA out of a cell, what would be the first thing you would have to do? The DNA is surrounded by a nuclear envelope — a double-thick layer of phospholipids — then outside of that is the cell's plasma membrane. You will have to use a substance that can break apart phospholipid membranes. Fortunately, this type of substance is readily available. We call it "soap." You use soap to clean greasy plates that have had fatty foods sitting on them. Soap can break apart lipid substances, even the lipid membranes of cells. Your skin doesn't just dissolve away when you wash yourself with soap because your cells are tightly bound together and the soap can't penetrate very far down.

After you split open all the phospholipid membranes, the organelles and the DNA can spill out of the cells. The organelles that are surrounded by a phospholipid membrane, such as the Golgi bodies, ER, and lysosomes, will also get broken down by the soap, leaving whatever proteins and enzymes were inside of them just floating around.

Now what would be the next thing you would need to get rid of in order to get pure DNA? Remember all those proteins that surround the DNA, and the histone spools that wind it up? You'd need to get rid of those. For this, you would need something that breaks down proteins. In this demonstration you will be using meat tenderizer to do this. Meat tenderizer contains enzymes called proteases that break down proteins. (Interestingly enough, the protease enzymes in meat tenderizer were extracted from pineapples or papayas.)

After the proteins have been stripped away from the DNA, you then need to gather it all together in one place. One single strand of DNA is ultra-microscopic. Even thousands of strands of DNA probably can't be seen without a microscope. You will be gathering together millions and millions of strands of DNA by using a chemical reaction between DNA, salt and isopropyl alcohol. The clump of DNA will look like a bunch of stringy white goo. So what can you do with your clump of DNA? Not much. You will probably dispose of it. But if you wanted to keep it, you could preserve it for months or maybe years by keeping it in a jar of alcohol.

What to do:

1) Put your living stuff (half a cup (125 ml) of it is enough) into the blender and add about a cup (250 ml) of cold water. Make sure to use cold water, not lukewarm water, as the temperature of the water can affect results. Add 1/8 teaspoon (1/2 ml) of salt. The salt will help the DNA clump together when you add the alcohol in step 7.

2) Blend on high for half a minute or so. The chopping action of the blender separates the cells from each other so the detergent will be able to penetrate every cell. In the case of plant cells, it will also chop up the tough cell walls, exposing the plasma membranes.

3) Put your strainer on top of the bowl and pour your soup into the strainer. This step is just to get rid of larger particles, such as fiber that did not get chopped fine enough. Then remove the strainer.

4) Add 2 tablespoons (15 ml) of dish detergent. Mix well, then let this mixture sit for about 5 to 10 minutes. (However, we got the experiment to work without letting it sit.)

5) Add a pinch of meat tenderizer to the soup and stir very gently. If you stir too hard you might break up the DNA too much and it will be harder to see. If you are using pineapple juice or contact lens cleaning solution, just add a few drops. (The enzymes in the meat tenderizer were extracted from either pineapple or papaya, which is why you can substitute these. Pineapple enzyme is called Bromelain. Papaya enzyme is called Papain.)

6) At this point you may want to transfer your soup into some smaller containers. You can use small, clear glasses, or you can use fancy lab equipment like beakers and test tubes, if you have them. The main goal is use something clear so that you can see the DNA precipitate out. However, if you want to leave all your soup in the bowl, the next step will still work.

7) Now add alcohol to the soup. Add enough alcohol so that you have the same amount of alcohol as soup. You will notice that the alcohol will float to the top.

8) Now you should see some white gooey stuff forming between the soup on the bottom and the clear alcohol on the top. If not, swirl it very gently and then wait. The DNA does not like to mix with alcohol. The salt you added in step 1 will also help the DNA to clump together (or more correctly, "precipitate out" of the solution). You can take a swab (or stick, or the end of a spoon) and pull out the stringy white blobs of DNA. Of course, there is RNA in the goop, as well as DNA. The cells also had a considerable amount of RNA in their cytoplasm. What you really have is a clump of nucleic acids.

STEP 1

STEP 3

STEP 7

STEP 8

ACTIVITY IDEA 7B: ART PROJECT: FIRST SESSION OF THE CELL "MINI-MURAL"

This project is divided into two sections because it is probably too much for one sitting, although students working at home with ample time and a long attention span could probably complete it in one long session. If you want to do it in one session, wait to begin until after the next chapter.

This drawing should be done mostly from memory (or as much as possible) so that it will be a record of just how much information the students have learned. By drawing "from scratch" and not copying it from any other source, the students' brains will be doing important processing of the information, creating procedural memories—something that reading and answering quiz questions just can't do. I found it helpful to my students if I demonstrated quickly on a white board approximately what they are supposed to sketch in each step.

You will need:

- two sheets of white card stock for each student (or substitute regular paper if necessary)
- tape
- pencils and erasers
- optional: compass or circle guides (I cut some circles from card stock, or had them freehand smaller circles.)
- a ruler
- optional: colored pencils or markers, or good quality drawing pens (Just pencil is okay, too.)

What to tell the students:

You will be making a detailed drawing of a cell using only what you've learned, not copying from any other pictures. You will be surprised at how much you know about cells! Your cell does not need to look like an illustration from a book. It doesn't need to be perfect. The goal is just to show the amazing amount of information you have learned about cells. Also, while you are drawing, your brain will be processing the information you are recalling from your memory, and sort of reorganizing in a new way so that you will be able to remember it even better.

How to do the drawing:

1) Tape the two sheets of paper together to make one large drawing surface. (Match edges carefully.) Put tape on only one side. Draw on the side without the tape.

2) Begin with lightly sketching where the plasma membrane will be. Fill the page as much as possible, but leave a little bit of space around the edge, perhaps an inch or so (2 or 3 cm).
VERY IMPORTANT: Remember to make your first lines ***very light*** and sketchy. This will allow you to erase them and make changes easily.

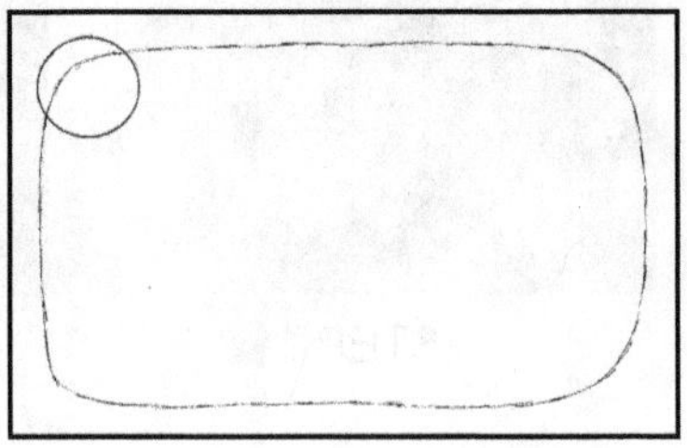

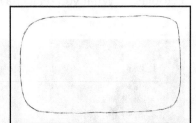

3) Now you will put in your first "inset." An inset is how illustrators show a close-up view of something inside a larger drawing. In this case, you need to show a close-up view of the plasma membrane. In the scale of your drawing, the phospholipids would be too small to see. The inset will allow you to draw the details of just a small portion of the membrane large enough to be able to see details. Use a compass or a circle guide to draw a circle in the upper left corner, overlapping the membrane. You can go ahead and make this circle dark, as you won't be putting anything over it. Then erase the pencil line inside it.

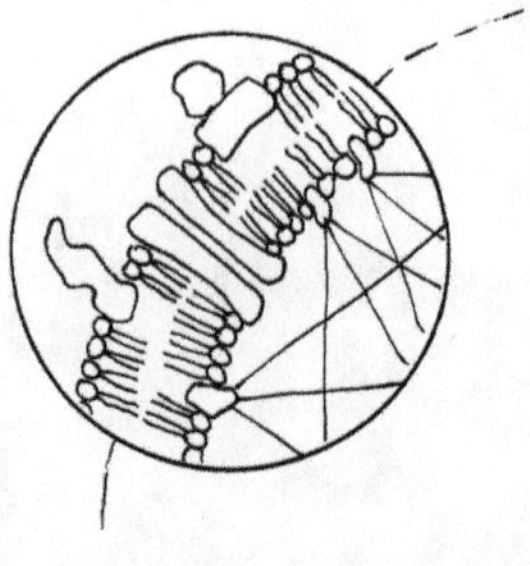

4) Inside this inset, draw a close-up of the phospholipid membrane. Show all the heads and tails lined up correctly (heads on outside, tails to the inside), and put in some portal proteins and some other membrane-bound proteins on both the inside and outside. You could also put in a lipid raft carrying a protein, or you could add some cytoskeleton attached to the proteins on the inside layer. Sketch in the parts lightly with pencil first, showing where you intend to put the layers of phospholipids and where the various proteins will go. Once you have it sketched out, then you can use darker pencil lines or pen lines to draw your final version.

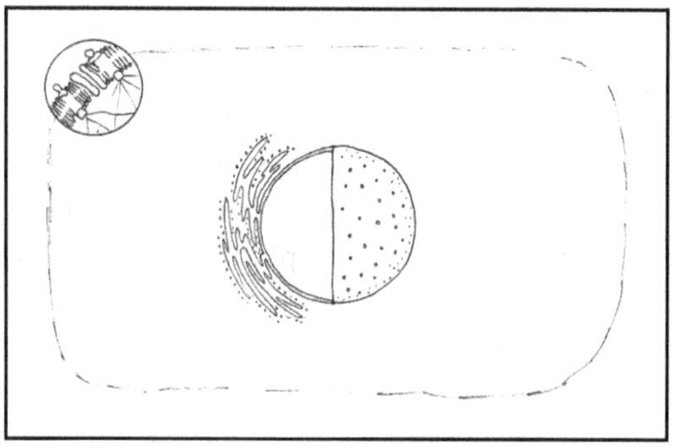

5) Next, sketch out where the nucleus will go. Put it in the center so the crack between the pages goes right down the middle. Make the nucleus about the size of your fist. Don't use a compass or circle guide for the nucleus. You want precise circles for the insets, but not for the natural cell shapes. Nuclei are not perfectly round in real cells, just sort of round. Draw a line down one side of the crack to divide the nucleus in half. The right side will be the exterior view and the left side will the interior view.

6) On the right side, draw tiny circles to represent the pores in the nuclear membrane. If you want to add a 3D effect, you could shade it using pencil (making it look round like a ball) and you could make the tiny circles get smaller and closer together toward the edges. You can go ahead and finish the right side with darker lines. The left side will be your "cut away view" showing what is inside the nucleus. We'll draw the inside in step 9.

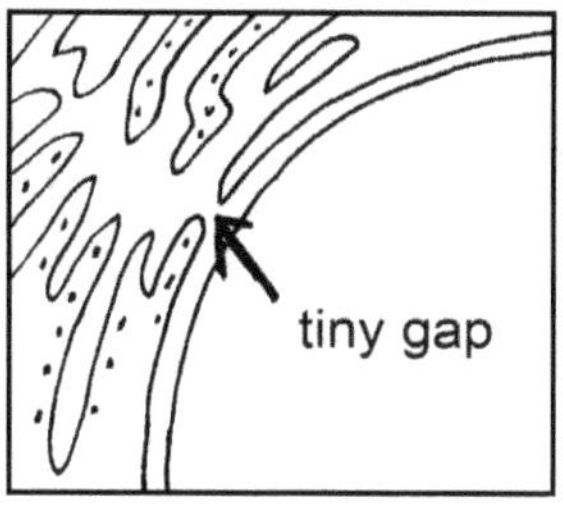

7) The nucleus has a double membrane so you will need to draw two lines going around the outside of the left semi-circle. The inner line will be solid, with no gaps. The outer line will have one or more tiny gaps where the endoplasmic reticulum will join to it. Remember, the ER is "continuous" with the nuclear membrane. A good way draw the ER is to put your pencil down at the top of the nucleus and don't pick it up until you make all your ER squiggles and finally reach the bottom. By not picking up your pencil you can be sure the membrane is continuous. After you finish the ER, put little dots all around it to represent the ribosomes. This side of the nucleus will have the rough ER. We will show the smooth ER on the other side. We will be putting a Golgi body in the space to the left, so you can also draw a few vesicles budding off the ER on that side, getting ready to drift off to the Golgi body.

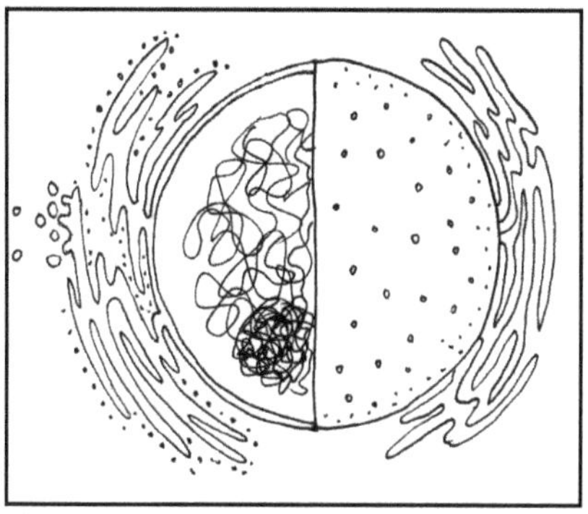

8) Draw the smooth ER on the other side of the nucleus. You can use the same technique of not picking up your pencil until you come back to almost the same point where you started. You could draw some vesicles budding off, but don't draw any ribosomes.

9) Draw DNA in the left semi-circle. You can just make scribbles because at this scale you would not be able to see any detail — just a bunch of DNA "spaghetti." Make one section of the DNA a bit more dense, to represent the nucleolus.

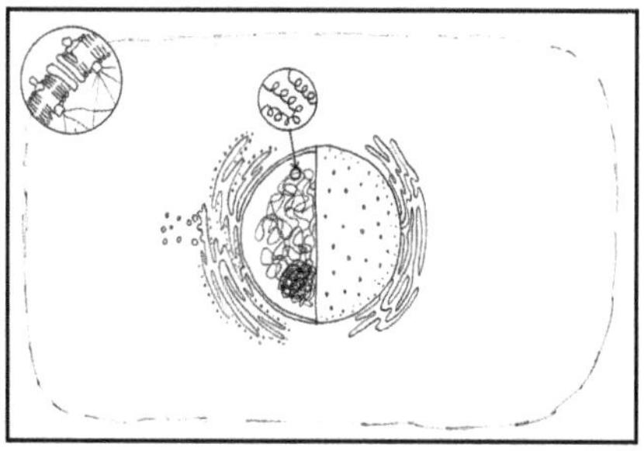

10) You will now add two more insets. These will be slightly different from the first one, as they will have arrows that point to smaller circles. Draw a very small circle around one section of your DNA spaghetti. Then make a large circle outside the nucleus. This larger circle should be at least an inch (2.5 cm) in diameter. Make an arrow pointing from the larger circle to the smaller one. The arrow indicates that the larger circle shows an up-close view of what is in the smaller circle. In this larger circle, draw the DNA as a line coiled around histone spools. (The arrow can point from the small circle to the large circle, if you prefer that.)

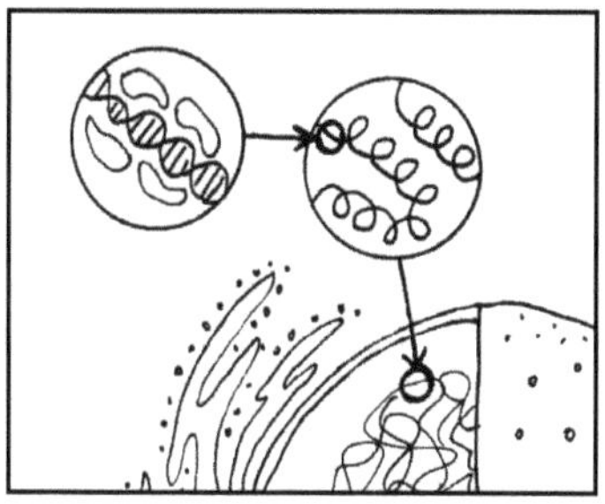

11) Now put a small circle around one tiny section of this coiled DNA. Again, draw a larger circle near by and then draw an arrow from the larger circle to the smaller one. In this second large circle, draw the DNA as a double helix. Don't forget to draw a few chaperone proteins surrounding it, too. These proteins can just be drawn beside the DNA, not covering it. You want the helix to be clearly visible. (The nucleus and rough ER are now finished. Lines can be made final, with pen or with darker pencil lines.)

12) Draw a Golgi body to the left of the rough ER. It can fill most of the space between the rough ER and the left side of the cell, as we won't be putting any other organelles in this space, just the Golgi. Lightly sketch in where the "pancakes" will be. Then as you trace around the shape making your final line, add some vesicles merging with the side facing the nucleus and some vesicles budding off the side facing the plasma membrane. You could also add some vesicles floating away from the Golgi, perhaps on their way to merge with the plasma membrane. It's up to you how 3D to make your Golgi. You can try to shade the pancakes, or you can leave it as a line drawing. It is also up to you if you want to add some enzymes inside the Golgi, working on modifying a protein.

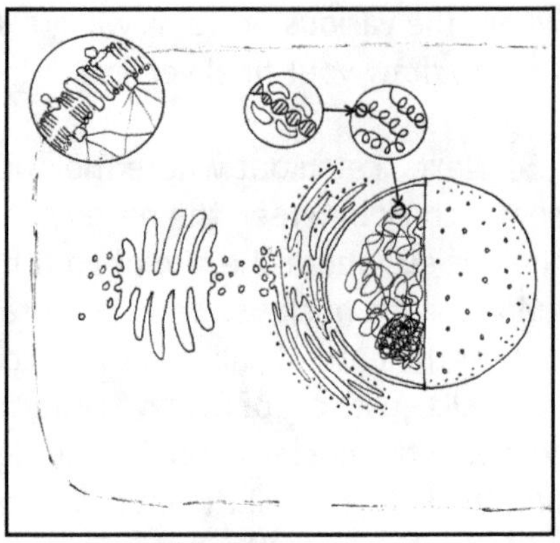

13) In the upper right corner of the cell, sketch an oval and a circle. (You can use a compass or circle guide to draw the circle, as it will be an inset.) The oval will become a mitochondrion and the circle will be used to show the electron transport chain. Draw a line across the middle of the oval. The top half of the oval will show the outside of the mitochondrion and the bottom half will show the inside. If you want to make your mitochondrion look 3D, you can shade the edges of the top half. Draw some lines at various places going partway across the top half. Micrographs of mitochondria usually show these lines, making the mitochondria distinguishable from other organelles. These lines are places where the inner patterns show through.

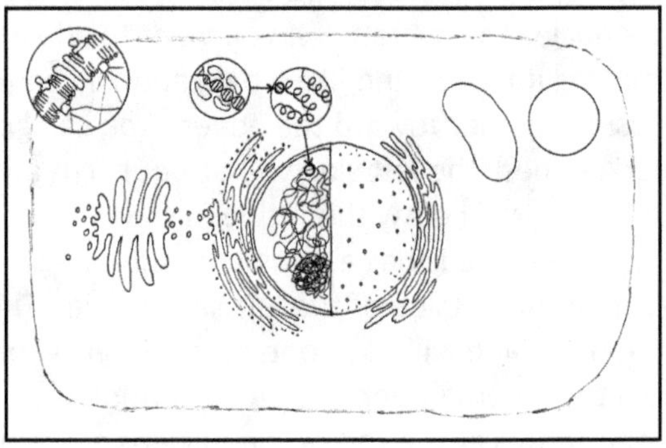

On the bottom half, draw the squiggly outline of the matrix. Draw a small circle around one tiny part of the matrix line, then draw an arrow from the large circle to the small one, just like we did with the close-ups of DNA. In the circle, draw a very simple schematic of the electron transport chain. You only need to indicate the membrane, three ion pumps, the "egg-beater" synthase machine, and little dots for protons above and below the membrane. You might want to make a lot more dots above the pumps than below, to indicate that the pumps are working to increase the number of protons above the synthase machine. You also might want to put a few (very tiny!) ATPs below the synthase machine.

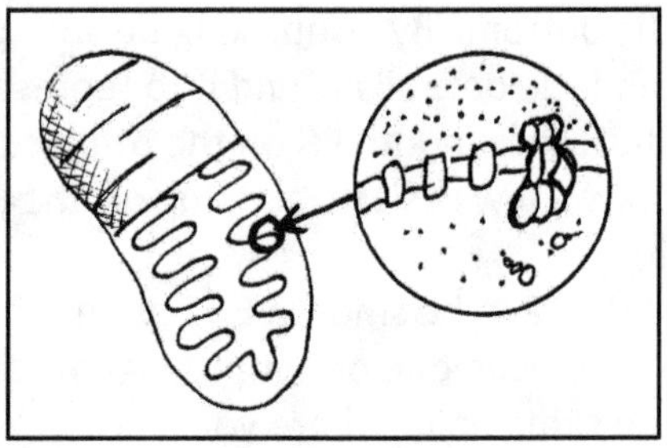

14) The last organelle for this drawing session is the lysosome. Draw an oval below the mitochondrion. Make it a little smaller than the mitochondrion. Making a dividing line to separate the outer view from the inside view. You can shade the outside view to make it look round. For the inside view, draw some small dots of various sizes and shapes to represent the enzymes found inside the lysosome. Make them tiny! At this scale, they still might be too small to see, but we are going to make them visible.

You can go over the organelles and insets with final lines, either pen or darker pencil.

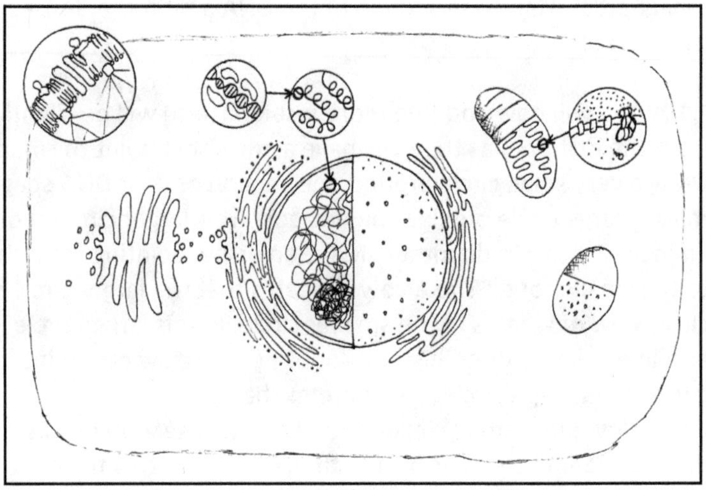

NOTE: DON'T DRAW FINAL LINES FOR THE PLASMA MEMBRANE YET! We still need to add some details to it. Strange as it may seem, the plasma membrane will be one of the last things we will fill in with final lines.

CHAPTER 8

ACTIVITY IDEA 9A: ART PROJECT: SECOND SESSION OF CELL "MINI-MURAL"

NOTE: Because this is a continuation of the project, not a new project, we won't start over with numbering the steps. Rather, we will pick up where we left off and continue with number 15.

15) Draw another small oval below the lysosome. This will be the peroxisome. Make a line dividing it in half and show the outside as a smooth, round shape. On the inside, make little dots along with some lines, representing the crystalline core that is full of enzymes. Of course, in a real cell there would be hundreds, or even thousands of peroxisomes, and also lots of mitochondria and lysosomes. But for clarity, we are putting only one of each in our cell. (Although you are welcome to put in a few more after our last step, if you have enough space.)

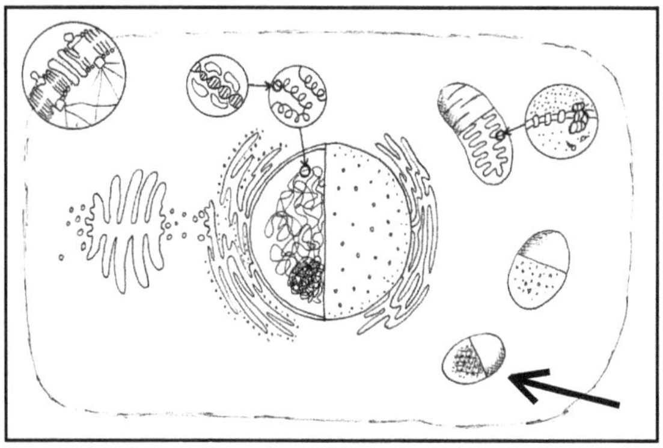

16) Ribosomes are very small. You already drew some ribosomes along the rough ER, but there are also "free" ribosome floating around in the cytoplasm. We will draw a close-up of one of these free ribosomes as it is in the process of making a protein chain. Draw a circle somewhere in the empty space below the Golgi body. Make this circle about the same size as the circles you drew for the DNA and the electron transport chain close-ups. Inside the circle show a ribosome with its two parts. Draw a piece of mRNA going in one side and coming out the other. Also show a finished protein chain coming out the other side. Then draw some tiny tRNAs, each bringing an amino acid to add to the protein chain. After you finish this inset, draw a few dots or tiny circles in that general area, to be in-scale ribosomes.

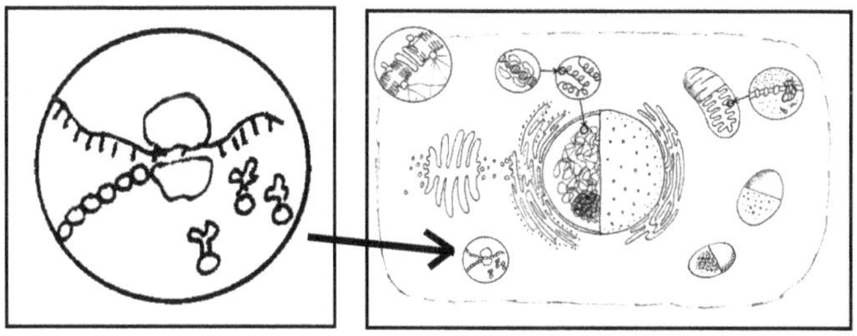

17) The last organelle is the centrosome. Draw a wiggly circle right below the nucleus and then put two barrel-shapes in side of it. (Remember, the two barrels are called the centrioles.) Ideally, they should be perpendicular. An easy way to do this is to draw the front barrel first, then put the other barrel behind it. Start each barrel by making the circle end first.

18) For this step, you will need to find four places in your drawing that have large enough blank spots so that you can add four more small insets. It doesn't matter where they go. Just find four places and draw four inset circles. They can be a little smaller than your other insets if your drawing is filling up. In these four circles, we will put close-up views of the "raw materials" that are floating around in the cytoplasm. In one circle, draw a few little hexagons. These will represent glucose molecules. In the second circle, draw a few circles with line going out from them. These will represent fatty acid molecules. In the third circle, draw a few ovals. These will represent amino acids. (If your ovals are large enough, you could even write the first three letters of some amino acids, such as "PRO" or "VAL" or "LYS.") In the fourth circle, draw a nucleotide. You'll remember that a nucleotide is made of a phosphate, a sugar, and a base. It's a piece of DNA ladder consisting of one rung and the side piece it is attached to. It looks a bit like a "T" lying on its side. If drawing a phosphate, sugar and base is too complicated, just draw some T's.

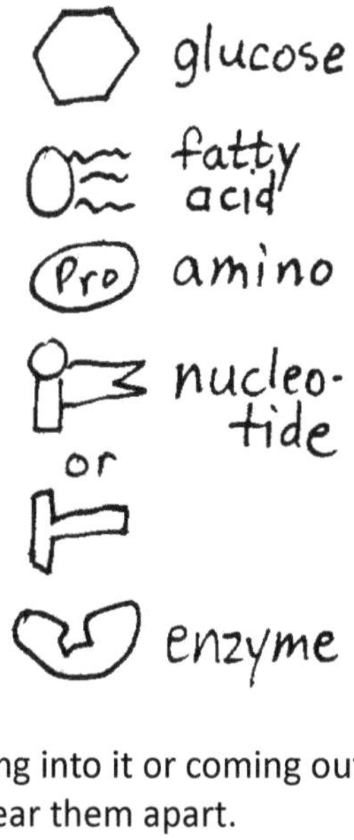

19) If you have one last spot available, draw one last circle. Inside this circle draw one or more enzymes. An enzyme can be drawn as an oval with a weird shape either going into it or coming out of it. Remember, enzymes are not nutrients. They are keys that put things together or tear them apart.

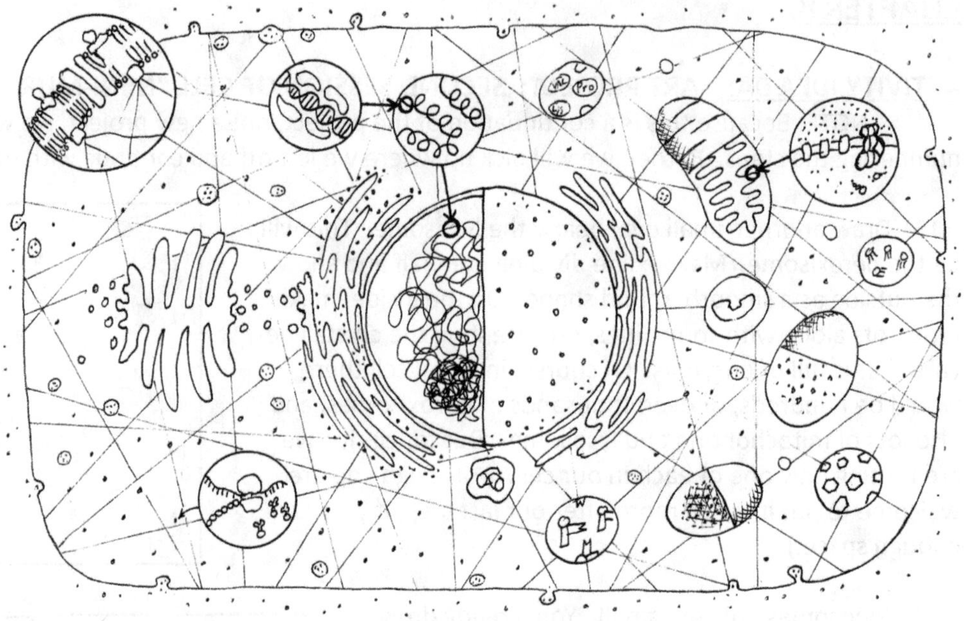

20) Now it's time to draw the final line for the plasma membrane. But first you need to sketch in some little details. Make a few places where pinocytosis or phagocytosis is going on. These would look like little indents in the membrane, some shallow and some deep enough that they are about ready to pinch off and become vesicles inside the cell. Draw some little dots to show the things that are being brought inside the cell. Then sketch in some areas where exocytosis is occurring. These would look like little bumps sticking out from the membrane. Once again, they would have little dots inside them to show the contents that are being exported out of the cell. Once you have these areas sketched in, you can then begin to trace around the membrane with your pen or dark pencil. Remember, you don't have to follow your sketchy lines exactly. If your sketchy lines don't look the way you want them to, you can erase them and move them.

21) Once you have your plasma membrane completed, draw in the cytoskeleton. The cytoskeleton lines should be fairly light so they won't obscure the other objects in the drawing.

Lay the ruler across the drawing (anywhere, at any angle) and draw a light line with your pencil. As you are drawing your line, if you run into any organelles, lift your pencil and don't draw over the organelle. Then move the ruler to another place and do this again. Of course, in a real cell, the microtubules would be running in front of them as well as behind them, but for clarity in our drawing we will leave them behind the organelles. Keep moving your ruler to different angles and adding more lines. Eventually it should look like a network of very light lines running all over the place behind your organelles. You may also want to add another inset that shows a close up view of a motor protein traveling along a microtubule. (Not shown on this sample.)

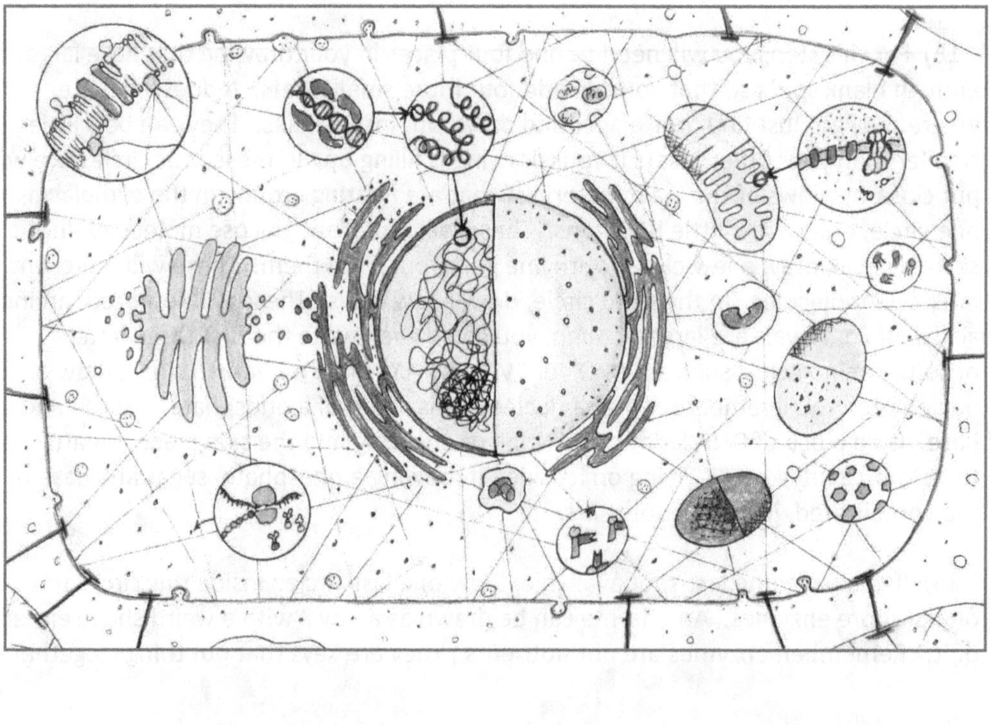

22) Add some desmosomes outside the plasma membrane. If you have space in the corners, you could draw in the edges of some other cells. Draw the little "plates" that are attached to the cytoskeleton just inside the membrane.

23) Now you can do any labeling you want to do. You'll have to make the words fairly small. If you want to add color, colored pencils are a good choice.

<u>ACTIVITY IDEA 9B:</u> GAME: CELL BINGO

The boards for this game will look a bit like the cell "mini-murals" the students drew. In fact, one option would be to play this game before finishing the mural.

Each player will need:

- A copy of the page with the cell squares
- A copy of the page with the blank squares
- About 24 tokens ("markers") to place on each square as clues are read (edible tokens are always popular!)
- A pair of scissors
- A glue stick (or other glue, if sticks are not available)

How to prepare:

Cut apart the cell squares so you have 25 individual square pieces. Glue these squares onto the page with the blank squares. (Don't cut apart the blank squares.) The cell corner pieces must go in a corner, the edge pieces must stay on an edge and the nucleus pieces must still form a circle, but other than that the players are free to put the pieces wherever they want to. Each player should create their own unique pattern.

How to play:

This is a bingo-type game. The only difference is that to win, you must fill both a vertical row AND a horizontal row. (A diagonal row would also be acceptable.) So winning patterns might look something like these:

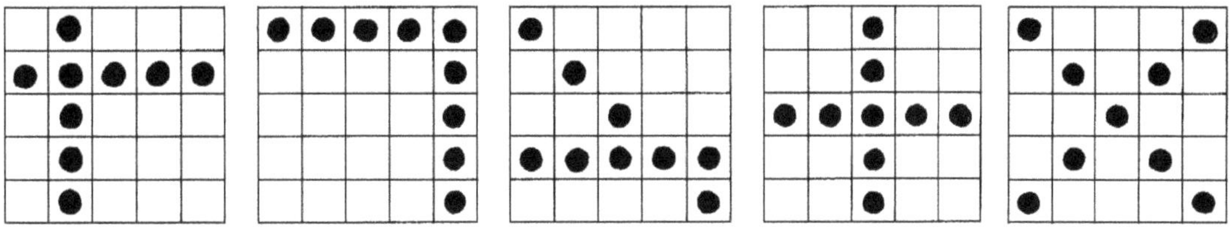

About the clues:

There are three pages of clues. This gives you several options.

- You can play the game three times, getting progressively harder each time.
- You can play just once, selecting the level most appropriate for your players.
- You can keep going even after you have a winner and play through the rest of the clues on that level so that all the clues are used (maximizing review). For example, if someone gets a bingo while you are on question 15 on the easy level, you can just pick up with clue 16 for the next game, then bridge over to the medium clues to finish that game. Then pick up with the rest of the medium clues for the next game.
- You can play all (or most of) the clues on the level if you let the rest of the players keep playing even after a bingo has been called.
- You can do the clues in order, or you can skip around and use the clues you feel are the best review questions for your players. (If you don't go in order, just remember to put a pencil mark next to the ones you've done so that you can keep track of them!)
- If you are using edible tokens, have the players "eat their bingo" and then continue playing.

RE-USABLE OPTION: If you want to use these cell pieces multiple times (for a permanent classroom set), copy the cell page onto white sticky-back label paper (the full sheet 8.5" x 11" labels). Then stick these to pieces of mat board or heavy cardboard. (You could also print the cell pages onto card stock, then use spray adhesive to adhere them to the mat board.) Use a sharp craft knife to cut apart the squares. Mark the backs of each set with a number or letter so that if the pieces get mixed up they are easily sorted again. Put each set in a small zip-lock plastic bag. When it is time to play, give each student a bag and have them arrange their 25 pieces into a square. The pieces will not be glued together, just set into place like puzzle pieces with straight edges.

CELL BINGO Easy Clues

1) This organelle generates energy in the form of ATPs.	MITOCHONDRIA
2) This is a dense area in the nucleus that contains the information for how to make a ribosome.	NUCLEOLUS
3) This organelle contains many digestive enzymes.	LYSOSOME
4) This organelle is often called the "post office" of the cell because it packages and labels things.	GOLGI BODY
5) This is the name of the watery fluid inside the cell.	CYTOSOL
6) These are sugar molecules that the cell breaks down to make ATPs.	GLUCOSE
7) This is called "rough" because it has ribosomes stuck to it.	ROUGH ER
8) This is how the cell takes in little "sips" of the fluid surrounding it.	PINOCYTOSIS
9) This organelle makes toxic hydrogen peroxide but then neutralizes it.	PEROXISOME
10) This structure is a central point for the organization the cytoskeleton.	CENTROSOME
11) These are like little "cardboard boxes" that the cell uses to transport things.	VESICLES
12) DNA is made of these.	NUCLEOTIDES
13) This is like the library of the cell. It contains all the information the cell needs.	NUCLEUS
14) These act like little connecting rods between cells.	DESMOSOMES
15) These things bring amino acids to ribosomes that are busy manufacturing proteins.	tRNAs
16) This organelle consists of a complicated network of smooth tubes. Its membrane is "continuous with" the outer membrane of the nucleus.	SMOOTH ER
17) This is what separates the cell from its environment.	PLASMA MEMBRANE
18) These are the little manufacturing units that take amino acids and string them together into long protein chains.	RIBOSOMES
19) These are the largest fibers of the cytoskeleton. Their structure is similar to drinking straws because they are cylindrical and hollow.	MICROTUBULES
20) These are what proteins are made of. There are 20 types.	AMINO ACIDS
21) These are the smallest fibers of the cytoskeleton.	MICROFILAMENTS
22) Your cells can use these to create and repair phospholipid membranes.	FATTY ACIDS
23) These proteins are "stuck" in the plasma membrane and serve quite a variety of functions, including as portals for letting things in and out.	MEMBRANE-BOUND PROTEINS
24) These proteins help join things together or break things apart.	ENZYMES
25) This is basically a very large empty vesicle.	VACUOLE

CELL BINGO Medium Clues

1) This is where the centrioles are located.	CENTROSOME
2) This has four layers of phospholipids around the outside.	NUCLEUS
3) This is basically the same thing as phagocytosis, just on a smaller scale.	PINOCYTOSIS
4) These are formed by the rough ER and used to transport polypeptides over to the Golgi body.	VESICLES
5) Your digestive system breaks down fats into these molecules.	FATTY ACIDS
6) This organelle has many functions, including storing calcium and manufacturing steroids. Part of its name means "network."	SMOOTH ER
7) These are attached to proteins that are bound to the inside of the plasma membrane. They help the cell maintain its shape or change its shape. They form the "skeletal" structure of a moving pseudopod.	MICROFILAMENTS
8) This organelle gets its name from the Latin word for "empty."	VACUOLE
9) This is where the electron transport chain is located.	MITOCHONDRIA
10) This organelle deals with several kinds of toxins produced by the cell.	PEROXISOMES
11) Glycolysis is when a cell starts harvesting ATPs from this molecule.	GLUCOSE
12) Most of this is consists of water molecules.	CYTOSOL
13) This organelle is responsible for labeling digestive enzymes (from the ER) so that they get transported over to the lysosomes.	GOLGI BODY
14) These are made of two pieces—a larger one and a smaller one.	RIBOSOMES
15) This contains lipid rafts.	PLASMA MEMBRANE
16) About 40 different kinds of these are found inside lysosomes.	ENZYMES
17) This is where ribosomes are made.	NUCLEOLUS
18) This is the smallest type of RNA we've learned about.	tRNA
19) This organelle pumps protons into itself.	LYSOSOME
20) This organelle has docking ports for ribosomes.	ROUGH ER
21) These are the highways along which motor proteins can travel.	MICROTUBULES
22) Some of these act as "flags" to identify the cell as part of the body.	MEM-BOUND PROTEINS
23) These connect skin cells to each other and give skin its stretchiness.	DESMOSOMES
24) The nucleus contains lots of these, all joined together into a twisted ladder shape.	NUCLEOTIDES
25) When proteins have been thoroughly digested, they break down into these simple molecules.	AMINO ACIDS

CELL BINGO Harder Clues

1) This is the only organelle that doesn't have any phospholipids in it.	RIBOSOMES
2) Although usually located towards the center of the cell, much of what this organelle does involves sending its products (such as steroid hormones) outside the cell, often to cells that are very far away.	SMOOTH ER
3) This organelle is full of chaperone proteins. Somehow they stay in place even though the organelle is in constant flux.	GOLGI BODY
4) Some of these require an acidic environment to work properly.	ENZYMES
5) During the last part of mitosis, the movement of these causes the cell to pinch in the middle.	MICROFILAMENTS
6) Tay-Sachs Disease is caused by a malfunction of this organelle.	LYSOSOME
7) This has an inner region called the matrix. The walls of the matrix contain ion pumps.	MITOCHONDRIA
8) Examples of these include: serine, proline, lysine, arginine, and valine.	AMINO ACIDS
9) Inside this are little "snipper" enzymes that can snip off the ends of protein chains being manufactured by ribosomes that are attached to it.	ROUGH ER
10) This is where glycolysis occurs.	CYTOSOL
11) This can be filled with waste and sent to merge with the membrane.	VACUOLE
12) This has an anti-codon.	tRNA
13) One of this organelle's main jobs is to break down long chains of fatty acids. It also can detoxify alcohol and other wastes.	PEROXISOME
14) This process forms small vesicles filled with water, minerals, and hopefully some food molecules.	PINOCYTOSIS
15) These were discovered by using antibodies stained with a fluorescent dye. The antibodies attacked and covered these, then the dye made them show up on a screen.	MICROTUBULES
16) This is where you would find histones with DNA wound around them.	NUCLEUS
17) This contains 6 carbon atoms and its basic shape is a hexagon.	GLUCOSE
18) This has thousands of sets of instructions for making just one thing.	NUCLEOLUS
19) These are formed by both smooth and rough ER and by the Golgi.	VESICLES
20) The structure of this is sometimes described as a fluid mosaic.	PLASMA MEMBRANE
21) These are basically long strings of carbon atoms.	FATTY ACIDS
22) Some of these act as receptors, receiving chemical messages from other cells.	MEMBRANE-BOUND PROTEINS
23) These are attached to the inside of the plasma membrane with a plate-like structure. The plate is then connected to the cytoskeleton.	DESMOSOMES
24) This is made of a base, a sugar, and a phosphate.	NUCLEOTIDES
25) This is made of two barrel-shaped things inside a blob of protein.	CENTROSOME

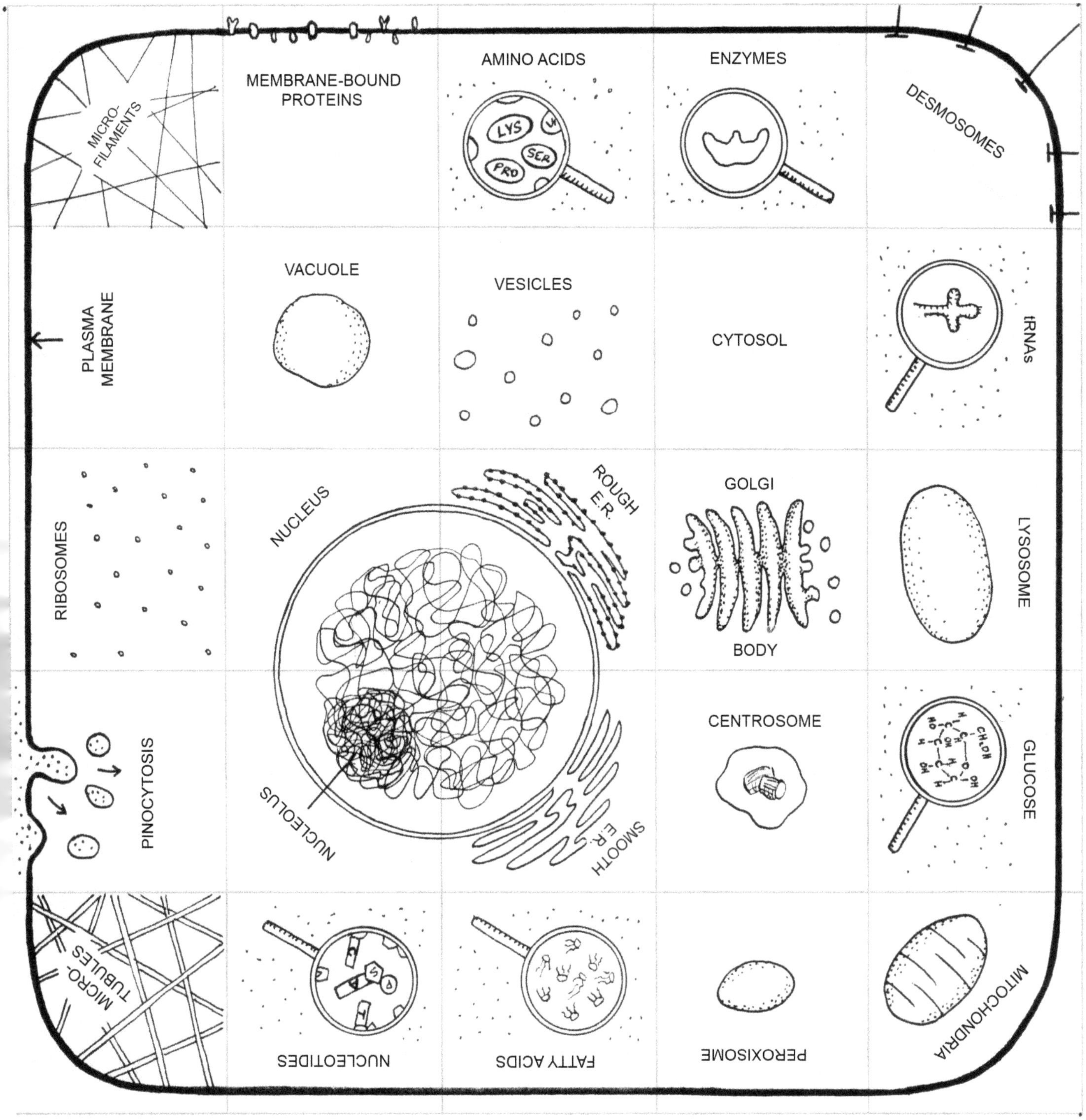
MICRO-FILAMENTS
MEMBRANE-BOUND PROTEINS
AMINO ACIDS
LYS
SER
PRO
ENZYMES
DESMOSOMES
PLASMA MEMBRANE
VACUOLE
VESICLES
CYTOSOL
tRNAs
RIBOSOMES
NUCLEUS
ROUGH E.R.
GOLGI BODY
LYSOSOME
PINOCYTOSIS
NUCLEOLUS
SMOOTH E.R.
CENTROSOME
GLUCOSE
MICRO-TUBULES
NUCLEOTIDES
FATTY ACIDS
PEROXISOME
MITOCHONDRIA

<u>ACTIVITY IDEA 8C</u>: TABLETOP RELAY RACE: "CATABOLISM RELAY" (includes the E.T.C.)

Was this chapter confusing? It's difficult to comprehend how all the parts (glycolysis, Krebs cycle, E.T.C.) go together to make one long process. This active game will help the students see the whole picture of how the processes involved in respiration fit together.

NOTE: I highly recommend watching the informal video I took the day I set this up for my class. It is posted on the Cells YouTube playlist.

You will need:

- 6 pairs of scissors
- 2 staplers
- some sheets of colored paper, and some sheets of white paper
- 2 medium sized cardboard boxes and something with which to cut it
- clear tape
- a black marker
- quite a few copies of the following pattern page on plain white paper
- the paper pumps parts and other small items from the E.T.C. relay race in chapter 4

What to tell the students:

This relay race will help you understand how all the parts of respiration fit together. You will act out what happens in a mitochondrion as we follow glucose molecules and see how they are chopped up and processed. Some of you will be the scissor-and-stapler enzymes. Others will transport acetyl-CoA through the Krebs cycle factory or run the pumps at the electron transport chain. You will see where those first little shuttles came from when we first played the E.T.C. relay.

How to prepare:

Below are posted some photos I took when I set this up for my class. I set up both sides of the tables with identical supplies, but color coded them, where appropriate, for a red team and a blue team. I used two 8-foot tables and one 6-foot table. (If you don't have tables, you could also set it up on the floor.)

STATION #1: DIGESTION

The person at this station will cut glucose molecules off the starch chain. The scissors represent enzymes such as amylase, which break apart starches.

I provided my players with pencils, too, and required them to "patch up" the glucose molecule after they cut it off, putting an O on one side and an OH on the other, at the places where the bonds were cut. (The enzyme that takes apart starches puts on the H and OH.) The glucose molecule was then handed to the person at the next station.

This station can be combined with station #2 if you have few players.

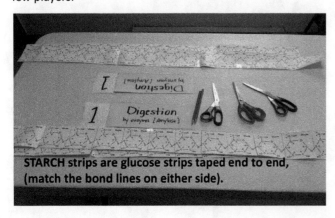

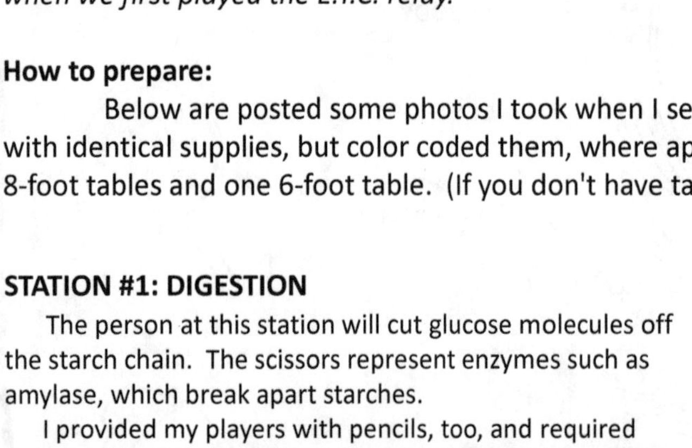

by enzymes (Amylase)
STARCH strips are glucose strips taped end to end,
(match the bond lines on either side).

STATION #2: GLYCOLYSIS

I used 10 cards to represent the 10 steps of glycolysis. I drew a number on each card and (during prep time) had them draw a funny cartoon "enzyme guy" under each number. To accomplish glycolysis, they must turn the cards over, <u>in order</u>. The turning of the card represented accomplishing that step. We did not "reset" the cards before the next round; the next player simply turned each card back over again. After turning the cards, the player took a scissor and cut a glucose in half to make pyruvates. (I did not worry about where it was cut-- there was no way to show the modification of the glucose molecule.) Pyruvates for the next step had been prepared ahead of time, and were already waiting in a bowl.

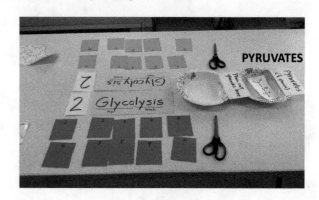

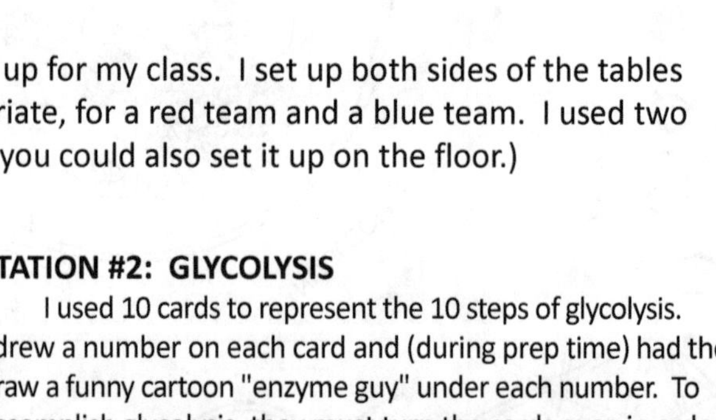

Copy the following pattern pages onto white paper. The number of copies you will make will depend on how many player you have in your group and how many times you want to run the relay. Make plenty!

Cut the glucose molecule strips and tape them end-to-end to make two very long starch molecules (one long strip for each team). My strips were about 10 feet (3 meters) long.

Cut two sets of 20 cards (10 for glycolysis, 10 for Krebs), for each team. Write a number on each side, but only draw a cartoon on one side so that it will be obvious whether or not a card has been flipped.

You can make your Krebs "factory" any way you'd like. In a pinch, you can even make do without a box.

Copy and cut out the CoAs and the pyruvates. I made about four dozen for my group and we used all of them. Technically, you need two pyruvates for every glucose molecule, but if this detail goes by the wayside and only one pyruvate gets handed over to the person at station 3, that's okay. (If you watch the video of my students doing these stations, they got really fast and efficient and it was not possible to keep track of how many glucose molecules were being processed. But they had a lot of fun and learned a lot —mission accomplished!)

How to play:

You will have to decide how many players to have at each station. You could even have one player do the whole process. I found that the kids in my group (ages 10 to 13) did not even care so much about the competition aspect. The challenge for them was to keep the processes going constantly, like a high speed assembly line. They wanted to keep getting faster and faster just to see how fast they could go.

I started by having each player at a station and doing the process once through. Then I had them all slide down a place and we did the process again. We slid down again and again until each player had a turn at each station. After that, my groups starting self-organizing and choosing their favorite stations. I just went along with it and observed. They showed great teamwork and became very efficient, finding ways to streamline and help each other.

The red letters show how the snipped glucose molecule needs to be "patched." (The H and OH come from a water molecule.)

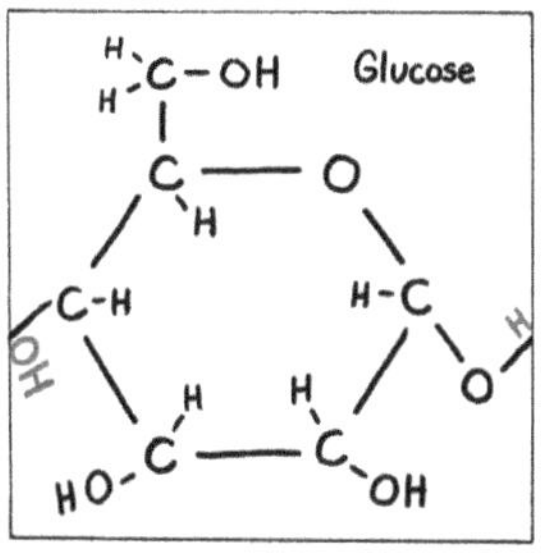

The dotted line shows where to snip in order to release a carbon dioxide. The other side gets stapled to a CoA.

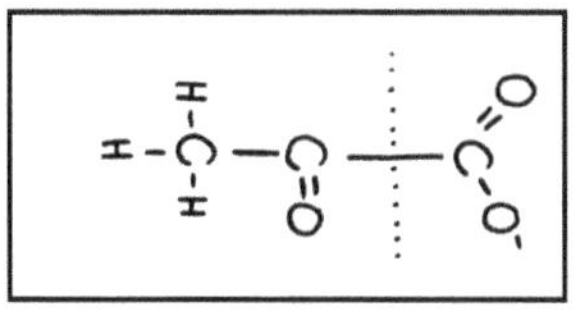

STATION #3: ENTERING THE MITOCHONDRIA

I used two blue stripes of masking tape to represent the membrane around the mito. The person running this station took a pyruvate, cut off the CO_2 on the end, and stapled a CoA to the 2-carbon part of the molecule, to make acetyl-CoA. They handed the acetyl-CoA to the person at the Krebs factory.

STATION #4: KREBS CYCLE

This station needs a "Krebs factory" and 10 cards to represent the 10 steps of the Krebs cycle (a.k.a. citric acid cycle). The paper acetyl-CoA went into a slot on the left side of the box. The player then turned over each card, in order, 1 to 10. After turning over the cards, the player put two electron tokens into the waiting NADH shuttle bus on the right side.

STATION #5: E.T.C. (Electron Transport Chain)

You already know what happens here. If you need to review, see instructions in chapter 4. and/or the video explanation in the video posted on the "Cells" YouTube playlist.

You can have more than one player working the E.T.C. since it is so complicated.

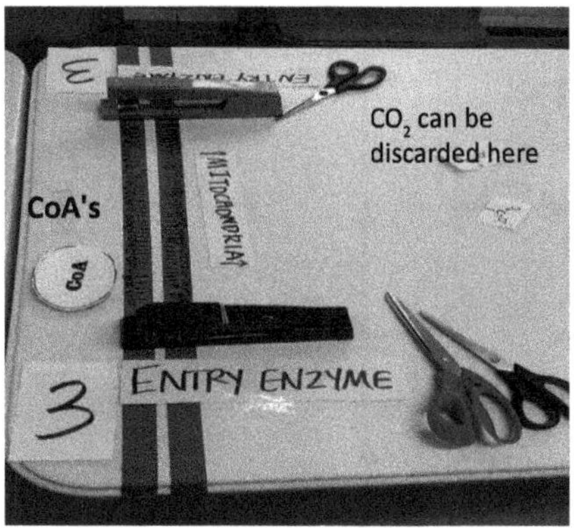

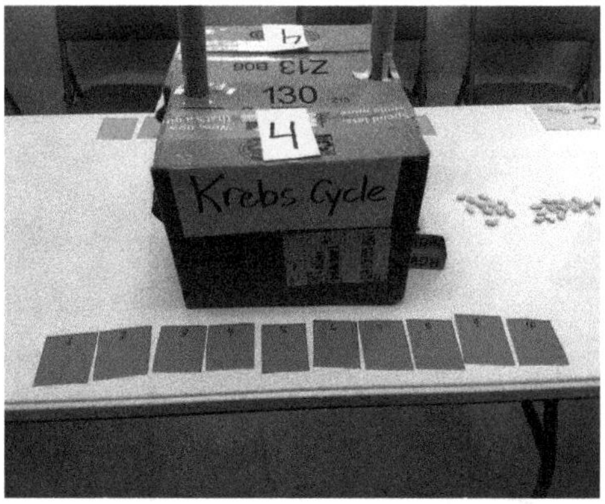

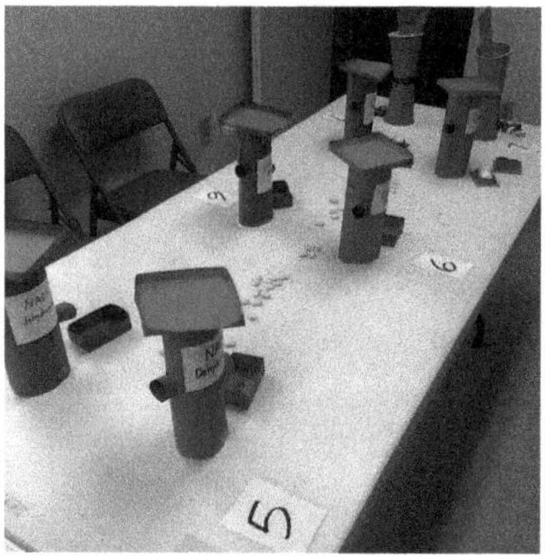

Coenzyme A molecule are very large and complicated—far too large to be shown hefre.

PYRUVATES

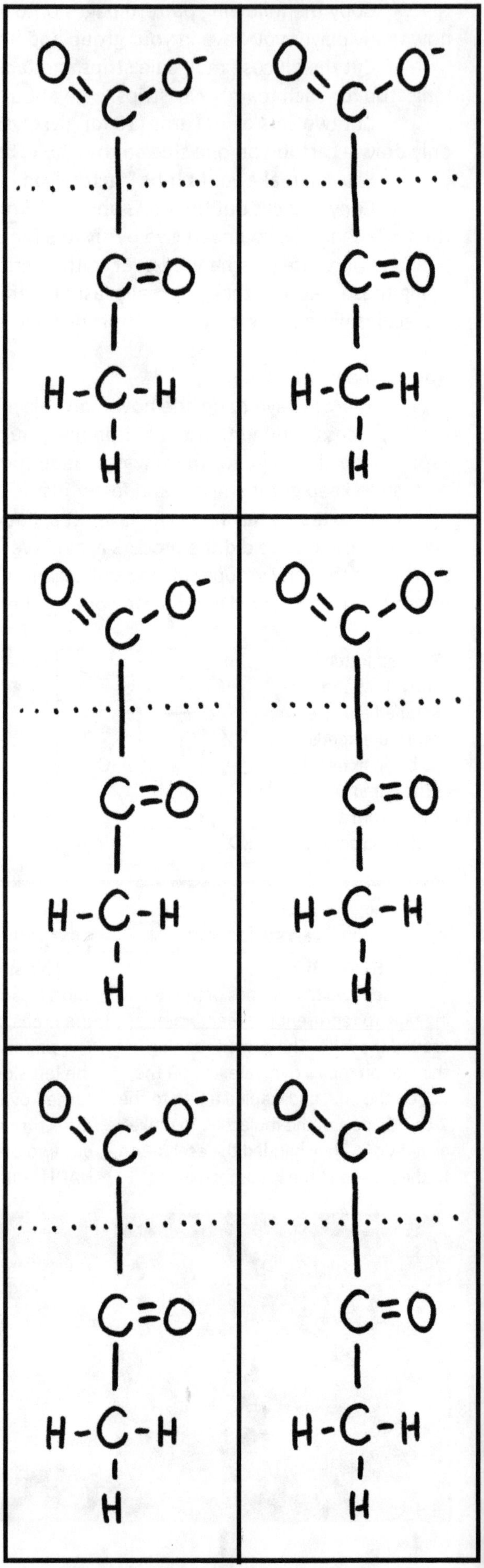

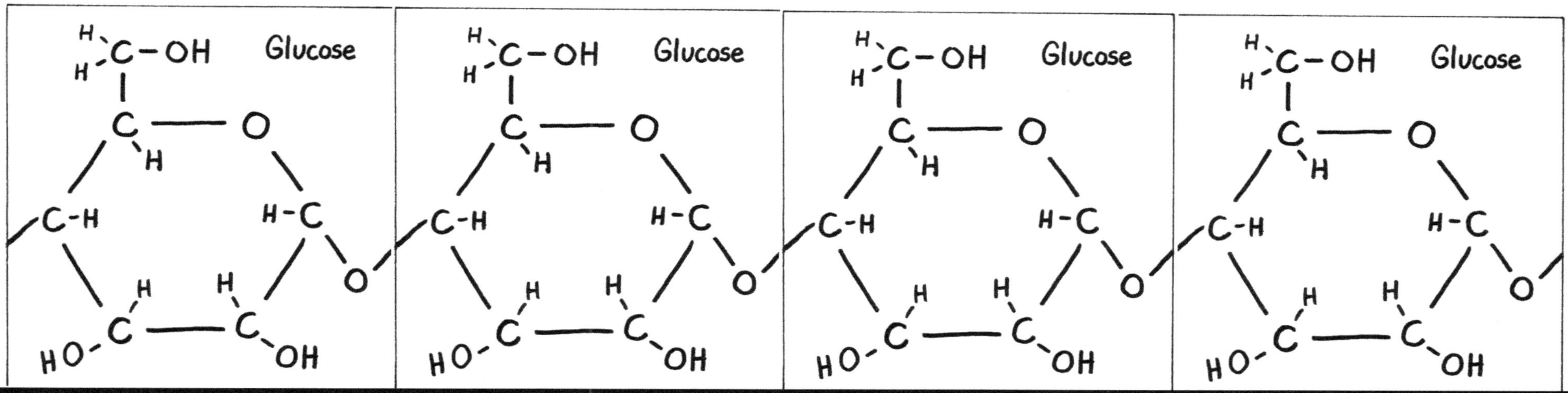
Glucose
Glucose
Glucose
Glucose

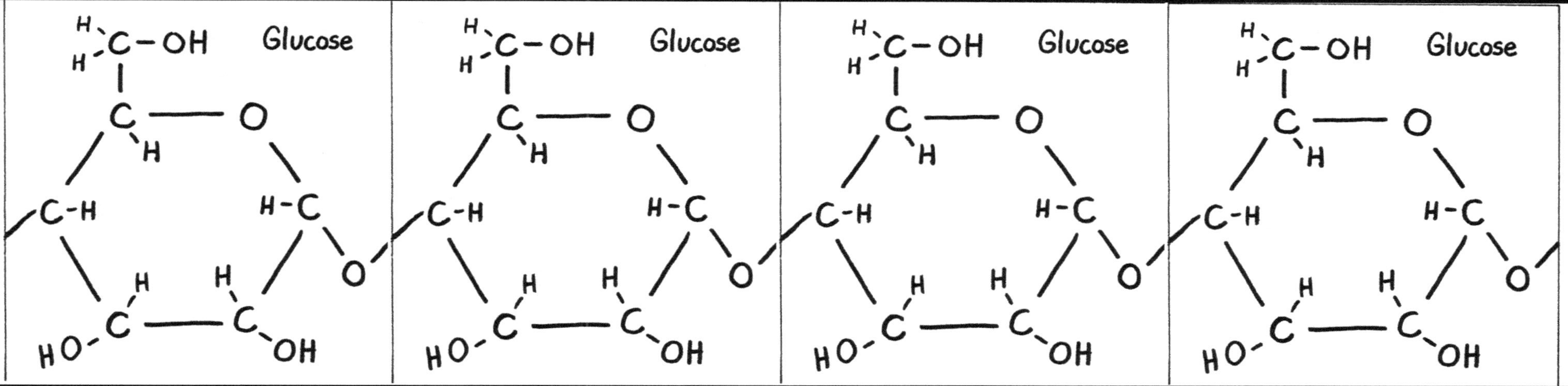
Glucose
Glucose
Glucose
Glucose

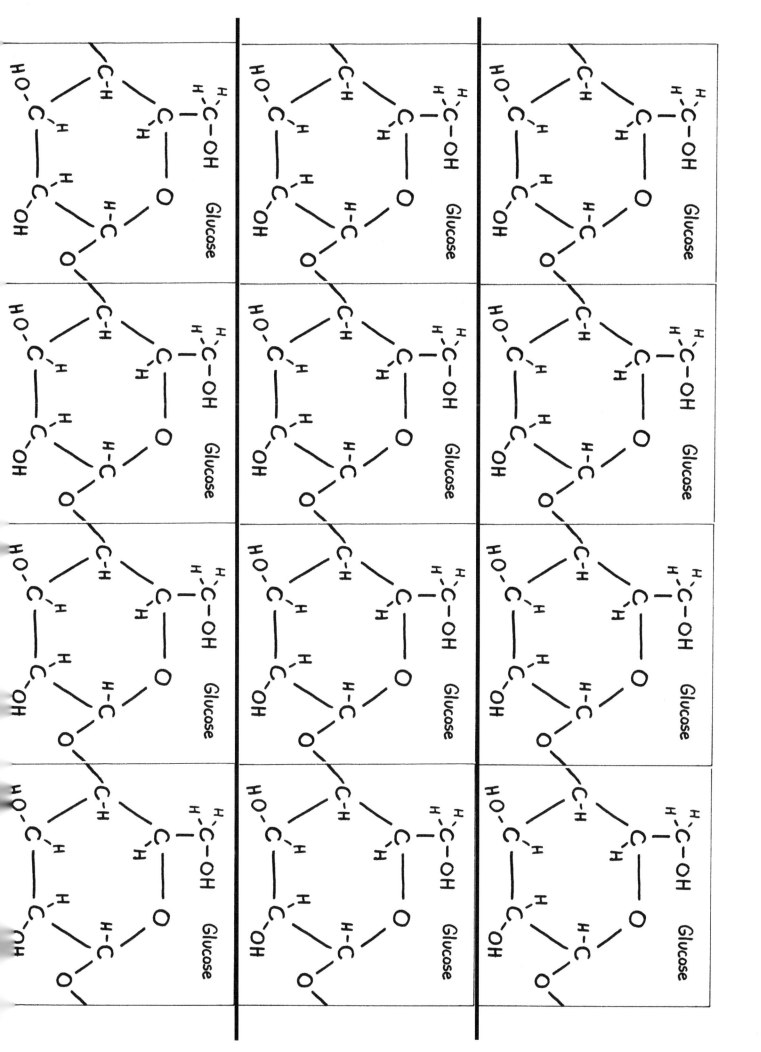

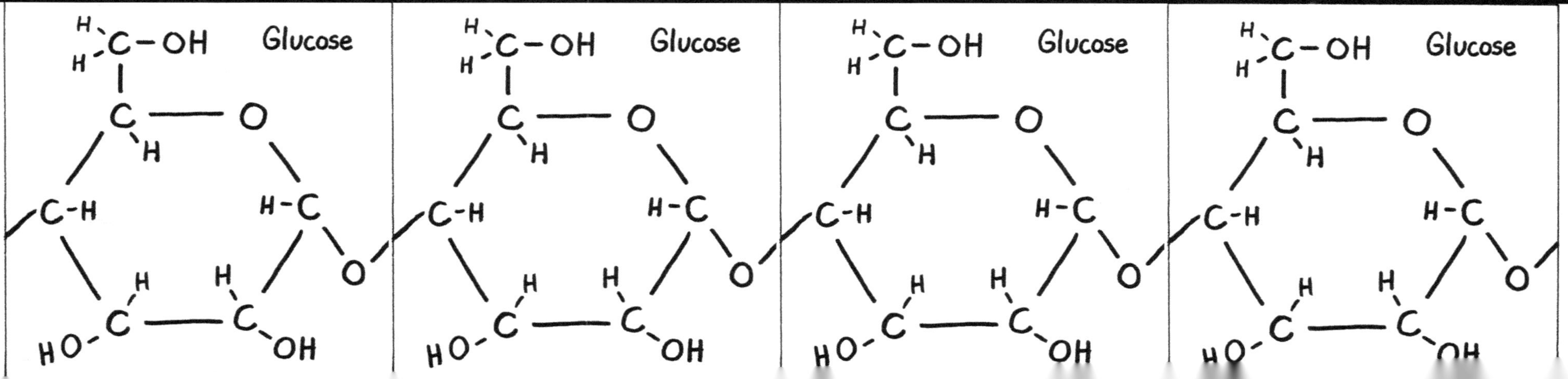
Glucose
Glucose
Glucose
Glucose

CHAPTER 9

Don't forget about the YouTube playlist videos! There are some really nice ones on mitosis and meiosis.

ACTIVITY IDEA 9A: VIRTUAL LAB about GEL ELECTROPHORESIS OF DNA

This lab procedure uses electricity and a gel medium to sort DNA into various lengths so that patterns can be seen. **http://learn.genetics.utah.edu/content/labs/gel/**

ACTIVITY IDEA 9B: ONLINE ACTIVITY: A matching game about karyotypes

You've seen karyotypes but didn't know the correct name for them. Karyotypes show you a complete set of chromosome with all the pairs matched and lined up in a row. (Remember, this only happens during mitosis or meiosis.) In humans there are 23 pairs of various lengths, with the last pair being the XY pair that determines gender. Here is an interactive online activity where you try to match chromosomes into pairs. They use micrographs of real human chromosomes.

https://learn.genetics.utah.edu/content/basics/karyotype/

ACTIVITY IDEA 9C: VIRTUAL LAB: PCR LAB (Polymerase Chain Reaction)

This procedure is used to make billions of copies of just one tiny piece of DNA. The tricks is to target one part of the DNA with "primers," then use enzymes and heat to switch various natural processes on and off.

http://learn.genetics.utah.edu/content/labs/pcr/

ACTIVITY IDEA 9D: ONLINE ACTIVITY: Meiosis mistakes

Here are some brief animations you can click on that show you possible scenarios for genetic mistakes during meiosis — mistakes that will cause the offspring to have genetics syndromes or disorders (such as Downs Syndrome). **http://learn.genetics.utah.edu/content/begin/traits/predictdisorder/**

ACTIVITY IDEA 9E: AUDIO VISUAL PRESENTATION on the 4 kinds of DNA

This is an online audio-visual presentation about inheritance of DNA and how it can be used to trace ancestry. **http://learn.genetics.utah.edu/content/extras/molgen/index.html**

ACTIVITY IDEA 9F: MAKE A MITOSIS FLIP BOOK

You will need:
- Patterns for flip book, printed onto card stock
- Scissors
- White glue (good PVA or craft glue, not white "school glue") or high quality glue stick
- Fine sandpaper (a belt sander is even better!)

Option 1: Use color pattern pages

If you would like to make a flip book from actual micrograph pictures, check out this address:
http://annex.exploratorium.edu/imaging_station/activities/flipbooks/flipbooks_mitosis.php

The pictures are from fruit fly embryo cells. There are many cells on each picture, not just one. The cells were stained with red and green, and the background is fairly dark, so the pictures have high saturated of color. The text says that fruit fly embryo cells multiply very quickly. The series shown in the flip book took about 10 actual minutes.

Option 2: Use black and white pattern pages

1) Print the following patterns onto card stock. The stiffer paper will make the book more flip-able.

2) Cut out the pages very carefully. The neater you cut, the better the book will work.

3) Decide whether your flip book will flip from the bottom or top. (Imagine you have a flip book in your hand and flip it. Was picture 1 on the bottom or top? Did you use your right hand or your left?) Stack the pages accordingly. Page 1 can be on the top or the bottom, as you choose.

4) As you glue the pages together, make sure the ends opposite the numbers are lined up as perfectly along the edges as possible. The natural inclination of the students will probably be to line up the edges where the numbers are. However, the spine of the book (where the numbers are) is not the crucial edge. The edge where your fingers flip the book is the edge you want to focus on. Line up the pages from this end so the book will flip smoothly.

Use only a TINY amount of glue. If any glue at all seeps out from the crack, you are using too much. Put on a tiny drop or two and spread them out with your finger.

5) Several blank pages are provided so that you can design a cover for the flip book, or add a few pages at the end to make it flip better. Before you put on a cover, decide whether you will be flipping the book with your left hand or your right hand. If you are going to flip with your left fingers, design the cover so that the spine of the book will be on the right of the cover. And vice versa for right hand.

6) After you have finished all the gluing, hold the book very firmly in one hand, and hold a piece of sand paper down on the table with the other hand, and pull the book across the sand paper in one direction. Continue sanding in this one direction until the edge is as smooth as possible. The smoother the edge, the better the book will flip. (A belt sander is faster and gives a better finish than sanding by hand, if you happen to have one. Just make sure to press gently so your whole book doesn't get ground away!)

When you flip your book, if there are pages that get skipped, this means that those pages are a little too short. You can do more sanding until all the pages are exactly the same length.

Option 3: Make one from scratch

Make several copies of the blank template page (onto card stock). Draw a simple cell on one the pages and mark it as page 1. Then place the second page on top of it and then hold them up to a window so you can see page 1 through page 2. Now make your second picture just slightly different from your first one, but don't stray too far from your first design. Mark this as page 2. Then do the same for pages 2 and 3, and so on. Make the changes very gradually. Don't glue the pages together until you have them all drawn.

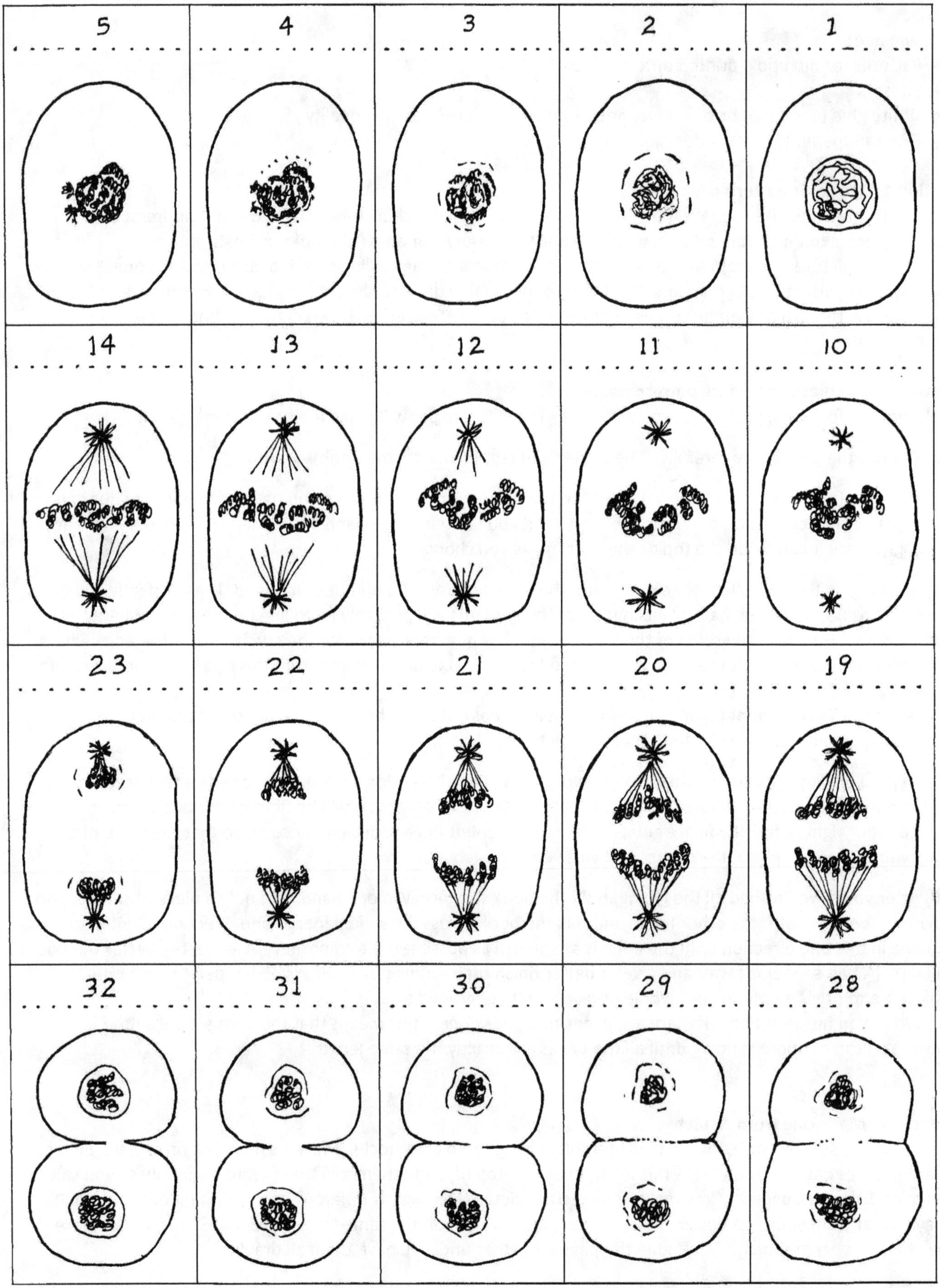
5
4
3
2
1
14
13
12
11
10
23
22
21
20
19
32
31
30
29
28

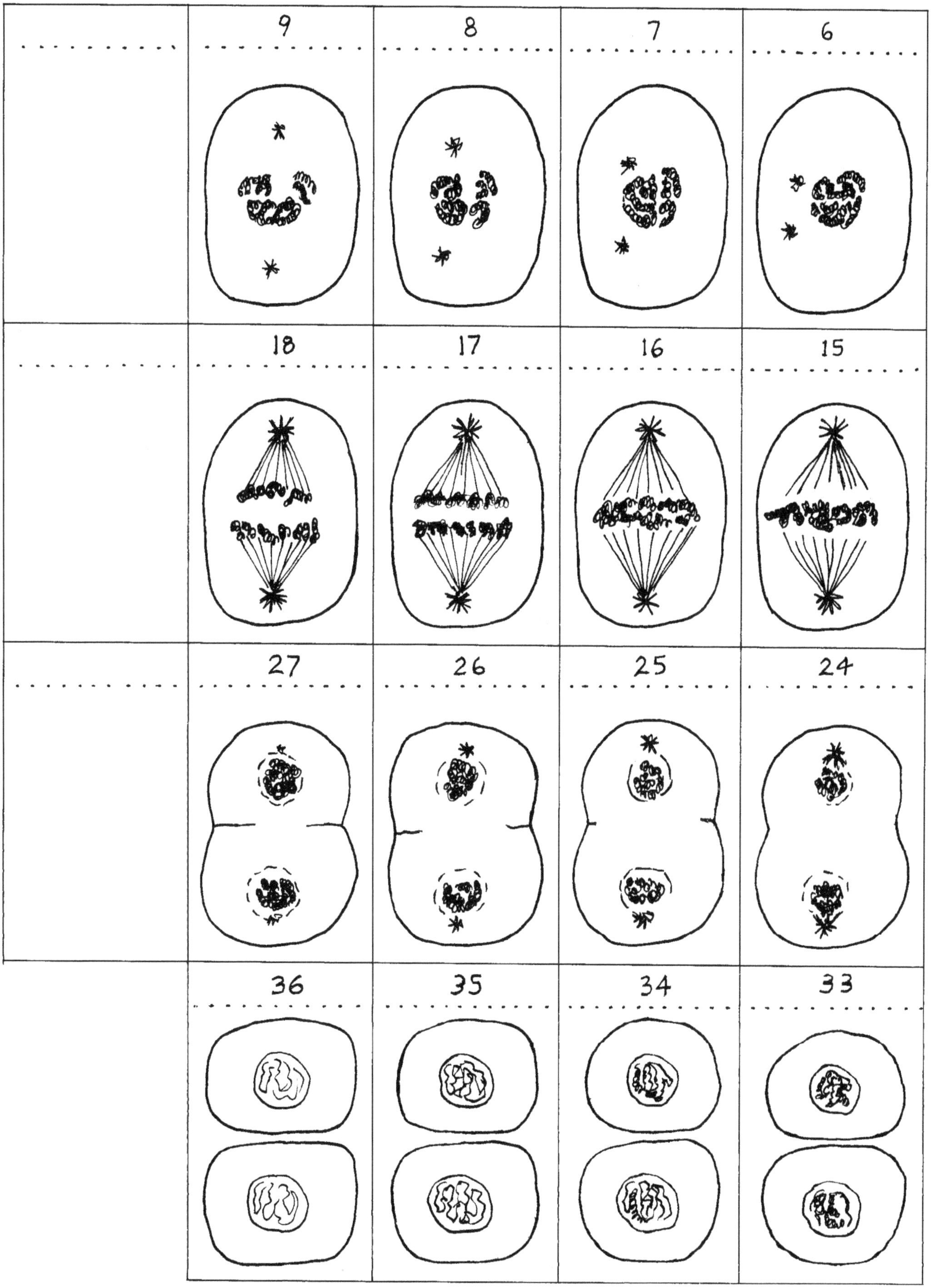
9
8
7
6
18
17
16
15
27
26
25
24
36
35
34
33

CHAPTER 10

ACTIVITY IDEA 10A: QUIZ GAME

This is a Jeopardy-style game but the answers don't have to be in question format. Most of the questions are taken out of chapter 10, although there are also review questions about cells in general, but disguised as questions about particular cells.

You will need:

- A playing board (see note below)
- Some kind of "markers" to put on the squares of the playing board as questions are answered
- The questions on the following pages

How to prepare:

If you are playing with a fairly large group you might want to make a version of the playing board that can be put on a wall. This is easily done with a piece of poster board and a marker. Or you can make it even larger, and use a sheet of paper for each square.

You will need some kind of "markers" such as squares of colored paper, or colored sticky notes, for keeping track of which team scores on which squares. You will need a color for each team. If you are playing with a board tacked to the wall you will obviously need your markers to be sticky.

Divide the group into teams. Two teams can work well if you let players take turns answering the questions. Depending on your group, you may want to have the player who is answering the question be able to ask advice of the other group members. This is a particularly good option if you have players that lack self-confidence.

You will note that there are two games available. It does not matter which you choose. They are designed to be approximately equal in difficulty.

How to play:

Choose a team to go first. They may choose any square on the board. They call out a cell type and the number, such as "Nerve Cells for 100" or "Bone Cells for 300." The teacher/adult then reads the question that corresponds to that square. If the correct answer is given, the team gets to put one of their colored markers on that square.

In the television show "Jeopardy" contestants continue to control the board as long as they give correct answers. This format won't work so well in this game and will result in cries of, "It's not fair!" or, "When will it be our turn?!" I recommend having the teams take turns calling out squares.

After the questions have all been answered, add up the scores to determine who won.

ACTIVITY IDEA 10B: ONLINE VIDEO ABOUT STEM CELLS

Check out this short but informative video on cells that make other cells (stem cells).

https://learn.genetics.utah.edu/content/stemcells/scintro

EPITHELIAL CELLS	SECRETORY CELLS	BONE CELLS	MUSCLE CELLS	NERVE CELLS	BLOOD CELLS
100	100	100	100	100	100
200	200	200	200	200	200
300	300	300	300	300	300
400	400	400	400	400	400
500	500	500	500	500	500

GAME ONE:

EPITHELIAL

100: In the word "epithelial" what does "epi" mean? OUTSIDE or OUTER
200: Epithelial cells divide using this process. MITOSIS
300: What are these hair-like structures in your intestines called? VILLI
400: What do you call the epithelial cells that form the surface layer of your skin? KERATINOCYTES
500: Endothelial cells in capillaries use this process to "drink" in nutrients from the bloodstream. PINOCYTOSIS

SECRETORY

100: This type of secretory cell makes mucus. GOBLET CELL
200: The main component of mucus is a protein called ______. MUCIN
300: The nucleus of a goblet cell is located where? Top, middle or bottom? BOTTOM
400: Mucin proteins expand after they are released from the cell. Their volume can expand by how much?
A) 10 times B) 100 times C) 600 TIMES D) a million times (Answer: C)
500: Where would you NOT find goblet cells? A) trachea B) skin C) intestines D) nose (Answer: B)

BONE CELLS

100: What does the root word "osteo" mean? BONE
200: Once an osteoblast is stuck inside the mineral "cage" it has built around itself, it is called an ___. OSTEOCYTE
300: The photographs of osteocytes in this chapter came from what animal? DINOSAURS
500: Which of these can an osteocyte NOT do? A) Live a long time B) Send signals to other osteocytes
C) Use mitosis to reproduce D) Control the mineral content of bone (Answer: C)
500: The spiderweb-like network of collagen made by osteoblasts is called the _______. MATRIX

MUSCLE CELLS

100: Muscle cells are also called muscle ______. FIBERS
200: Which muscle protein is also found in a cell's cytoskeleton— actin or myosin? ACTIN
300: Which molecule plays a critical role in the action of myosin and actin?
A) ATP B) glucose C) phosphate D) water (Answer: A)
400: Muscle fibers have a special membrane around them called the ___________.
A) plasma membrane B) sarcolemma C) sarcoplasmic reticulum D) myolemma (Answer: B)
500: Inside the muscle fiber's many mitochondria, ATPs are recycled by a tiny machine called ___.
(Hint: This machine looks like an egg beater.) ATP SYNTHASE

NERVE CELLS

100: The branch-like things on the cell body of a neuron are called __. DENDRITES
200: In the brain, neurons are surrounded by cells that protect them, hold them in place and nourish them.
These cells are called: A) Schwann cells B) glial cells C) epithelial cells D) matrix cells (Answer: B)
300: Which of these cells is NOT found in the brain?
A) astrocyte B) oligodendrocyte C) Schwann cell D) microglia (Answer: C)
400: There is a tiny gap between neurons, called the __. SYNAPSE
500: What goes across the tiny gap between neurons? CHEMICALS, or NEUROTRANSMITTERS

BLOOD CELLS

100: What is the correct name for a red blood cell? ERYTHROCYTE
200: Where are blood cells "born"? BONE MARROW
300: How long can a red blood cell live? ABOUT 3 OR 4 MONTHS
400: Which of these is NOT found in a red blood cell?
A) nucleus B) mitochondria C) ribosomes D) all of these E) nucleus and ribosomes (Answer: D)
500: Red blood cells are full of this oxygen-carrying molecule. HEMOGLOBIN

GAME TWO:

EPITHELIAL

100: This epithelial cell makes a pigment called melanin. MELANOCYTE
200: This epithelial call lies at the bottom of your epidermis. BASAL CELL
300: These are long, thin strands of protein which gradually fill up skin cells. KERATIN
400: This epithelial cell in the villi of the intestines is responsible for pulling nutrients out of food. ENTEROCYTE
500: This epithelial cell in the villi of the intestines makes defensins to fight germs. PANETH CELLS

SECRETORY

100: How many types of sweat glands do you have? 2
200: All types of secretory cells use this organelle to manufacture their proteins. RIBOSOME
300: Which organelle packages mucin proteins into vesicles? GOLGI BODIES
400: What ingredient in mucus causes it to be sticky? SUGARS
500: Which of these is NOT produced by a secretory cell? A) blood B) milk C) saliva D) tears (Ans: A)

BONE CELLS

100: Bone cells are arranged in a circular pattern called an ___ OSTEON
200: Which mineral would NOT be found in bone?
A) iron B) magnesium C) calcium D) phosphorus (Answer: A)
300: The branch-like "thread feet" of bones cells are called__. FILOPODIA
400: What kind of bone cell breaks down the matrix to release minerals? OSTEOCLASTS
500: How long can an osteocyte live? SEVERAL DECADES (at least 20 years)

MUSCLE CELLS

100: How many nuclei are in a muscle fiber? A) none B) 1 C) 2 D) many (Answer: D)
200: The bundles inside muscle fibers are called:
A) bundles B) myofibrils C) myosins D) sarcolemmas (Answer: B)
300: When calcium atoms bind to myosin proteins, the muscle fiber will do this. CONTRACT (GET SHORTER)
400: Which would a muscle have more of—mitochondria or Golgi bodies? MITOCHONDRIA
500: What does the sarcoplasmic reticulum do? A) Store calcium ions B) Manufacture proteins C) Attach the muscle cell to a capillary D) Control diffusion of water molecules into cell (Answer: A)

NERVE CELLS

100: "Dendrite" is Greek for ____. TREE
200: The long, thin middle part of a neuron is called the ______. AXON
300: In peripheral nerves (such as the nerves in your arms and legs), the cells that wrap around the neurons' axons and provide insulation are called __________. SCHWANN CELLS
400: Schwann cells are full of what? A) mitochondria and ATP B) Golgi bodies and vesicles C) cholesterol and fat D) ribosomes and proteins (Answer: C)
500: Another name for the insulating wrapping provided by Schwann cells is the ___.
(Hint: two words) MYELIN SHEATH

BLOOD CELLS

100: What is the correct name for white blood cells? LEUKOCYTES
200: Which type of lymphocyte makes antibodies? B-CELLS
300: Which type of white blood cell is responsible for itching and swelling? BASOPHIL
400: Which type of white blood cell "feels" the surface of cells looking for the ID flag (MHC 1)? Natural Killers
500: Why would a detective analyze a neutrophil taken from a blood sample at a crime scene?
TO DETERMINE GENDER OF CRIMINAL OR VICTIM

<u>ACTIVITY IDEA 10C</u>: CRAFT: MAKE ANTIBODY REFRIGERATOR MAGNETS

What to tell the students:

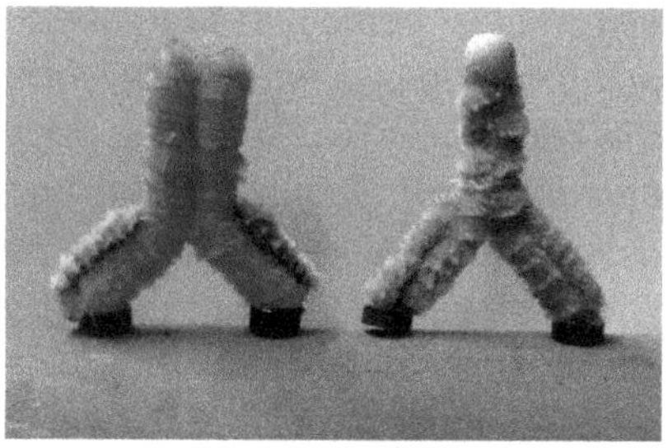

This craft takes us back to chapter 3, where we learned that the cytoskeleton was discovered using antibodies that were tagged with a molecule that could glow. Back then we did not know what an antibody was. Now we know that B cells make antibodies to fight infections. B cells will make antibodies to fight any foreign substance that comes into the body even if it is not a virus or a bacteria. Scientists can take bits of tubulin protein from human cells and inject them into mice, and the B cells of the mice will make antibodies against them. The antibodies can then be collected using very sophisticated equipment. The antibodies are then tagged with the fluorescent molecules. In this activity you will make a Y-shaped antibody that will attack... refrigerators! Or anything else that will hold a magnet.

You will need:

- chenille stems
- hot glue gun (or very strong all-purpose craft glue)
- small magnets
- Optional: a large bead (glow-in-the-dark if available)

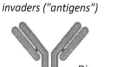

The tips of the Y stick to invaders ("antigens")

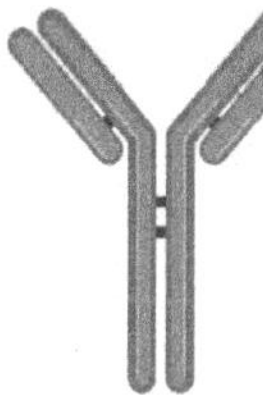

Diagrams of antibodies often look something like this.

What to do:

NOTE: Antibodies consist of four parts, two identical long parts (called the long chains) and two identical short parts (called the short chains).

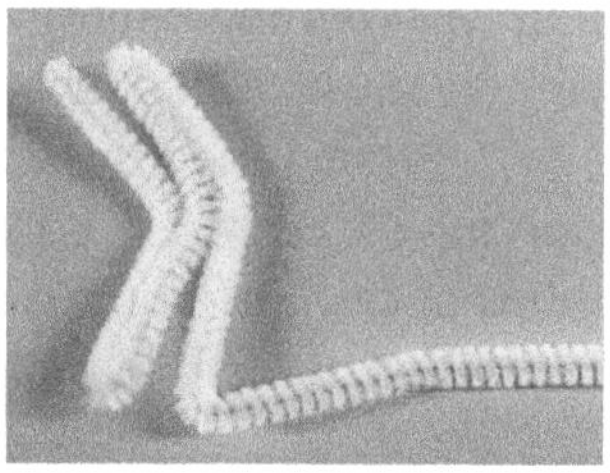

1) Make the long parts first. (They look like half of a Y.) Twist two chenille stems of the same color together at the ends to make one very long stem. Bend the end to make half a Y, then reinforce this up and down a few times you will have a "core" around which to twist the rest of the stem.

2) Twist the rest of the stem tightly around and around this core to thicken it. Once this is finished, make another exactly like it. (If you find a way that works better for you, that's great. The exact method isn't critical.)

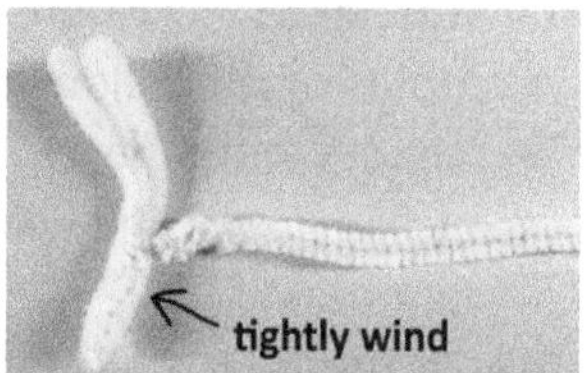

3) To make the shorter part, use just half a chenille. Bend the end back and forth to make a straight "core" that is the length of the top part of the long parts you just made. Wrap it tightly to thicken it.

4) Use hot glue to attach the two longer pieces so that they form a Y.

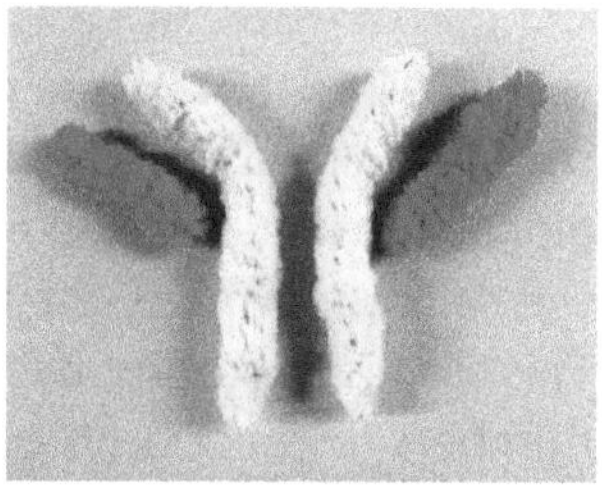

5) Use hot glue to attach the shorter pieces right under the sides of the top of the Y.

6) Use hot glue to attach magnets to the tips of the Y.

7) If you would like to simulate a tagged antibody, you can hot glue a large bead to the base of the Y. (See sample photo at top, antibody on right.)

More info:

You saw a diagram of a real antibody in this chapter. They really are shaped like the letter Y. The tips of the top of the Y have a unique shape that will fit into a matching shape like a puzzle piece. Most antibodies never meet the shape they match. Your body makes millions of randomly shaped antibodies that it will never use. However, one in a million will be a match for a germ that enters your body. If this match is found, your immune system has a way to signal that B cell to start cloning itself so that this useful antibody can be mass produced.

The end of the base (stick part) of the Y can also stick to things. Notably, they can stick to basophils. (Our cartoon basophil had antibodies for hair.) It can also hold onto fluorescent molecules that glow in the dark.

ACTIVITY IDEA 10D: GAME: "SUPPLY CHAIN" A game about protein transport in neurons

What to tell the students:

This game puts together many of the bits of information you have been learning throughout this course. You will see the entire process, from beginning to end, of how a neuron makes and delivers some of its neurotransmitter chemicals. The electrical signal that travels through a neuron consists of a sudden flow of sodium atoms into the axon. When this flow reaches the ends of the axon, it causes calcium channels to open up and allow calcium atoms to flow in. The calcium atoms trigger the waiting vesicles to do exocytosis and release their neurotransmitter chemicals. The chemicals will travel across the tiny gap between neurons and will cause the next neuron in the line to begin its own electrical signal. This game board shows only one neuron, but in reality, it would be part of a series of neurons.

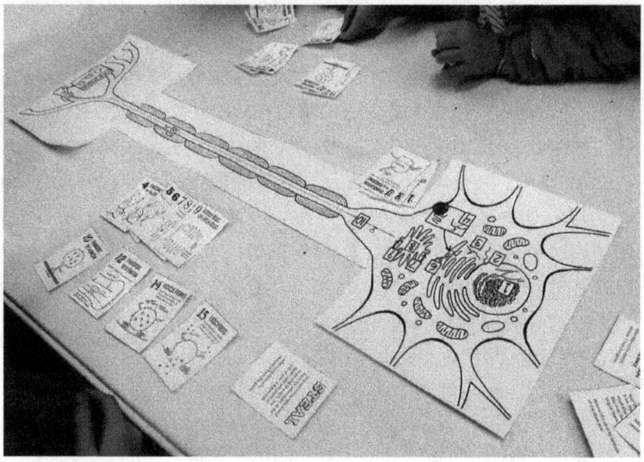

NOTE: As of the writing of this edition, there is a website where you can see a short animation of the processes described on cards 13, 14 and 15 (as well as info on the other steps, too).

https://nba.uth.tmc.edu/neuroscience/m/s1/chapter10.html

Scroll down to section 10.7. Also, section 10.8 will give you some information about the recycling of both vesicles and neurotransmitters after the exocytosis process. (Eventually, vesicles wear out and must be completely broken down and recycled by lysosomes.)

When I used this animation in my class, I added the analogy of cocking an air rifle or a flintlock rifle. The docking of the vesicle is like loading the gun, the priming of the vesicle is like cocking the gun, and the sudden influx of calcium atoms is like pulling the trigger. (The entry of calcium was caused by the electrical signal when it reached the terminal knobs. The signal caused the calcium portals to open and allow calcium atoms to enter.)

You will need:

- copies of the following pattern pages printed onto card stock (game board can be regular paper)
- scissors and tape
- a token (anything will do, even a coin) to place on the numbered squares on the neuron board

How to prepare:

1) Copy the pattern pages onto card stock. Cut out all the cards. Assemble neuron board.

2) IMPORTANT: Before playing, have each team lay out a set of all 15 cards in order so they can see the complete chain. Read each card out loud, in order, adding any comments you feel are important as you go along. The large labeled neuron should be helpful in visualizing where the events on the cards are happening in the cell. You can use the following list of questions if you find them helpful to jog the students' memories. This step is important because once they start playing the game, their brains will prioritize looking at the numbers (by necessity) and they'll tend to ignore the pictures and words on the cards.

3) Make sure the cards are shuffled REALLY well. This is very important. In fact, you might want to scan through them quickly and make sure there are not any sets clumped together.

LIST OF REVIEW QUESTIONS YOU MIGHT WANT TO ASK WHILE THEY ARE LOOKING AT THE BOARD:

1) What was the initial source of the DNA in the nucleus? (the father and mother)
2) Why does the DNA in the nucleus look like tangled spaghetti? (It not doing mitosis.)
3) How is the membrane around the nucleus different than the outer membrane? (It is a double layer.)
4) What might those blank ovals be, near the mitochondria? (lysosomes)

5) The motor protein that is carrying the vesicle AWAY from the nucleus would be what type? (kinesin)
6) What type of motor protein would be required to take a vesicle or a mitochondria back to the cell body? (dynein)
7) What is missing around the DNA? (many proteins that control which parts are opened)
8) What organelle manufactured the protein enzymes that are working inside the Golgi body? (ribosome)
9) What organelle manufactured the proteins that surround the DNA? (ribosomes)
10) What organelle manufactured the proteins that are embedded in the vesicle that contains the neurotransmitter chemicals? (ribosomes)
11) Where did the information come from to manufacture the protein that are embedded in the walls of the vesicle that is being transported? (from the DNA)
12) What organelle manufactures neurotransmitters that are made of proteins? (ribosomes)
13) What organelle manufactured the motor proteins? (ribosomes)
14) What type of cytoskeleton fiber is the motor protein traveling on? (microtubule)
15) If your cells need to manufacture new phospholipids, what is the ultimate source of the raw materials such as fatty acids? (the food you eat)
16) What are those lumpy things on either side of the axon? (Schwann cells) What shape are these cells if you unrolled them? (flat) What do they do? (Provide insulation, like the rubber around an electrical cord.)
17) Is the neuron in the brain or in the rest of the body? (It is a peripheral nerve, outside of the brain or spinal cord, because of the Schwann cells. Brain neurons do not have Schwann cells, they have oligodendrocytes.)
18) Would the DNA in the Schwann cells be the same as the DNA in the neuron? (Yes , the DNA is identical but different parts of it are being used or expressed.)

How to play with two teams, with 1 to 3 players on a team:

The object of the game is to complete your series of supply chain cards before the cells completes its supply chain. The cell's progress will be marked by moving a token on the neuron from number 1 down to number 15. Moving the cell's token down one space will mark the beginning of a turn. Therefore, you have only 15 turns to complete your series of cards. If the cell gets to 15 before you complete your series, the cell wins the game. If you complete yours before the cell does, you win the game. Both teams can win, or only one team can win. Even if one team wins before 15 is reached, keeping playing until either the other team wins or the cell wins.

BEFORE YOU START MAKE SURE THE CARDS ARE SHUFFLED REALLY WELL.

1) Deal 4 cards to each team. At the end of every turn you will always have a maximum of 4 cards. The only exception is that you may hold one TRADE or STEAL card off to the side, in addition to your 4 regular cards.

2) On each turn, both teams will draw 3 cards. If there are 3 players on a team, each player will draw a card. If there is one player on each team, they will draw all 3 cards. If there are 2 players on a team, they can take turns drawing one or two cards, but since the cards are for the whole team it really won't matter who draws how many.

Teams should alternate who gets to move the token and draw cards first (though in the grand scheme of things it won't make or break the game if you lose track of who turn it is). The only time it will matter is if both teams want to use a TRADE or STEAL card on the same turn. The team who moved the token and drew first gets to use their TRADE or STEAL card first.

3) A turn begins by moving the cell's token to the next boxed number. (As I suggested, teams will take turns doing this.) So the first turn begins by moving the cell's token to the number 2. Now each team draws 3 cards.

4) If either team draws a TRADE or STEAL card, they have the option of using it before anything else happens on that turn. If they draw both a TRADE and a STEAL card, they must return the STEAL card to the bottom of the draw pile and draw a replacement card. You can only hold one TRADE card **<u>or</u>** one STEAL card. If both teams have either a TRADE or STEAL card, the team that drew first gets to go first. You don't have to use your card on that turn; it can be set aside for future use. As soon as it is used it is removed from the game.

A TRADE card allows you to force a trade with the other team. You may take one of the cards in their hand (but not any cards from completed sets) but you must give them one in return. A STEAL card permits you to take a card without giving one in return.

5) After using a TRADE or STEAL card, it is removed from the game.

6) During each turn, the team looks at the cards in their hand and checks to see if they have any complete sets of three numbers. The numbers are printed in similar fonts to make this easy to see. If so, they take that set out of their hand and set it out in front, along the edge of the neuron, so it is obviously "out of play."

7) At the end of each turn, whether not any sets were achieved, the teams always have to discard so that they only have a maximum of 4 cards left in their hand (with the exception of a TRADE or STEAL set off to the side).

8) When the NEURON GETS A FREE TURN card is drawn, it is not counted as one of the 3 drawn cards. This card will cause the neuron to immediately advance to the next number. Once used, this card is removed from the game. The team that drew this card can draw a replacement, to make a total of 3 drawn cards on their turn.

9) The turns continue like this: a) advance the token on the neuron board, b) each team draws 3 cards, c) trades or steals are accomplished, d) a set of 3 may be taken out of hand put into line of finished sets, e) teams must discard to pare down to a maximum of 4 cards at the end of the turn. If they end up with zero cards at the end of a turn, that's okay, they will draw three more cards at the beginning of the next turn.

How to play solo:

1) Use only two sets of numbers (1-5) instead of 3. Take out the TRADE and STEAL cards.

2) Otherwise rules will be the same, advancing the token on the neuron board at the start of each turn, then drawing 3 cards. Even playing solo, you still might lose to the neuron!

1 TRANSCRIPTION

RNA polymerase copies a section of DNA

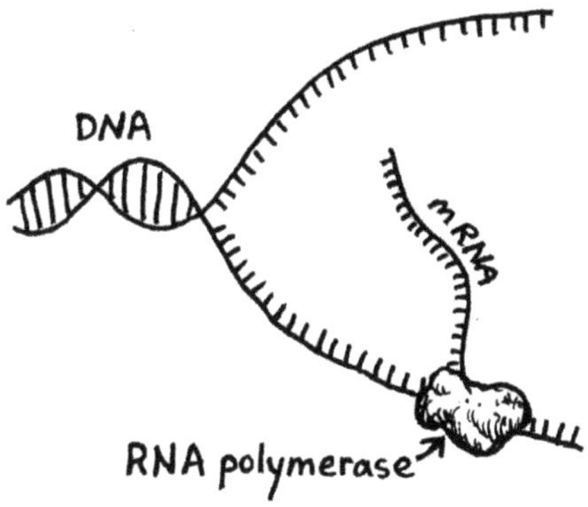

2 Messenger RNA EXITS NUCLEUS

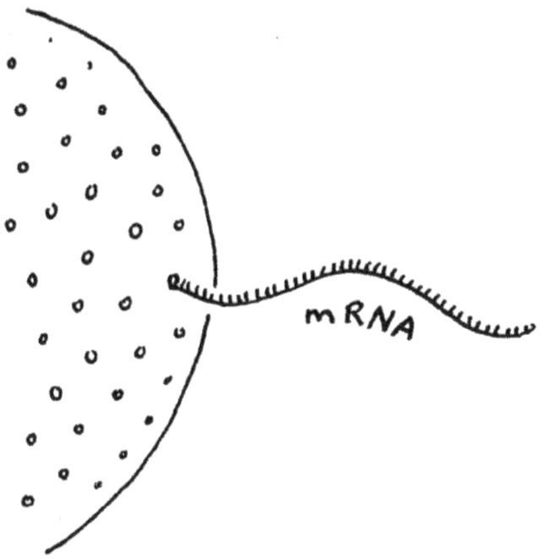

3 TRANSLATION in a RIBOSOME

The first part of the protein says, "Take me to the E.R.!"

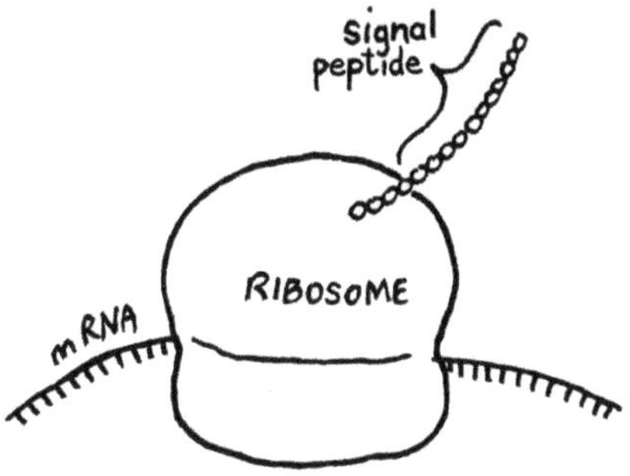

4 RIBOSOME DOCKS on the E.R.

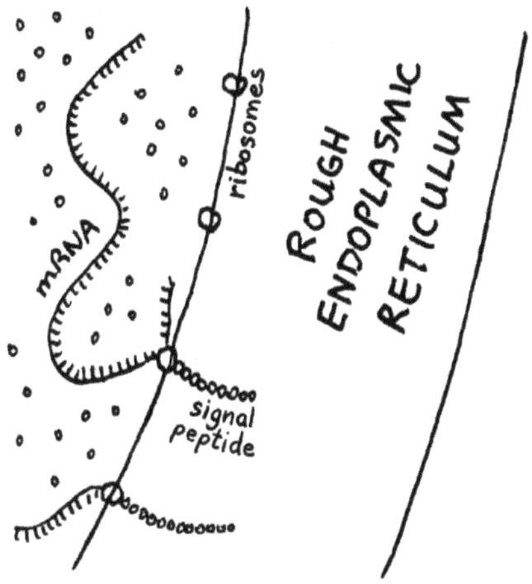

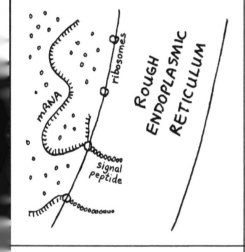

5 PROTEIN MADE INSIDE THE E.R.

The protein will be a neurotransmitter.

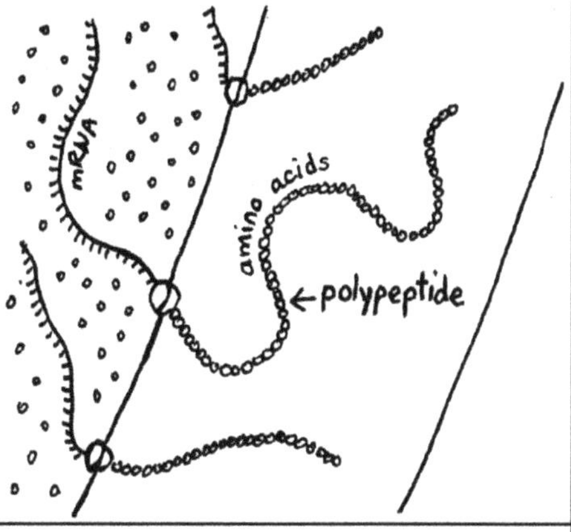

6 VESICLE BUDS OFF FROM E.R.

Vesicle contains neurotransmitter

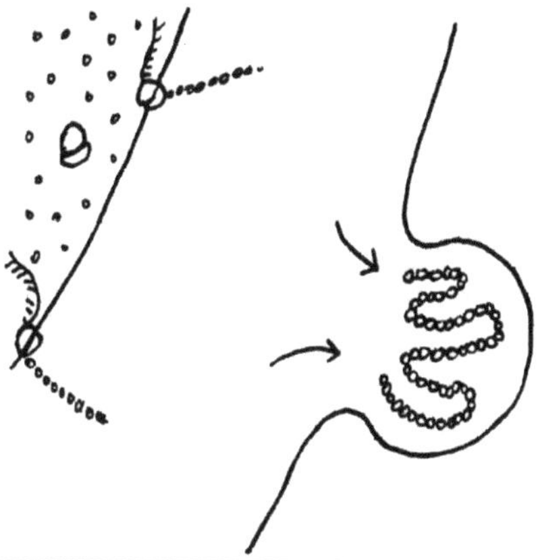

7 VESICLE MERGES INTO GOLGI BODY

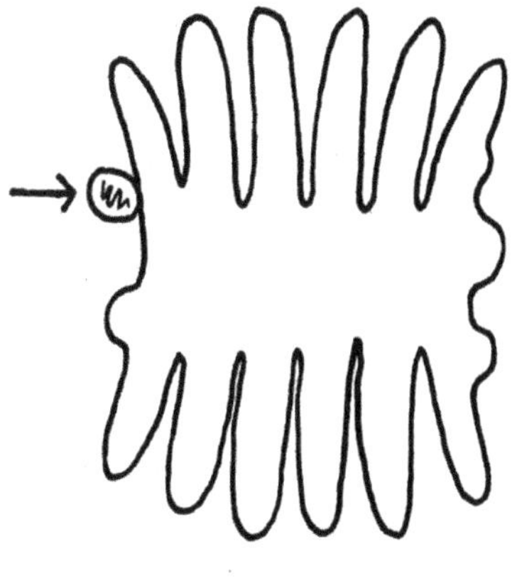

8 GOLGI MODIFIES PROTEIN

Golgi enzymes make changes such as folding or adding sugars and phosphates.

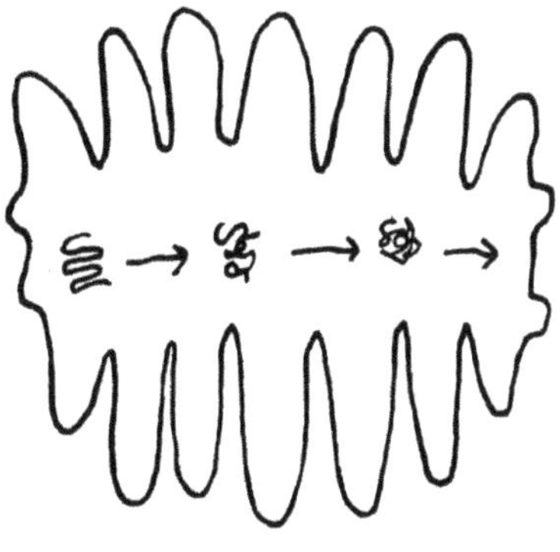

9 VESICLE BUDS OFF FROM GOLGI

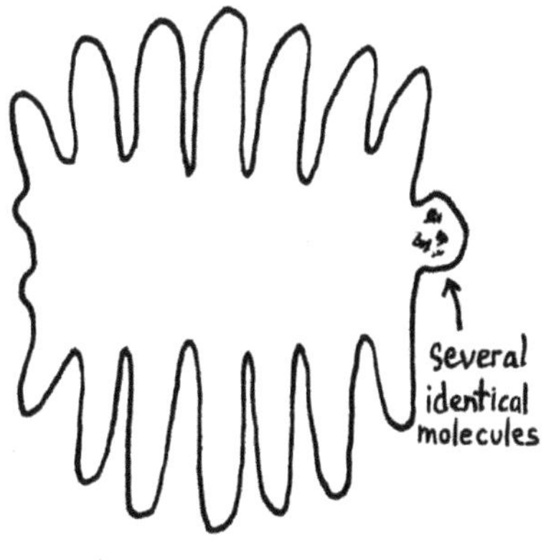

MAKE THREE COPIES ON CARD STOCK

10 TRANSPORT BY A KINESIN

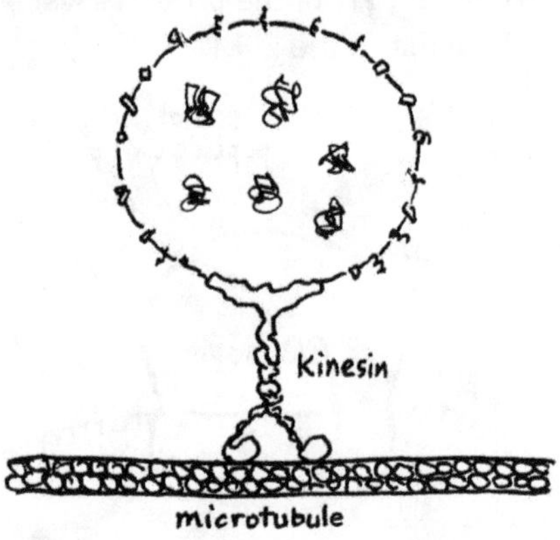

11 TRANSPORT BY A KINESIN

Vesicle is handed off to another kinesin

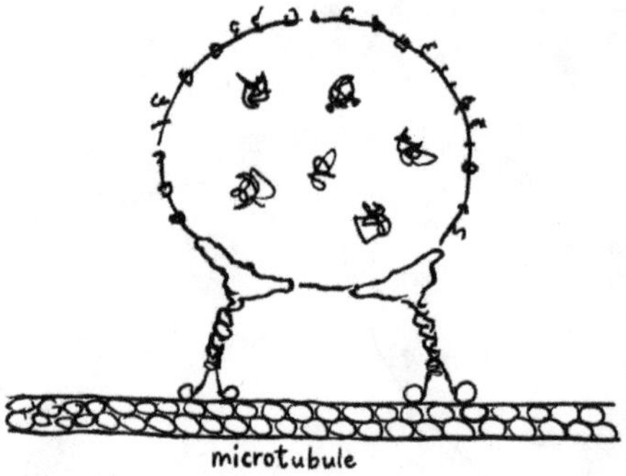

12 TRANSPORT BY A KINESIN

Vesicle is handed off to a final kinesin

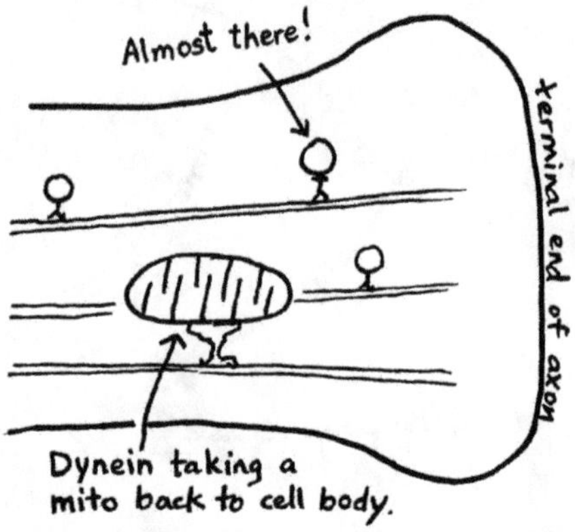

13 VESICLE DOCKS

There are docking proteins on vesicle and membrane

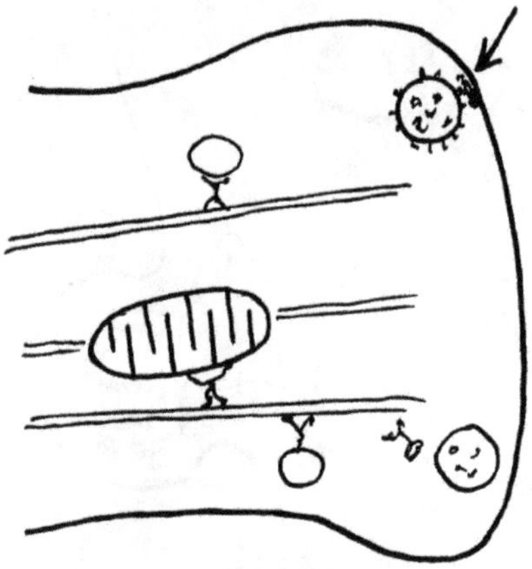

14 VESICLE PRIMED

Docking proteins change shape and get ready to receive calcium ions.

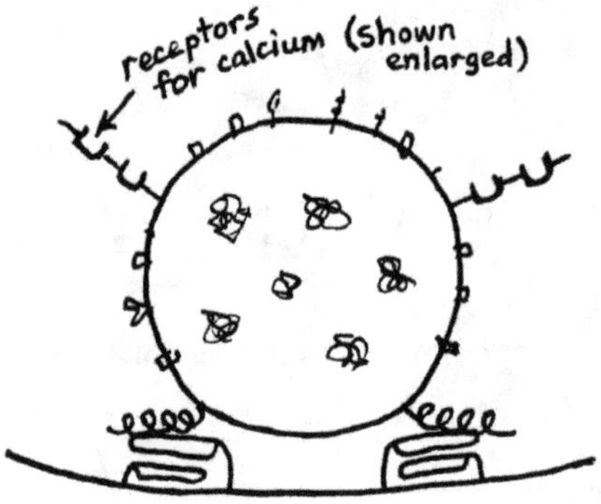

15 EXOCYTOSIS

Docking proteins change shape and get ready to receive calcium ions.

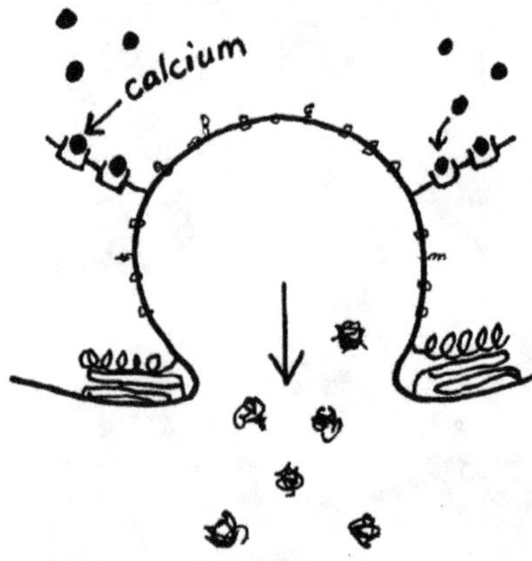

STEAL

Your team may take a card from the other team at the beginning of a turn. The stolen card will count as one of your 3 draw cards.

After using this card, remove it from the game.

TRADE

Your team may take a card from the other team but must also give them one. You may decide which card to give them.

After using this card, remove it from the game.

NEURON GETS A FREE TURN

After this card is played, remove it from the game.

MAKE THREE COPIES ON CARD STOCK

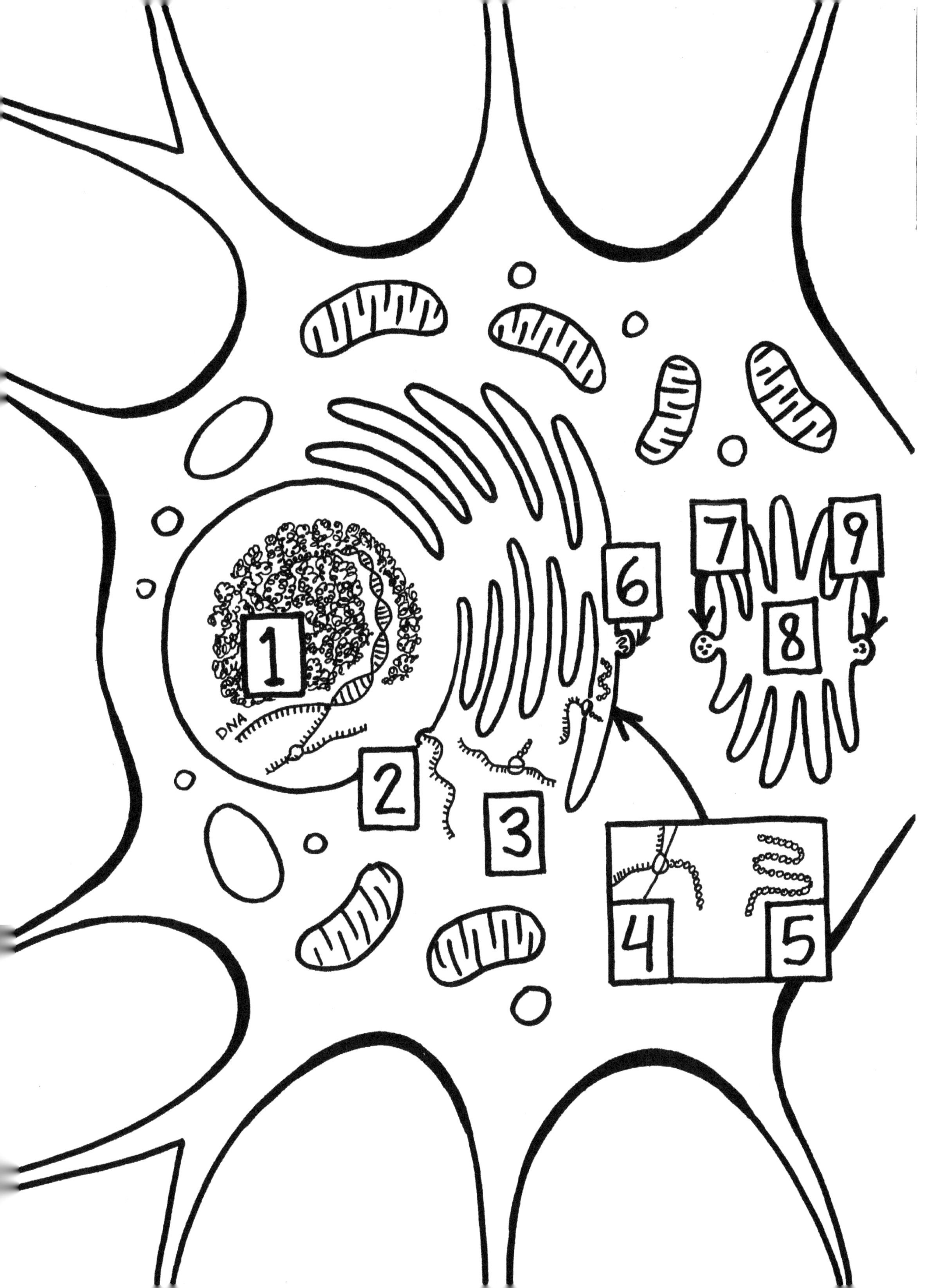
1
DNA
2
3
4
5
6
7
8
9

11
One motor protein is handing off the vesicle to another.

This bump is called the hillock.
10

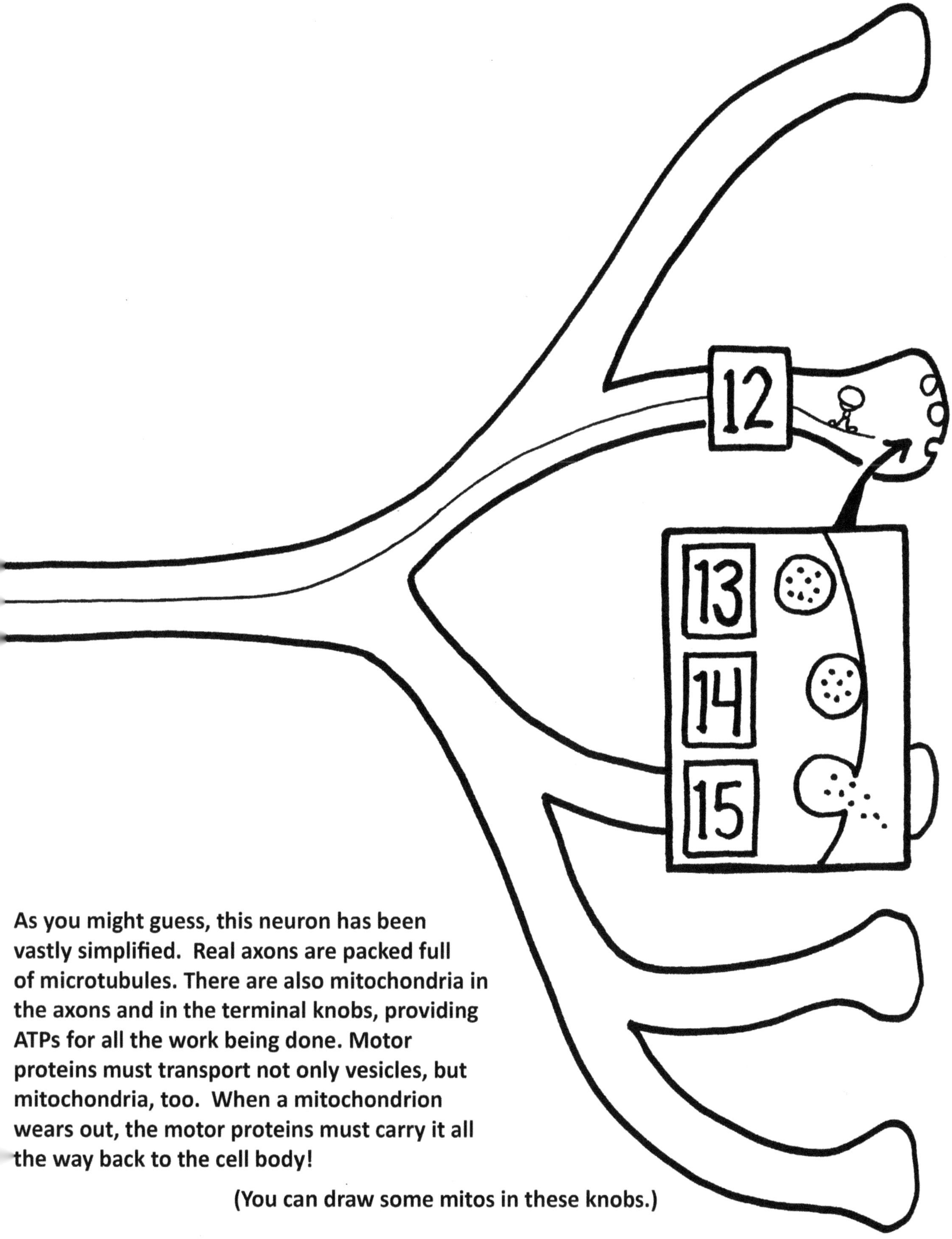

As you might guess, this neuron has been vastly simplified. Real axons are packed full of microtubules. There are also mitochondria in the axons and in the terminal knobs, providing ATPs for all the work being done. Motor proteins must transport not only vesicles, but mitochondria, too. When a mitochondrion wears out, the motor proteins must carry it all the way back to the cell body!

(You can draw some mitos in these knobs.)

BIBLIOGRAPHY

Books for young people that I used for reference and comparison:

Animal Cells; The Smallest Units of Life by Darlene R. Stille. Published in 2002 by Compass Point Books, Woodbury, NY. ISBN 978-0-7565-1761-8 (www.cshlppress.com)

Cell Scientists; From Leeuwenhoek to Fuchs by Kimberly Fekany Lee. Part of the "Mission: Science" series. Published in 2009 by Sally Ride Science (www.sallyridescience.com). ISBN 978-0-7565-3964-1

Enjoy Your Cells by Fran Balkwill and Mic Rolph. Published in 2002 by Cold Spring Harbor Laboratory Press, Minneapolis, MN. ISBN 0-87969-584-6

Cell Communication; Understanding How Information Is Stored and Used in Cells by Michael Friedman and Brett Friedman. Published by The Rosen Publishing Group, Inc., 2005. ISBN 1-4042-0319-2

Eukaryotic and Prokaryotic Cell Structures; Understanding Cells With and Without a Nucleus by Lesli J. Favor, PhD. Published by The Rosen Publishing Group, Inc., New York, NY, 2005. ISBN 1-4042-0323-0

Cell Regulation; Understanding How Cell Functions, Growth, and Division are Regulated by Lois Sakany. Published by The Rosen Publishing Group, New York, NY, 2005.

Textbooks used by my consultant to check facts:

Biology of the Cell by Sylvia S. Mader. Published by McGraw-Hill, 1993. ISBN 0-697-15098-4

Biology of Genetics and Inheritance by Sylvia S. Mader. Published by McGraw-Hill,1993. ISBN 0-697-15099-2

Molecular Biology of the Cell, 5th edition by B. Alberts, A. Johnson, J. Lewis, M. Raff, K. Roberts and P. Walter. Published by Garland Science, Taylor and Francis Group, LLC., 2002. 978-0-8153-4105-5

Molecular Biology of the Cell, 3rd edition by B. Alberts, D. Bray, J. Lewis, M. Raff, K. Roberts, and J. Watson. Published by Garland Publishing, Inc., 1994. ISBN 0-8153-1619-4

Samples of on-line references I used: (I did not list all the Wikipedia articles I consulted.)

Detailed info about proteins:
http://www.ncbi.nlm.nih.gov/bookshelf/br.fcgi?book=mboc4&part=A388 (ebook)
http://www.zoology.ubc.ca/~berger/b200sample/unit_8_protein_processing/golgi/lect28.htm

Motor proteins and microtubules
https://www.ncbi.nlm.nih.gov/books/NBK22572/
https://www.ncbi.nlm.nih.gov/books/NBK9932/
https://www.ncbi.nlm.nih.gov/books/NBK26888/
https://www.youtube.com/watch?v=jGmz4xVP50M
https://www.ncbi.nlm.nih.gov/pmc/articles/PMC5362366/

Role of sugar moelcules:
https://www.ncbi.nlm.nih.gov/books/NBK453034/

Info about protein synthesis, ER and Golgi bodies
http://users.rcn.com/jkimball.ma.ultranet/BiologyPages/P/ProteinKinesis.html

Translocation of proteins into the rough ER:
http://www.cytochemistry.net/cell-biology/rer2.htm

Protein production and transport in neurons:
https://web.williams.edu/imput/synapse/pages/I.html
https://nba.uth.tmc.edu/neuroscience/m/s1/chapter10.html

Animation of how sodium potassium pump works:
http://highered.mcgraw-hill.com/sites/0072495855/student_view0/chapter2/animation__how_the_sodium_potassium_pump_works.html

Cell parts:
http://www.wisc-online.com/Objects/ViewObject.aspx?ID=AP11403
https://en.wikipedia.org/wiki/Mitochondrial_matrix

Nuclear pores:
http://www.genetik.biologie.uni-muenchen.de/research/parniske/nucleoporins/index.html

Nulceosomes and histones:
http://scienceblogs.com/transcript/2006/08/nucleosome_binding_sites_1.php

Chromatin:
http://micro.magnet.fsu.edu/cells/nucleus/chromatin.html

DNA binding proteins:
http://www.ncbi.nlm.nih.gov/pmc/articles/PMC2726711/

DNA handedness
https://www.the-scientist.com/features/left-handed-dna-has-a-biological-role-within-a-dynamic-genetic-code-67558

Protein structure:
http://www.ncbi.nlm.nih.gov/books/NBK26830/
https://en.wikipedia.org/wiki/Alpha_helix

Intermediate filaments and desmosomes:
http://www.cytochemistry.net/cell-biology/intermediate_filament_intro.htm
https://alg.manifoldapp.org/read/fundamentals-of-cell-biology/section/301523e1-6d5d-488f-8dce-e73b46200340

Capillaries:
http://education.vetmed.vt.edu/Curriculum/VM8054/Labs/Lab12b/Lab12b.htm

Glycolysis and the Krebs cycle:
http://www.sparknotes.com/testprep/books/sat2/biology/chapter6section1.rhtml
http://www.daviddarling.info/encyclopedia/C/citric_acid_cycle.html
http://biology.clc.uc.edu/courses/bio104/cellresp.htm
http://www.elmhurst.edu/~chm/vchembook/612citricsum.html

Digestion of fats:
http://www.annecollins.com/digestive-system/digestion-of-fats.htm

Reproduction of Golgi bodies:
http://books.google.com/books?id=Eb1TaY6P8HwC&pg=PA187&lpg=PA187&dq=how+to+golgi+bodies+replicate&source=bl&ots=vJXxVxslbD&sig=vhY-Pxqrvl1PisTH3aOh0c0mOJs&hl=en&ei=sgBXTcTlF4X6lweczNWUBw&sa=X&oi=book_result&ct=result&resnum=5&sqi=2&ved=0CD0Q6AEwBA#v=onepage&q&f=false

NK cell maturation:
http://researchnews.osu.edu/archive/nkstages.htm

Printed in the USA
CPSIA information can be obtained
at www.ICGtesting.com
LVHW080753100923
757503LV00013B/250